W. T. Lippincott

University of Arizona
Tucson, Arizona

Devon W. Meek

The Ohio State University
Columbus, Ohio

Frank H. Verhoek

The Ohio State University
Columbus, Ohio

Experimental General Chemistry

SECOND EDITION

 *SAUNDERS GOLDEN SUNBURST SERIES*

W. B. SAUNDERS COMPANY
Philadelphia, London, Toronto

W. B. Saunders Company: West Washington Square
Philadelphia, Pa.   19105

1 St. Anne's Road
Eastbourne, East Sussex BN21 3UN, England

833 Oxford Street
Toronto, M8Z 5T9, Canada

Experimental General Chemistry   ISBN 0-7216-5788-5

Last digit is the print number: 9   8   7   6   5   4   3

PREFACE

This is a laboratory text for a college general chemistry course. It is designed to show some ways in which chemists synthesize desired compounds, and how they obtain information by experiment and then use the collected data to draw conclusions about the properties and behavior of matter. It is planned to help students learn not only to perform essential laboratory operations, but also to observe, to correlate and interpret data, and to think and act in laboratory situations.

The 35 experiments include portions of a variety of topics, including separations, identification, stoichiometry, qualitative and quantitative analysis, coordination chemistry, chemical equilibrium, electrochemistry, thermodynamics, organic chemistry and biochemistry. They were selected to illustrate both modern and traditional techniques, and to provide teachers and students with a large number of options in choosing experiments to be used and in determining the amount of time and study to be given to any one experiment.

Approximately one-third of the experiments consist of several parts, any or all of which may be assigned or elected. This provides the opportunity for individualized assignments and for some students to spend more or less time on a given experiment than others do. Thus, one student may complete three or four parts of a single experiment, while others in the same laboratory are working on parts of several different experiments. For example, many students, having learned a new and moderately complicated technique, such as measurement of vapor pressure, especially like to continue to use this technique in more challenging ways; other students prefer to move on to other experiments rather than repeat a learned activity. Options such as these in several variations are included in numerous combinations of experiments.

Some experiments relate to the chemistry of everyday life. For example, in Experiment 33 students can synthesize at least four medicinally useful compounds; in Experiment 34 they can identify amino acids obtained from the palms of their hands; in Experiment 7 they can discover for themselves some relationships between color, wavelength and energy of electronic transitions. Other experiments illustrate exciting modern chemistry, such as the synthesis and properties of ferrocene (Experiment 6), the structure of ionic crystals from x-ray diffraction data (Experiment 26) and the electronic spectra of coordination compounds (Experiment 25).

Because lecture and textbook presentations often emphasize the results rather than the details of experiments, each experiment begins with an introduction, giving background information needed to bridge the gap between lecture presentation of the topic and laboratory study of a small segment of it.

Many experiments require more than one laboratory period for completion. Because each experiment msut be done carefully and new techniques constantly learned, it is not expected that all the experiments can be completed in a one-year course. With the large number provided, it is possible to select a different group of experiments in successive years or to give students a choice of experiments during one year.

Among the techniques included are: use of the analytical balance; paper, column and thin-layer chromatography; spectroscopic and spectrophotometric methods; qualitative analysis of cations, anions and simple organic substances; precipitation, ion exchange and solvent extraction quantitative separations; colorimetric, gravimetric and volumetric analysis, including acid-base, precipitation, oxidation-reduction and complexometric titrations; measurement of vapor pressure, vapor density, enthalpy change, free energy change, and equilibrium, reaction rate, and electrochemical data; certain aspects of inorganic and organic synthesis.

All experiments are designed for students with little previous laboratory experience, but students are expected to work carefully and to learn good laboratory habits. The apparatus and instruments used are, with few exceptions, inexpensive and simple to work with. All experiments require a written report, and there is a section dealing with record-keeping in the laboratory notebook. Because the evaluation of numerical data is essential in many experiments, special sections on errors of measurement and their treatment, use of significant figures and graphical representation of data are provided (Experiment 35).

Most of these experiments have been used in the general chemistry laboratories at the Ohio State University for several years. They have been scrutinized by thousands of students and hundreds of instructors. Pre-laboratory films prepared to accompany certain of these experiments also have been used for several years at Ohio State. They have been found to increase significantly the speed and efficiency with which students learn to use the many experimental techniques called for.

No laboratory experiments that have been developed at a large university can be considered to be the work of three people alone. To the students and teaching assistants, to those who have shared their experiences through the literature, and especially to our colleagues at Ohio State, Arizona and elsewhere who have used or examined this laboratory program, we acknowledge with grateful appreciation our debt for assistance and advice.

W. T. Lippincott

Devon W. Meek

Frank H. Verhoek

TO THE STUDENT

Chemistry, like all other sciences, is built on experimentation. This means that the knowledge accumulated by chemists is obtained or inspired by experiment. In order to appreciate the ecological, medical, and bio-chemical problems confronting us all in this technological age, and to appreciate science more fully, one needs to become actively involved in the process of collecting and interpreting data. No description of an observed phenomenon, no development of a theory, no classification of facts has complete meaning until a person has experienced for himself the difficulties in obtaining a reliable experimental result, understood the implications of an assumption made in developing a theory, or verified the results of another worker.

It is impossible for anyone to repeat even a small percentage of the chemical experiments that have been performed during the past 200 years. Thus, in planning an introductory course, most chemical educators elect to present the important principles of the science in the lecture portion of the course and to illustrate some of these principles with appropriate laboratory exercises. In addition to illustrating principles in a quantitative manner, we have attempted in this laboratory program to incorporate some new techniques, manipulative skills, and ingenuity into the experiments and methods of experimentation.

For example, Experiment 12 illustrates the methods by which vapor pressure data of a liquid can be obtained and the way these data can be used to determine the heat of vaporization of the liquid. In this case, a value for a property—specifically, the heat of vaporization of the liquid—is obtained without actually measuring that property. Since it is often difficult or impossible to measure certain properties of matter, such techniques are employed frequently. Experiment 6 involves the synthesis of ferrocene, $Fe(C_5H_5)_2$, and a study of its oxidation to the ferrocenium ion, $Fe(C_5H_5)_2^+$, and the reduction back to ferrocene. Ferrocene was prepared in 1952 for the first time; its unusual "sandwich" structure and chemical properties stimulated intense research into a new class of organometallic compounds that now encompasses almost all of the transition metal elements and very recently has been extended to "sandwich" compounds of the lanthanide and actinide elements. By judging the relative colors of mixtures containing ferrocene and ferrocenium ion, you will be asked to arrange the reducing and oxidizing agents in a relative order. Experiment 33 illustrates the transformation of a common organic chemical into several different compounds either by one reaction step or by a sequence of steps. A series of reactions originating from one compound is often used by modern biochemists and by organic and pharmaceutical chemists for the synthesis of a complicated new chemical or a drug.

As mentioned above, several of these laboratory experiments are sequential. For example, in Experiment 2 you separate a mixture and determine its quantitative composition. Then in Experiment 3 you identify qualitatively the organic and inorganic salt components of the mixture. In Experiment 9 you prepare and standardize a sodium hydroxide solution and then use the solution in Experiment 10 to determine the equivalent weight of an unknown acid. Thus, both Experiments 9 and 10 must be performed carefully to obtain the correct equivalent weight. Similarly, Experiments 29, 30 and 31 involve three different quantitative analyses of the same unknown.

Many of the experiments have a quantitative basis; consequently, you will generally be asked to determine a value or to make calculations based on your data. Thus, careful efficient work is required throughout the laboratory program. Chemicals generally will be in the laboratory one week before an experiment is scheduled and for one to two weeks afterward, so that a student may proceed at his own pace. However, experience has shown that certain deadlines for experiments must be established if the work is to be completed during the term. The instructor will remind you of these deadlines.

No chemist can afford to begin an experiment without having planned it. Similarly, students of chemistry must be prepared *before* entering the laboratory. This means that you must study and *understand* the principles and experimental procedures before going to the laboratory. Your instructor will help you through the difficult aspects of an experiment if you are prepared, but he cannot be of much help if you do not study the experiment before the laboratory period.

Experimental data are recorded as a permanent record in a laboratory notebook in an organized and legible manner. You should study the experiment before class and prepare the necessary data tables in your notebook in which to record the observations *as you collect them.* Never record numbers or observations on a loose piece of paper with the intent of transferring the data to your record book later. Time spent before class in organizing your data collection and recording steps is invested wisely.

A report of the experiment should be prepared as described in the following section or as modified by your instructor. Careful analysis and interpretation of the data are just as important as making the observations. As you collect the data and as you correlate the results, be aware of instances in which quantitative techniques are required, sources of error, and whether the inherent precision of the measuring device or of the investigator is the determining factor to accuracy in a given experiment. During the first six experiments you will gain experience with the types of problems encountered in collecting and evaluating data. Because the experiments have a quantitative emphasis, at this point it is important to study "Evaluation of Experimental Data" carefully; include a treatment of errors and an evaluation of the precision in your reports of subsequent experiments.

Some experiments give you a choice of projects, any or all of which may be undertaken; others are open-ended so that you can pursue an interesting aspect of the problem as a small individual research problem. Discuss your proposed extensions with the instructor and obtain his permission before proceeding. **NEVER ATTEMPT TO CONDUCT AN UNAUTHORIZED EXPERIMENT.**

The experiments in this program are intended to challenge you. Whereas the lecture material gives you an overview of chemistry, these experiments should give you a good idea of the methods by which a chemist obtains information by experimentation and how he uses the data to formulate conclusions about the properties and behavior of matter. All the chemical information that you will be reading, discussing and studying during the year has been accumulated from data collected in experiments and from the interpretation of those data. This laboratory program will help you gain experience in carefully performing experiments, learning chemical techniques and recording your observations. Also, it should serve as a guide for relating the data to those obtained in other experiments, questioning the significance of the data and creatively interpreting the results.

SAFETY AND LABORATORY RULES

Accidents in a chemical laboratory usually result from improper judgment on the part of the victim or one of his neighbors. *Learn and observe the safety and laboratory rules listed below.* If an accident to you or your neighbor does occur, summon help immediately from the laboratory instructor.

1. *Wear safety glasses.* Because the eyes may be permanently damaged by spilled chemicals and flying broken equipment, be sure to wear safety goggles or safety glasses *at all times* in the laboratory. If you get anything in your eye, report it immediately to your instructor. In laboratories equipped with an *eye wash fountain*, learn its location and how to use it the first day in the laboratory. Note that routine eye-washing with a contact lens in place will not clear a splashed chemical from the eye; wash, remove your contacts and wash again.

2. *Locate safety equipment.* During the first laboratory period familiarize yourself with the location of the safety features of the laboratory, including the *safety shower, fire extinguisher, fire blanket, laboratory first aid kits* and *first aid room.* The safety shower should be used if your clothing catches on fire or if a corrosive chemical is spilled on you in quantities that cannot be easily flushed away at laboratory faucets.

3. *Don't cut yourself.* Cuts and burns are the most common injuries occurring in chemistry laboratories. Cuts can be prevented by following a few simple rules:
   a. When inserting glass tubing or thermometers into rubber stoppers, *always* use glycerin (available on the shelf) or soapy water as a lubricant on both the glass tubing and the hole. *Always* protect your hand by wrapping the glass tubing with a towel.
   b. Fire-polish all sharp edges of broken glass.
   c. Discard cracked or broken glassware immediately.
   d. Never heat a graduated cylinder with a burner flame.

4. *Wash chemicals from skin.* If you receive a chemical burn from acid, alkali or bromine, immediately wash the burned area with *large quantities* of water. Ask another student to summon the laboratory instructor.

5. *Be careful with flames.* A lighted gas burner can be a major fire hazard. The burner should be burning only for the period of time in which it is actually utilized. Carefully position it on the desk away from flammable materials and overhanging reagent shelves. Before lighting your burner, make sure that flammable reagents (such as acetone, benzene, ether and alcohol)

on neighboring desks are well separated from your burner. Be careful not to extend your arm over a burner while reaching for something.

When heating a test tube over a burner or carrying out a reaction in one, never point the test tube toward your laboratory neighbor or yourself.

Long or tousled hair, especially if covered with hair spray, is a major hazard to the wearer in the laboratory. Keep long hair tied back so that it cannot fall forward into a flame. Sprayed hair, long or short, presents a large surface area of flammable plastic to the oxygen of the air, causing almost explosive combustion if ignited. Beards, too, present a hazard to their owners. *Keep hair away from flames.*

6. *Wear suitable clothing.* Clothing is a protection against spilled chemicals or flaming liquids. Dress which exposes large areas of bare skin is a laboratory hazard, and open-toed shoes or sandals are an invitation to maimed feet. *Footwear must be worn in the laboratory at all times.*

7. *Smell cautiously.* Many chemicals used in the laboratory are toxic. If you are instructed to smell a chemical, do so by pointing the vessel away from your face and carefully fanning the vapors toward your face with your hand and sniffing gently. Never taste a chemical unless specifically directed to do so.

8. *Don't pipet by mouth.* Always use a rubber bulb or rubber tubing connected to an aspirator to fill your pipet. Never fill the pipet by using your mouth.

9. *Assemble safe apparatus.* Makeshift equipment and poor apparatus assemblies are the first steps to an accident. Always assemble an apparatus as outlined in your instructions.

10. *Pour acid into water.* Never pour water into concentrated acid. Concentrated acids and bases may be diluted by pouring the reagent into water while stirring it carefully and constantly.

Never add concentrated acid to concentrated base or vice versa.

11. *Use the hood.* Any experiment involving the use of or production of objectionable (i.e., poisonous or irritating) gases must be performed in the hoods.

12. *Read the label.* Read the label carefully before taking anything from a bottle. Many chemicals have similar names, such as sodium sulfate and sodium sulfite; it is obvious that use of the wrong reagent can spoil an experiment or, in some cases, cause a serious accident.

13. *Leave reagents on the shelves.* Do not carry reagent bottles to your desk. This is a matter of courtesy to the other students in the class, and it minimizes the possibility of contamination of the reagent. Obtain the required quantities of chemicals from the reagent shelf by taking clean test tubes or beakers to the reagent area. Take the amount you need and return the bottle to the shelf. Make it a practice not to take much more material than is required for the experiment, because many chemicals are quite expensive.

14. *Don't contaminate reagents.* Do not insert spatulas or pipets into reagent bottles. Remove a solid reagent from the stock bottle by pouring it onto a clean watch glass or filter paper by gentle rotation of the bottle. To transfer a solid material to a test tube, first place the solid on a creased narrow strip of paper. Then insert the paper into the test tube and shake the solid down the crease by gentle tapping with a finger.

Never put a stopper on the table or working surface where it can pick up dirt or another chemical; hold it in the hand while removing reagent. A stopper with a flat head is conveniently removed by grasping the stopper between the two middle fingers, with the *palm of the hand facing upward.* Hold the stopper in this position until you return it to the bottle.

Never return excess materials to reagent bottles. Pour them down the sink drain if they are water-soluble and nontoxic, and flush with water. Water-insoluble and toxic materials should be placed in the waste jars provided for this purpose. If you are not certain how to dispose of excess chemicals, consult your instructor.

15. *Insoluble solids or liquids that do not dissolve in water go into waste cans.* Place paper, glass and similar items in the proper waste cans, *not in the sinks.* Water-insoluble liquids such as benzene and dichloromethane, go into capped waste-solvent containers.

16. *Perform only authorized experiments. Anyone attempting to conduct unauthorized experiments will be subject to immediate and permanent expulsion from the laboratory.*

17. *Plan ahead.* Carefully study each experiment prior to the laboratory period in which the experiment is to be performed. To use the laboratory time efficiently, you must have a clear plan of action, and all data tables must be outlined in the laboratory notebook before performing the experiment.

18. *Keep your work space orderly.* Never place coats, books, and other belongings on the laboratory bench where they will interfere with the experiment and are likely to be damaged. Place tall items, such as graduated cylinders, toward the back of the workbench so they will not be overturned by reaching over them.

19. *Clean up for the next student.* At the end of each laboratory period, wash and wipe off your desk top. Be sure the gas and water are turned off. *Return all special equipment to the stockroom.* Be sure to put everything back into your locker drawer, and also *be sure that your locker drawer is locked before you leave the laboratory.*

THE LABORATORY NOTEBOOK

The importance of keeping an up-to-the-minute record of what you are doing in the laboratory cannot be overemphasized.

A section entitled Notebook and/or Report is included with each experiment to assist you in organizing the recording of observations and data. When completed, this section should become a portion of the notebook and report.

The authors recommend that each student provide himself with a spiral-bound notebook of the size specified by the instructor (8 inch × 10 inch is convenient). A loose-leaf notebook is not acceptable, and some instructors may require a bound notebook. The spiral-bound type is convenient because it folds back on itself when open and requires less space on the laboratory bench. All data and comments should be entered *directly in the notebook, never on a loose sheet of paper* with the intention of copying it into the notebook later. Scraps of paper or loose sheets can get lost between laboratory periods; notebooks are much less likely to be misplaced. A Report Sheet is included with each experiment, and that should provide a neat summary of your notebook data for the experiment.

Enter the work carried out in each laboratory period under that day's date, and add nothing to these entries at a later date. If you discover on Wednesday that something additional should have been recorded on Monday, write it under Wednesday's date, with a note that it should have been a part of Monday's record. Never erase anything in a notebook. If a mistake has been made, cross out the erroneous material with a single line *so that it can still be read,* and rewrite the correct statement in place of it. Even this should not be done except on the day the error is made; if the error is discovered at a later date, simply make an entry on that date (October 14, for example) to the effect that "the material in the second paragraph under the date of October 12 is incorrect and should be ignored; the corrected version follows," and record the correct information under the date of October 14. The material in quotation marks indicates the informal style in which a notebook may be kept; it is usually written in the first person; e.g., "Although the procedure suggested that about 7 ml of 2 M HCl would be required to neutralize the base present, I had to add 9.5 ml before the indicator changed color. The greater volume apparently did no harm, because the precipitate formed readily, as it was supposed to."

WHAT IS TO BE RECORDED IN THE NOTEBOOK?

The criterion of an acceptable laboratory notebook is that the record should be so complete that a second person, with the notebook at hand, could repeat your experiments exactly as you did them, and would know the reason you selected the particular procedure you followed. This means that you should carry out the following steps: (1) Record all numerical data in such a way that the reader knows to what the numbers refer. A three line entry that appears as

$$1.2463$$
$$0.8297$$
$$0.4166$$

for example, is meaningless to a second person (and will become meaningless to you, too, after a few weeks have passed). But an entry such as the following

Wt. of crucible and salt	1.2463 g
Wt. of empty crucible	0.8297 g
Wt. of salt	0.4166 g

is perfectly clear (provided that the salt is described elsewhere). The weights given should have been recorded at the balance; take the notebook there, and copy the balance reading directly. (2) Other observations should be indicated. If the solution turns red when acidified, record this information. If the precipitate is gelatinous rather than crystalline, write it down. Nothing is too trivial to record; if there is any doubt whether something should or should not be recorded, the answer is always "Record it!" (3) The notebook must include the calculations based on the data. Many laboratory workers find it convenient to use the right-hand pages for recording procedures, observations, and data, and to use the left-hand pages, opposite the data, for calculations based on those data. Again, no calculation is too trivial to record; if you are to weigh out 0.10 mole of KCl, show on the left-hand page your addition of the atomic weights of potassium and chlorine to find the weight of 1 mole and your multiplication by 1/10 to determine the weight required. (4) Write chemical equations for all reactions that occur. (5) Sometimes, in the experiments, you are instructed to display graphically the results of your experiments or calculations. You would normally paste these graphs on one of the pages of your notebook. However, your instructor may request that you hand them in as part of your report; if so, make a free-hand sketch in the notebook of the way the graph looked, and paste the graph in the notebook when the report is handed back to you. (6) The procedures you followed in carrying out the experiments would normally be part of the notebook record. In most of the experiments under discussion here, however, detailed procedures are given in the laboratory manual. There is no need to copy these into the notebook; simply refer, in the notebook, to the instructions in the manual, but again in sufficient detail so that the reader will know exactly what you did; e.g., "Following the procedure outlined in the paragraphs below Figure 9-1 on p. 109, I set up two plastic funnels with ashless filter paper, and . . . ." (7) Record discussions of the work with your neighbors or the instructor: "I showed the precipitate to Mr. Blank, telling him what I had done, and he suggested that I add 4 ml of . . . ."

It is not easy to keep a satisfactory notebook. One naturally desires to keep it clean. This stands in the way of recording at any instant the observed data or thoughts in connection with the work. The notebook must be available for use at the time the observations are made; if this availability causes the notebook to become spotted, regard the spotting as a small price to pay. But this does not mean that a notebook need be *illegible* as well as spotted. Neatness in keeping a notebook will pay off in understanding what you did when you refer to the notebook at a later date. Make it a habit to tabulate data whenever possible (be sure to label column headings), separate the text into paragraphs, numbered if necessary, and use as much space as is needed.

THE REPORT

In all cases, the essential features of the experiment and its results should be summarized at the end of the notebook record. Often this summary, plus the notebook record, will be sufficient to serve as a report, if this is required by the instructor. The essential features of a report include answers to the six questions: What did you do? Why did you do it? What were the results? What conclusions did you draw from these results? How valid or precise are these conclusions? How could you improve the precision or definiteness of the conclusions? A two-page report is usually ample, and a shorter one often is adequate.

COMMON LABORATORY EQUIPMENT

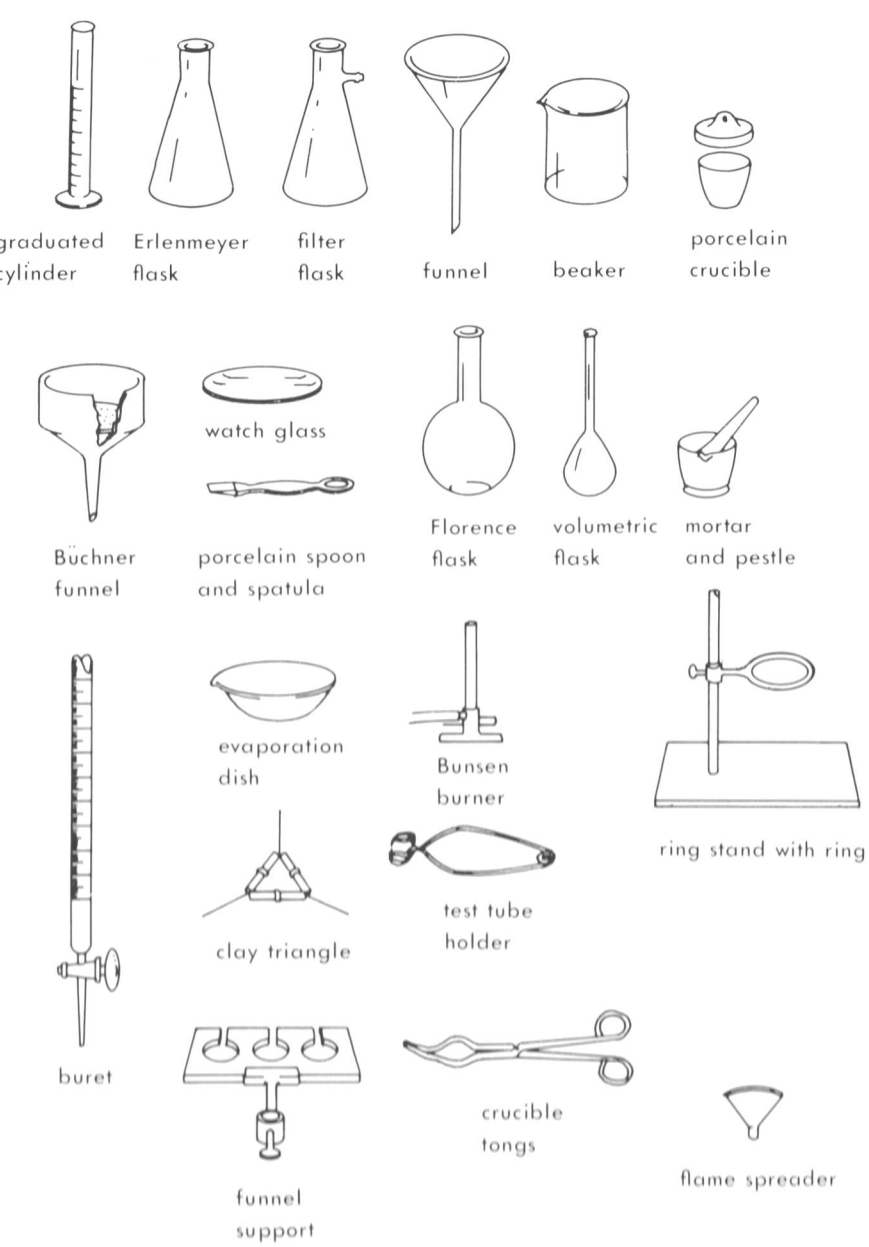

Figure C.1. Laboratory Apparatus.

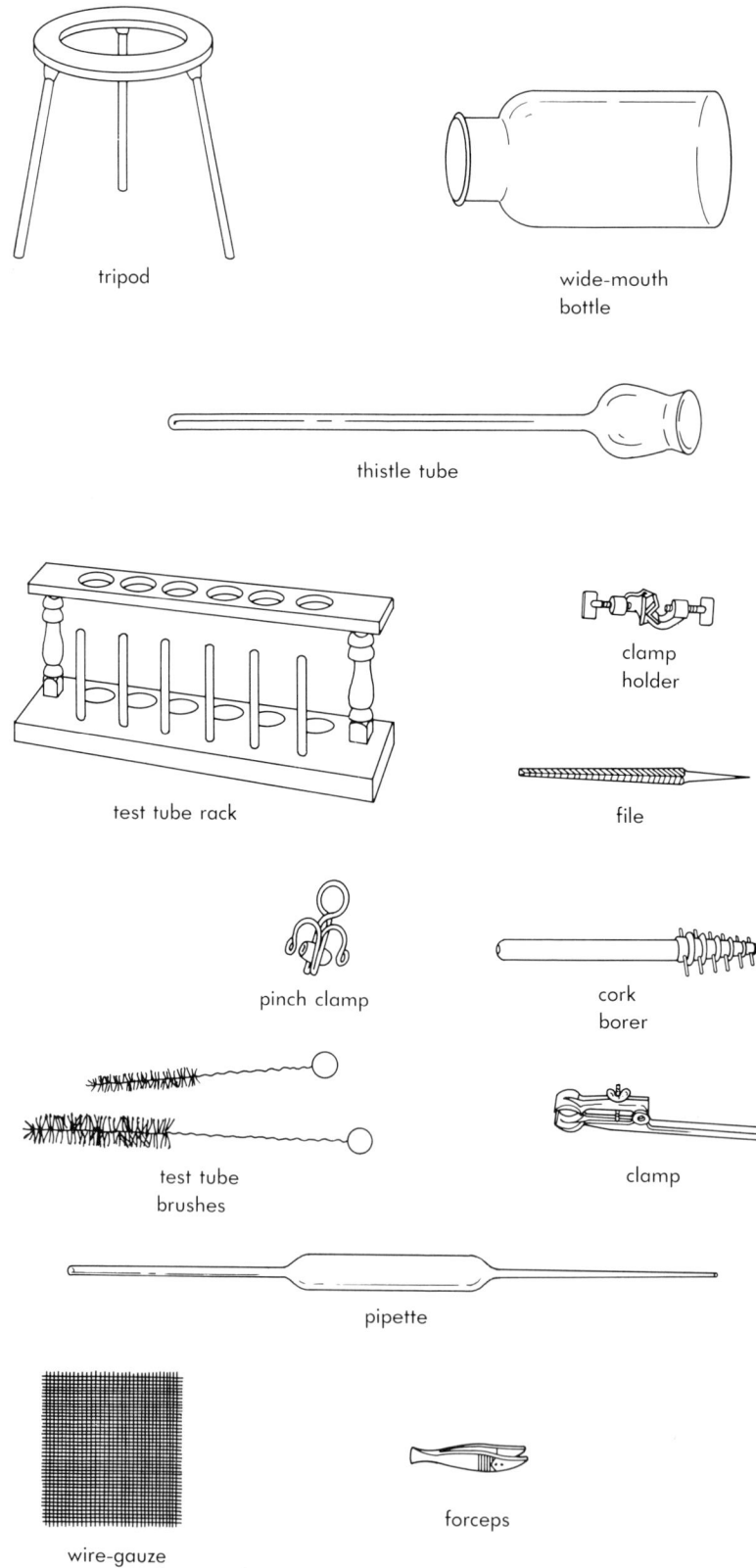

Figure C.2.   Laboratory Apparatus.

CONTENTS

THERMODYNAMICS

INORGANIC SYNTHESES

QUANTITATIVE ASPECTS OF COORDINATION CHEMISTRY

SOLID STATE

INORGANIC QUALITATIVE ANALYSIS

QUANTITATIVE ANALYSIS OF SOLUTIONS

ANALYSIS OF A MIXTURE OF SALTS

ORGANIC AND BIOCHEMISTRY

FOUNDATION

USE OF THE SINGLE-PAN
ANALYTICAL BALANCE

OBJECTIVE

To learn to weigh precisely and to compare the reliability of weighings of the same objects made on balances of different sensitivities.

EQUIPMENT NEEDED

Single-pan analytical balance with a sensitivity of 0.0001 g.; porcelain crucible and lid; crucible tongs; triple-beam balance with a sensitivity of 0.01 g.

REAGENTS NEEDED

None

INTRODUCTION

The measurement of mass is one of the fundamental operations performed in any chemical laboratory. We use balances as mechanical devices for determining the mass of an object; they are constructed in several styles and sizes, and they can measure a variety of mass ranges. Laboratory balances vary from rather rough measuring devices (platform or triple-beam balances) sensitive to 1 or 0.1 g. to precision analytical instruments sensitive to 1 part in 10^{-7} g.

Single-pan analytical balances with a sensitivity of 0.0001 g. are now common. For objects weighing approximately 10 g. this sensitivity means there is an uncertainty in the mass measurement of 1 part in 100,000.

Pictures and specific operating instructions for a typical single-pan balance are given in Appendix 1. All single-pan analytical balances employ the principle of substitution weighing, and utilize an optical lever mechanism to record the small increments of weight ranging from 0.1 mg. to as much as 0.1 g. The balances are simple to operate because no weights are transferred by hand. Instead, the weights are located inside the balance housing and are normally never seen by the operator. The operator manipulates the weights by turning a series of dials on the outside of the balance. Associated with each dial is a scale that records the mass of the weights manipulated. When the weighing is completed, the mass of the object on the pan is read directly from the scales of the balance.

In order to weigh masses accurately to within 0.0001 g., it is necessary to use exacting techniques to insure the safety of the instrument and to obtain maximum reliability and reproducibility in the measurements.

Before using the balance, one should be familiar with the parts of the balance and their functions, the factors that influence the readings, the correct method for weighing materials, and the care of the instrument. (Textbooks in quantitative analysis should be examined for details on the use, theory, and construction of analytical balances. See references 1 and 2.)

Errors in Weighing

Moisture. Condensation of water vapor from the atmosphere onto dry samples, glassware, and precipitates may add several milligrams to the weight of the sample. Precipitates should be weighed rapidly if they are open to the air; a preferable procedure is to weigh the dry sample in a stoppered bottle that has been stored in a desiccator.

Temperature. All weighing should be done at room temperature. Hot objects, such as heated crucibles (Exp. 4), set up convection currents, which tend to push up on the pan of the balance and reduce the apparent weight.

Static Electricity. A static charge can be set up on the surfaces of glass vessels when they are rubbed with a dry cloth, especially in the low humidity conditions of cold, dry winter days. The charge may cause the balance to behave erratically. This effect usually does not present a major problem except with very large vessels.

Buoyancy. According to the principle of Archimedes, a large container will be buoyed up by the air in the balance case. The true weight of the object (i.e., its weight if weighed in a vacuum) will be greater than the weight obtained in air by the net buoyancy on the object. The buoyancy effect on the weights partially cancels the buoyant effect of the air on the object. The net buoyancy on the object being weighed is the difference between the buoyancy experienced by the object (because of its displacement of a volume of air) and that experienced by the weights. The larger the volume of the object and the smaller its density, the greater will be the buoyancy effect. In most cases, buoyancy effects are not significant because the volume and density of the object being weighed are such that they result in an insignificant buoyancy correction. When objects are weighed in large containers, one usually takes the difference between the weight of an empty container and the weight of the container with the object. In such cases, the buoyancy effect of the large container cancels out.

Care of the Balance

Analytical balances are expensive, precision-made instruments and do not function properly if abused. *Each student is responsible for cleanliness, knowledge of proper operating steps, and adherence to the following rules whenever he uses the balance.*

1. Never move the balance. Check to see that the balance is level, as indicated by the leveling bubble. If it is not level, ask the instructor for assistance.
2. Be sure that the beam arrestment control is in the arrest position before placing the object to be weighed on or removing it from the pan.
3. Always move the arrestment control gently. Avoid a sudden jar, which might damage the knife edges.
4. Always close the balance case before adjusting the weights. Air currents must be avoided.
5. Do not overload the balance. The balances in the laboratory are designed for maximum loads ranging from approximately 100 to 200 g., as stated by the manufacturer. (See your instructor for specific operating instructions for the analytical balance that you are using.)
6. Never weigh any chemical or moist object directly on the balance pan. Clean up any material that is spilled on the pans or within the balance case *immediately*, making sure the beam arrestment control is in the arrest position.
7. Never place a hot object on the balance pan; its apparent weight will be incorrect because of convection currents set up by the rise of heated air.
8. Avoid using the fingers to handle objects to be weighed. Do not touch the balance pan or the pan support. Care must be taken to prevent weight changes caused by absorption of moisture or oil from the hand.
9. When weighing is completed, return the arrestment control to the full-arrest position, remove all objects from the pans, and close the balance case.

Steps in Weighing

The procedure for weighing consists of four major operations: Checking the zero point to be sure the balance is properly adjusted; placing the object to be weighed on the pan; setting the weights by using the appropriate knobs; and reading the mass of the object from the appropriate dials and scales.

1. *Checking the zero point.* This must be done each time the balance is used. (Start this step only after instructions from the laboratory instructor.)
   a. Unload the pan. Clean it if necessary, using a camel's-hair brush.
   b. Close the windows.
   c. Set all weights to zero.
   d. Slowly turn the beam arrestment control to the fully released position.
   e. When the optical scale is at rest, use the zero adjustment knob to set the scale 0-line to coincide with the pointer.
2. *Placing the object on the pan.* The object to be weighed may be placed on the pan only when the balance is completely arrested. Whenever possible, use a pair of tongs or tweezers to avoid transmitting moisture and oil from the hands to the object to be weighed.

After introducing the object to be weighed, close the window immediately.

3. *Setting the weights.*
   a. Set the beam arrestment control in the partial-release position.
   b. Estimate the weight of the object and begin dialing the weights, removing the large increments of weight first. The procedure for obtaining the final weight setting varies with the type of balance used. Obtain the specific instructions for your balance from your instructor.
   c. After setting the 1- and 10-g. weights, return the beam to the arrest position.
   d. Fully release the beam and allow the optical scale to come to rest. Remove your hands from the balance and balance table.

4. *Reading the mass.*
   a. Read, making any adjustments of the micrometer or digital control that are necessary.
   b. *Arrest the beam,* and remove the object; return the weight settings to zero; then close the doors. Check to see if you have accidentally spilled any material on or in the balance; if you have, clean it up before leaving.

Since a major portion of your laboratory results will depend upon your ability to make rapid and valid measurements of mass on the analytical balance, it is essential that you learn to use the instrument as soon as possible. Your instructor will show you how to use the balance during the first and second laboratory periods. You will then be asked to determine the mass of an object. If you experience difficulty in making the weighing, ask your instructor for additional help.

NOTEBOOK AND REPORT

Before going to the laboratory, carefully read the steps in weighing and the rules for the care of the balance.

Prepare a data table in your laboratory notebook in which to record the results of your weighings on the two different types of balances.

PROCEDURE

Determine the mass of crucible and lid on triple-beam balance. Determine on the laboratory triple-beam balance the mass of a porcelain crucible and its lid separately and then together. Does the total weight equal the sum of the component weights? Repeat the complete weighing sequence four times and record the values. Determine the average deviations (p. 443) to obtain a measure of the precision of this type of balance.

Determine the mass on an analytical balance. Using the general rules and sequences of weighing steps listed previously and the specific operating instructions for your type of analytical balance, determine the mass of the crucible, its lid, and the combination five times on the single-pan analytical balance. From these data, calculate the average deviations, and compare the precision with that of the triple-beam balance. Ignoring all sources of error except the limitations of the balances, indicate the reliability of the determinations of mass from each of the balances.

QUESTIONS—PROBLEMS

1. Setting the zero point of the analytical balance is, in effect, adjusting the beam of the balance so that its pointer rests in the center of the optical scale. When a weighing is completed, the beam is brought back to its zero point position. What error will be introduced if you fail to set the zero point of the balance before making a reading?

2. Supposing the micrometer scale of the analytical balance were broken but the other scales were still usable, what would be the reliability of the mass measurement made on the balance under these conditions? Supposing the 1-g. control were broken, what reliability could be obtained? Explain your answers.

3. Under what situations might it be more appropriate to weigh something on the triple-beam balance than on the analytical balance?

4. The density of an object is its mass divided by its volume. In determining the density of an object, suppose the volume were determined first and found to be 23.2 ml. Could the mass be determined on the triple-beam balance without decreasing the reliability of the density determination? Justify your answer.

REFERENCES

1. Fisher, R.B. and Peters, D.G.: *A Brief Introduction to Quantitative Chemical Analysis.* W.B. Saunders Co., Philadelphia, 1969.
2. Skoog, D.A., and West, D.M.: *Fundamentals of Analytical Chemistry.* 2nd. edition. Holt, Rinehart, and Winston, New York, 1969, pp. 63-119.

Experiment 1
REPORT SHEET

Name _____ Section No. _____

Data

A. Weighings on Triple Beam Balance

Trial

	1	2	3	4	5	Average Value	Average Deviation
Crucible							
Lid							
Sum of Above							
Crucible and Lid							

B. Weighings on Analytical Balance

Trial

	1	2	3	4	5	Average Value	Average Deviation
Crucible							
Lid							
Sum of Above							
Crucible and Lid							

Answers to the Questions—Problems

SEPARATION TECHNIQUES AND IDENTIFICATION

THE SEPARATION AND QUANTITATIVE DETERMINATION OF THE THREE COMPONENTS IN A MIXTURE

OBJECTIVE

To illustrate the separation of the components of a mixture on the basis of their differences in solubility in different solvents.

EQUIPMENT NEEDED

One 250-ml. suction filter flask and trap bottle; 3-inch Büchner funnel; filter paper to fit Büchner funnel; clamp and iron ring stand; stirring rod; balance (accurate to 0.001 g.); aspirator; two beakers (approximately 100 ml.); one 125 ml. Erlenmeyer flask; medicine dropper or 3″ of 8-mm. glass tubing; rubber hose; flexible flat metal spatula or rubber policeman; burner; desiccator; plastic squeeze bottle.

REAGENTS NEEDED

Unknown mixtures containing one organic and two inorganic components, ~6 g./student; dichloromethane; distilled water; 3 M nitric acid, HNO_3.

INTRODUCTION

When two or more substances that do not react chemically with each other are present in a sample, the result is a mixture in which each component of the mixture retains its own physical and chemical properties. The separation of mixtures and the qualitative and quantitative determination of components are problems often facing a chemist. For example, in chemical research, in industrial manufacturing and in testing laboratories, the separation and identification of components of mixtures on a routine basis can often mean the difference between success and failure for a new process or between life and death in hospitals.

Each component of a mixture is an element or a compound which has fundamental properties such as physical appearance, solubility, density, melting point, and boiling point. A pure substance will melt at a definite temperature, and if the external pressure is fixed, it will boil at a different definite temperature. Also the substance has a specific solubility: a certain

13

quantity of the pure component will dissolve in a fixed amount (e.g., 100 ml.) of a given solvent, at a constant temperature.

Some of the physical methods often used by a chemist in the separation of mixtures are *decantation, filtration, crystallization, distillation,* and *sublimation.* We shall use filtration and crystallization in this experiment.

We shall be concerned with the separation of a mixture of three solids into its components, and with the determination of the percentage composition of the mixture. The separation can be accomplished by taking advantage of the large differences in solubility of a compound in various solvents. The percentage composition of the mixture will be determined by weighing a sample of the mixture, by carefully separating the mixture into its components, and by collecting and weighing the components.

Each component of the mixture can be further analyzed qualitatively to identify the specific ions or molecules present. Such specific tests and procedures are given in Experiment 3. If your instructor has assigned Experiment 3, be sure to save each component of your mixture in a labeled container to use later.

TECHNIQUES

Since the objectives of this experiment are to separate the components and to determine quantitatively the percentage composition of the components, filtration and weighing are important procedures to be accomplished carefully. Review the procedure and operating instructions for the analytical balance given in Experiment 1 before starting this experiment.

Suction Filtration. Separation of a solid from a slurry of solid and liquid by filtration is a frequent task in the chemistry laboratory. The process can often be speeded up by applying suction to draw the liquid (the *filtrate*) through the filtering medium (commonly porous paper, *filter paper*). A convenient set-up is shown in Figure 2-1. A circle of filter paper of

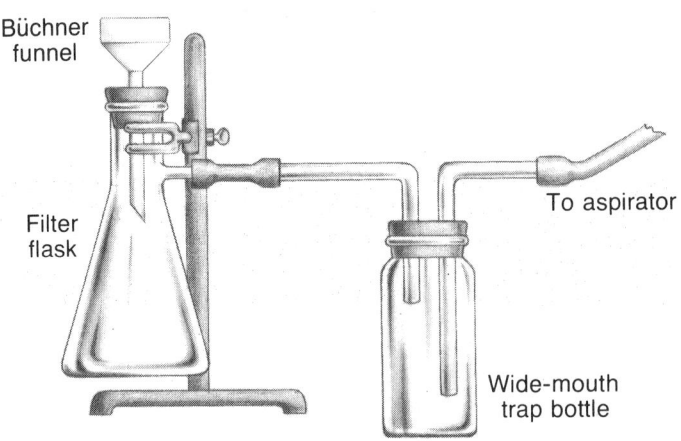

Büchner funnel

Filter flask

To aspirator

Wide-mouth trap bottle

FIGURE   2-1

Set-up of apparatus for suction filtration.

a size that will lie flat on the bottom of the Büchner funnel and cover all the holes in the filter plate holds the solid back while the liquid is pulled through. The paper is moistened after placing it in the funnel. Note that the slanted end of the funnel is turned away from the sidearm on the suction flask so that emerging liquid drops are not drawn into the sidearm. The trap bottle prevents water from the aspirator from being sucked back into the filtrate in case a sudden decrease in water pressure should occur. The water for the aspirator should be turned on before liquid is poured on the filter, and filtration should be stopped by disconnecting a hose or removing a stopper, not by turning off the aspirator. (Why?)

NOTEBOOK AND REPORT

Your notebook record of this experiment should include, besides the usual statement of purpose, a brief outline of procedure, *i.e.*, which steps are to be accomplished in sequence and which are to be undertaken while waiting for filtration or other slow steps, calculations of the percentage composition, answers to the questions, and a table to show the raw data collected. These data should show weights of empty and filled containers, net weights of the components, and volumes of solvents used. It is always advisable to prepare a skeleton table in your notebook before you go to the laboratory, so you will have an organized manner in which to collect the data and so that your notebook record will be intelligible to you and your instructor at a later date.

PROCEDURE

A. Before coming to the laboratory, complete the chart in the Report Sheet by referring to a Chemical Rubber Company *Handbook of Chemistry and Physics*, Lange's *Handbook of Chemistry*, other chemistry handbooks, and to your text.

B. *Examination of your Mixture*

1. Physical Appearance. Record in your notebook any aspects of your unknown sample that can be described, such as color, odor, size and shape of the crystals, and heterogeneous or homogeneous character.

Record physical properties of your unknown sample.

2. Solubility. Solubility is defined as the quantity of one substance that will dissolve in another substance (usually a liquid) at a given temperature, forming a mixture that is both transparent and uniform in appearance. The resultant solution may be called a *homogeneous mixture.*

The separation scheme to be used is based on the fact that certain compounds are very soluble in organic solvents such as benzene or dichloromethane whereas they may be essentially insoluble in water. Also, other compounds may be quite soluble in water but not in dichloromethane or benzene. Your unknown sample will be a mixture containing (1) an organic compound that is soluble in dichloromethane but insoluble in water, (2) a simple, soluble inorganic salt containing one cation and one anion, and (3) a metal carbonate that is insoluble in water and reacts with acid solutions.

Examine the separation scheme below carefully and prepare a step-wise outline in your notebook to be followed in the laboratory.

Scheme to Separate the Three-Component Mixture

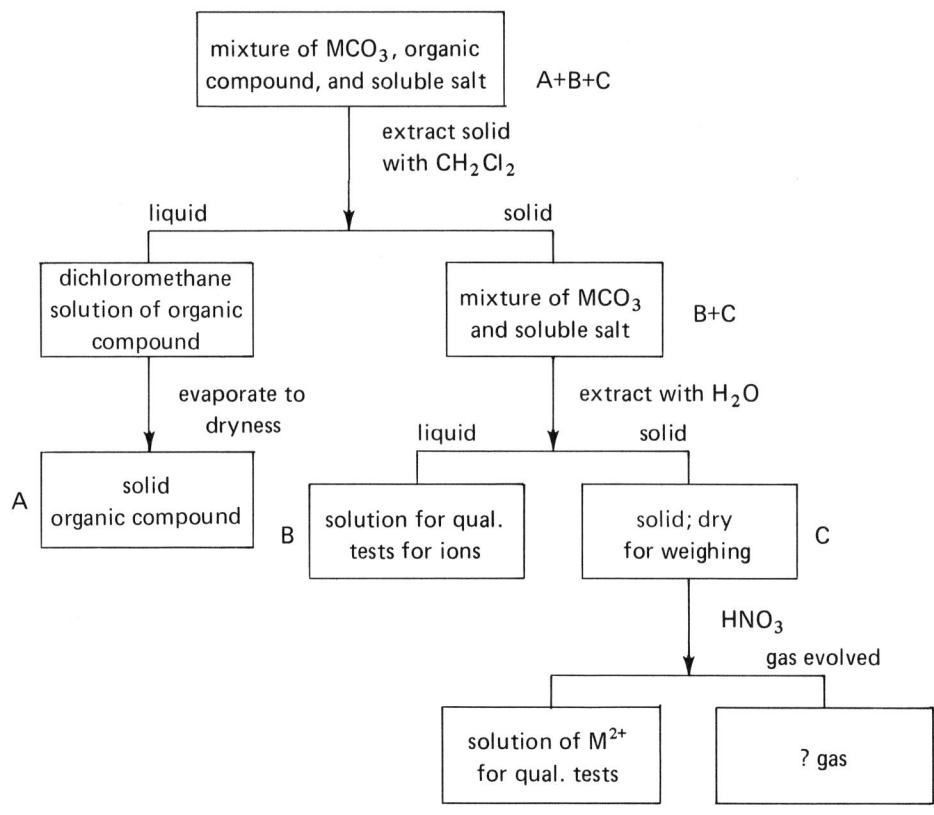

3. *Extraction with Dichloromethane.* Weigh about 6 grams (to 0.001 g.) of your unknown sample, and transfer the solid to a 100-ml. beaker. Add 25 ml. of dichloromethane and stir the mixture for 3 minutes, being **Weigh sample and** careful not to spill or splash out the solution. Obtain a piece of filter paper **treat with CH₂Cl₂.** to fit the Büchner funnel and assemble the suction filtration apparatus as shown in Figure 2-1. Place the filter paper flat in the funnel and wet it with a little dichloromethane. Turn on the aspirator *just enough* to suck the wet **Filter with gentle** filter paper snugly against the Büchner funnel. CAUTION: If the suction is **suction.** very hard, you may tear a hole in the filter paper or you may pull the finely divided solid materials into the pores of the filter paper, thereby making the filtration go abnormally slowly. Decant most of the liquid in the beaker onto the filter as you maintain a very gentle suction. Add 25 ml. of dichloro- **Repeat.** methane to the solid, most of which should have remained in the beaker, and stir for 3 minutes. Again pour the liquid onto the filter paper and collect the filtrate in the filter flask.

Weigh (to 0.001 g.) a 100-ml. beaker and then pour the filtrate from the filter flask into the beaker for evaporation. The dichloromethane can be evaporated fairly rapidly by directing a stream of air across the surface of the beaker as shown in Figure 2-2. As dichloromethane vapors are somewhat toxic, the evaporation should be performed in a hood if possible. The stream of air can be a compressed air line source that is filtered with a loose cotton plug (Figure 2-2A) or it can be created by a vacuum pump or an aspirator **Evaporate the** (Figure 2-2B). Evaporate the dichloromethane while you perform the next **CH₂Cl₂ filtrate.** extraction. Remove all or nearly all of the dichloromethane. Record the

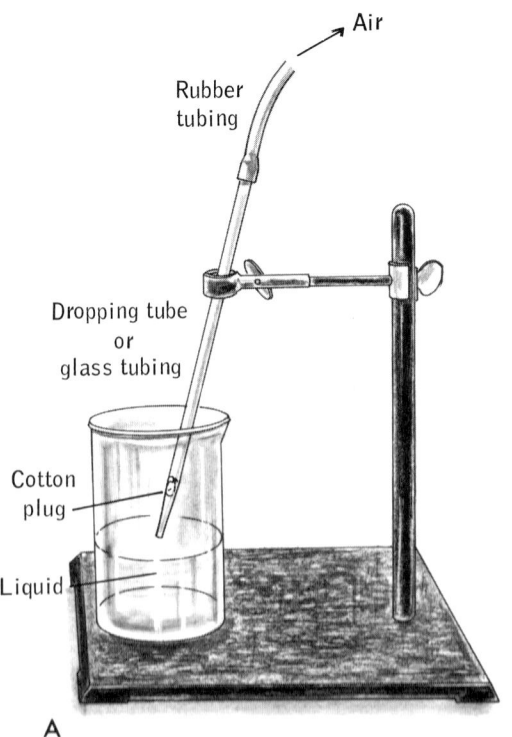

Air

Rubber
tubing

Dropping tube
or
glass tubing

Cotton
plug

Liquid

A

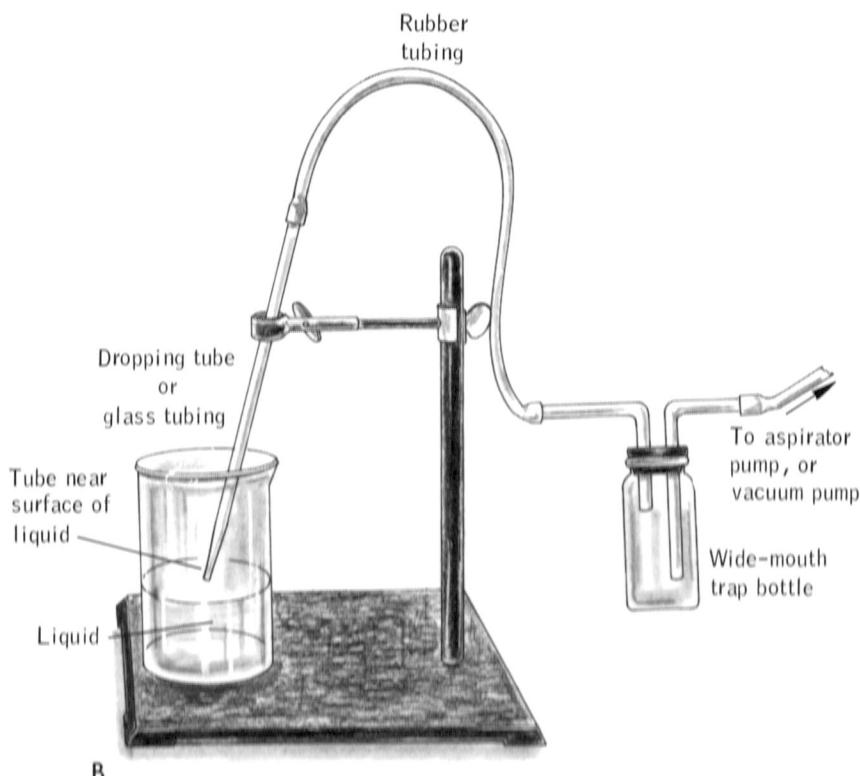

Rubber
tubing

Dropping tube
or
glass tubing

Tube near
surface of
liquid

Liquid

To aspirator
pump, or
vacuum pump

Wide-mouth
trap bottle

B

FIGURE 2-2

Illustration of apparatus to evaporate the dichloromethane; (A) for an air stream; (B) for a
vacuum.

colors and appearances of the solid and solution. At the end of the present laboratory period or at the next one, weigh the beaker that had contained the dichloromethane solution to determine, by difference, the weight of the solid that was extracted into dichloromethane. Is the residue a crystalline solid, or does it have a viscous, gummy appearance? If it is gummy or an oil, obtain some suggestions for recrystallization solvents from your instructor.

Treat material insoluble in CH$_2$Cl$_2$ with H$_2$O.

4. Extraction with Water. The solid residue on the filter paper should be scraped off and added to that in the first beaker. Add 25 ml. of *distilled* water and stir the mixture moderately for 3 minutes. Assemble the filtration apparatus, as shown in Figure 2-1, swirl or stir the mixture of solid and liquid, and pour onto the filter paper using a glass rod to guide the flow, as shown in Figure 2-3. Do not allow the level of liquid in the filter funnel to come higher than one-half inch from the top of the funnel. Collect the filtrate and transfer it to a 125 ml. flask. Then transfer the remaining solid in the beaker to the filter by washing with 10-15 ml. of water from the squeeze bottle as shown in Figure 2-4. Using a rubber policeman, a flexible metal spatula, or some other thin, flat object, gently lift the filter paper from the funnel and dry the solid on the filter paper in the desiccator until the next laboratory period. Use the aqueous filtrate for the qualitative tests in Experiment 3. Record the colors and appearances of the solid and filtrate.

Filter, collecting the filtrate.

Save filtrate for Expt. 3.

Dry the solid; obtain its weight.

At the next laboratory period weigh the dry filter paper containing the remaining solid, then carefully scrape the dry solid into a clean 100-ml. beaker, and reweigh the filter paper to determine (by difference) the amount of the mixture that was not soluble in either dichloromethane or water.

How can you easily obtain the weight of the substance that was soluble in water?

5. Dissolving the Remaining Solid in Nitric Acid. After you have obtained the weight of solid C, add a few ml. of water to make a paste with solid C. Then carefully add 10 ml. of 3 M nitric acid; control the addition rate so that no foam or bubbles overflow the beaker. After the nitric acid is completely added, gently heat the beaker until essentially all of the material has dissolved and the bubbling has stopped. If the solution is not fairly transparent, you should filter it before performing the specific ion tests in Experiment 3.

Dissolve weighed solid in HNO$_3$.

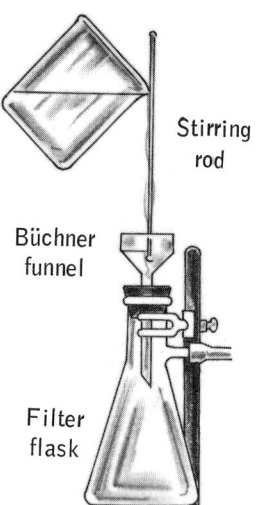

Stirring rod

Büchner funnel

Filter flask

FIGURE 2-3

Filtration by suction. Transfer of the liquid to the funnel.

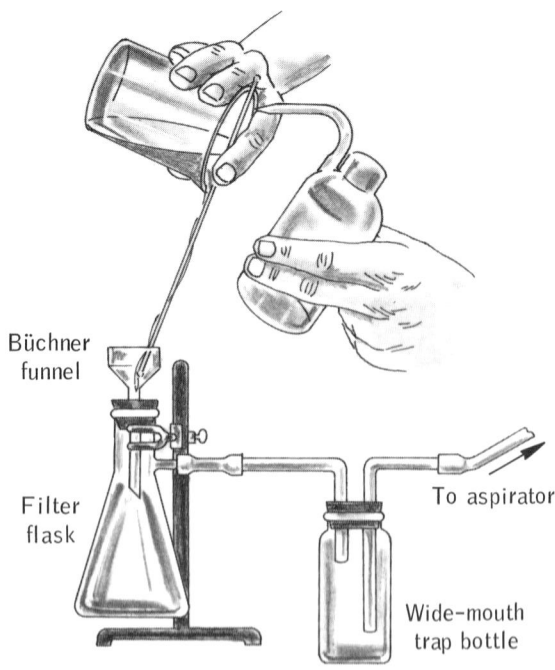

FIGURE 2-4

Filtration by suction. Transfer of remaining solid to the funnel by washing.

QUESTIONS

1. How would you separate iodine from sodium chloride? Sodium carbonate from p-dichlorobenzene?

2. From the data in part A of your Report Sheet, what relationships exist between the melting points of the substances and their solubility in water? How are the melting points related to the solubility in organic solvents such as ethyl alcohol?

3. Do the compounds that are insoluble in water dissolve in nitric acid because they are colored? What type of test or experiment would you propose to examine the validity of this statment?

4. What chemical inference can you make from the gas evolution observed in Procedure B.5?

Attach sheets with answers to assigned questions.

Experiment 2
REPORT SHEET

Name _____ Section No. _____

A. *Handbook Data*

Substance	Melting Point	Boiling Point	Does the Solid Sublime?	Solubility in Water	Solubility in Ethyl Alcohol
sodium chloride, NaCl					
iodine, I_2	113.5°C				
p-dichlorobenzene, $C_6H_4Cl_2$					
sodium carbonate, Na_2CO_3					
lead carbonate, $PbCO_3$					
dichloromethane, CH_2Cl_2					
acetonitrile, CH_3CN					

B. *Your Mixture*

Physical Appearance

Which components of your unknown mixture, on basis of physical appearance, are soluble in which solvents?

Weights of the Components A, B, and C.

Percentage Composition of the Mixture in Terms of Components A, B, and C.

QUALITATIVE IDENTIFICATION OF THE COMPONENTS OF A MIXTURE

OBJECTIVE

To identify the organic component of the mixture, to become familiar with qualitative analytical tests for specific ions, and to identify the inorganic ions present in the 3-component mixture.

EQUIPMENT NEEDED

250-ml. beaker; thermometer; six 16 X 150 mm. test tubes; 100-ml. beaker; rubber bands; six capillary melting-point tubes; micro spatula; watch glass; burner; cobalt glass and platinum (or Nichrome) wire for flame tests; centrifuge.

REAGENTS NEEDED

Small bottles of each compound in Table 3-1; small dropper reagent bottles containing 0.1 M nitrate solutions of iron(III) (Fe^{3+}), cobalt (Co^{2+}), nickel (Ni^{2+}), copper (Cu^{2+}), and magnesium (Mg^{2+}) ions; dropper bottles containing 0.1M solutions of ammonium chloride (NH_4Cl), sodium chloride (NaCl), potassium chloride (KCl), potassium bromide (KBr), and potassium iodide (KI); dropper bottles containing 0.01 M solutions of sodium chloride (NaCl), calcium chloride ($CaCl_2$), potassium nitrate (KNO_3), and potassium chloride (KCl); dropper bottles containing 1 M solutions of potassium thiocyanate (KNCS) and ammonium thiocyanate (NH_4NCS); sodium fluoride (NaF); concentrated nitric (HNO_3), hydrochloric (HCl), hydrobromic (HBr), and sulfuric (H_2SO_4) acids; 3 M HCl; 3 M HNO_3; 6 M HCl; 4 M sodium hydroxide (NaOH); concentrated and 2 M ammonia ($NH_3 \cdot H_2O$) or (NH_4OH); 1 M barium nitrate ($Ba(NO_3)_2$); 10% barium chloride solution; 1% solution of dimethylglyoxime in alcohol; EDTA-NaOH solution; 0.1% thiazole yellow solution; lime water ($Ca(OH)_2$); 0.1 M sodium sulfate (Na_2SO_4); and 0.1 M silver nitrate ($AgNO_3$); diphenylamine-sulfuric acid solution; sodium tetraphenylborate; saturated solution of silver sulfate; unknown mixtures.

INTRODUCTION

After separating a mixture into its major components on the basis of some physical property such as solubility in different solvents, melting point, or vapor pressure, chemists often need to identify specifically each component. In this experiment we shall be interested in determining the identity of the organic component (*i.e.*, the component that was soluble in dichloro-methane) and the identity of the two components that were inorganic salts.

A pure compound has a definite melting point and boiling point (Note 1). Thus, the investigator can often eliminate 90-95% of the possible compounds by simply comparing the melting point or boiling point of the substance with a list of known melting points and boiling points. For solids, determination of the melting point is an easier experiment. On the basis of the experimental melting point and the fact that your compound will be one of those in Table 3-1, suppose that you are able to narrow the number of choices to two or three. Then how could you proceed to obtain a specific identification?

Chemists often resort to one or more of the following types of informa-tion: quantitative elemental analysis to obtain the percentage composition of the compound for a match with the calculated values; infrared and/or ultra-violet absorption spectroscopy to examine specific structural features of the molecule; and mixed melting point. In this experiment you are to utilize data from each of these methods to identify the organic component of your unknown mixture. One chemist may not perform all of these determinations himself; instead, he (or she) often does some of the measure-ments and then relies on another person to perform a specialized determina-tion, *e.g.*, the quantitative percentage composition of the compound. As a way of illustrating this division of labor, you will determine the melting point of the unknown compound (within an acceptable range) and then your instructor will provide an *experimental* percentage carbon and the principal infrared absorption peaks for your compound.

The inorganic components of the mixture will be identified by taking advantage of qualitative "wet-chemistry" tests that are specific for the ions.

TECHNIQUES

1. **Melting Point.** The melting behavior of a compound is often a simple test of its purity, because an impure compound melts over a range of temperatures and at a lower temperature than does the pure compound (with exceptions). Consequently, the chemist is frequently called upon to

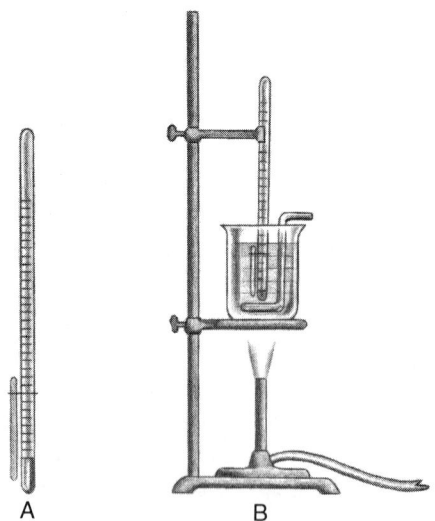

FIGURE 3-1

Simple apparatus for determining melting points.

A   B

determine the melting point and melting range of compounds he prepares in the laboratory.

To determine a melting point, a small quantity (0.0001 to 0.0002 g.) of the powdered compound to be examined is placed in a melting point capillary, and the capillary is fastened to a thermometer by use of a rubber band or a slice of rubber tubing (Fig. 3-1A). The thermometer and capillary are then inserted to the immersion mark into a beaker containing a liquid that is heated gradually with constant stirring, and the material in the capillary is watched carefully (Fig. 3-1B).

FIGURE 3-2

Melting point tube, Thiele.

The thermometer is read at the first sign of collapse of the solid material in the capillary, and again at the exact temperature at which the last of the solid disappears. For substances with unknown melting points, it is often convenient to make a preliminary determination by rapidly heating the bath liquid to obtain an approximate melting point, letting the bath cool a few degrees, and heating the bath much more slowly for the accurate determination of the melting point of the compound, using a second capillary. In place of the beaker and stirrer, a Thiele melting point tube may be used (Fig. 3-2). Heating the side arm with a small flame produces convection currents in the liquid, which heat the thermometer slowly and evenly.

To fill the melting point capillary, crush a small portion of the crystalline sample on the watch glass with a spatula, scrape the powder into a mound, and scoop it into the capillary by pushing it against the spatula. Shake the material down by gently tapping the capillary, supported in an upright position, on a hard surface. Add material until you have a column equivalent to one-half to three-fourths the length of the thermometer bulb tightly packed in the bottom of the capillary.

2. **Mixed Melting Point.** The mixed melting point is particularly useful for identification of organic compounds, as the technique is simple and the determination is rapid. The technique simply involves mixing (by grinding thoroughly) approximately an equal amount (~10 mg.) of the unknown compound with each of 3-4 known possible compounds that have melting points close to the unknown and determining the melting points of the resultant 3-4 mixtures. If the unknown is the same as one of the known

compounds, the mixed material will still be a pure substance and it will melt at the characteristic melting point of the compound. However, if the mixed material contains two different compounds, the melting point of the mixture usually will be lower than for either of the individual pure components. That is, an impurity depresses the melting point of a compound. Thus, by comparing the melting points of the different mixtures with the melting point of the pure unknown compound, one may determine which added compound gives no depression of the melting point.

PROCEDURE

1. Melting Points

For practical reasons the organic compound to be identified in this experiment will be limited to those in Table 3-1. Consequently, you should be able to identify the compound with a thermometer that is accurate to one or two degrees. Depending on the time allotted for this experiment and your instructor's directions, an alternative procedure to a complete calibration[1] would be to determine the melting point of your compound, and then check the accuracy of your thermometer by choosing one of the known compounds in Table I that melts close to your unknown and that has a narrow melting point range.

Set up the apparatus shown in Figure 3-1 or obtain a Thiele melting point tube and determine the melting point of your unknown organic compound, as described under Techniques (p. 24). Remember that it is often faster to make an approximate determination of the melting point by heating the m.p. bath rapidly with a first sample in a capillary until the solid melts, letting the bath cool a few degrees, and then determining the melting point of the compound in a second capillary by heating the bath much more slowly, i.e., about one degree per minute. Use a new capillary melting point tube for each measurement. After you have determined the melting point, examine Table 3-1 for all of the compounds with melting points within five degrees of your compound. Obtain very small amounts (enough to cover the end of a small spatula) of those compounds, grind approximately an equal amount of your unknown with each of these compounds, and then determine the "mixed melting points" of these mixtures. Recalling that any different compound will depress the melting point of your unknown compound, you should now have nearly a unique identification of the organic component of your original three-component mixture from Experiment 2.

2. Analytical and Infrared Data

Now ask your instructor to give you the percentage carbon and the major infrared peaks for the organic compound in your unknown. Compare the carbon percentage with the calculated value for your compound and with calculated %C values of all the compounds that you used for the mixed

---

[1] Instructions and procedure for calibration of a thermometer are given in Appendix IX.

melting point determinations (Note 2). Is the % carbon consistent with your identification based on the results of the mixed melting points? Read the discussion about the use of infrared peaks in Experiment 32 and Appendix VIII-A. From the infrared peaks determine the possible structural or functional groups in your compound and compare these with the structural formula of your compound in Table 3–1. Do the infrared peaks confirm your other data? On the basis of the melting point, mixed melting point, percentage carbon, and infrared peaks, there should be only one possibility for the organic compound, so you have arrived at a positive identification.

3. Qualitative Tests on the Aqueous Filtrate, *i.e:*
Component B

We shall be interested in determining the chemical species that were extracted with water, *i.e.,* the ions that were quite soluble in water at room temperature. Take 1 ml. of the filtrate (obtained by extraction with water) in a test tube for each of the specific tests given on pages 27 to 32. Since the unknown contains one soluble simple salt, it should be the source of only one cation and one anion. When you have identified one of each, no further tests are necessary.

4. Qualitative Tests on the Acid Solution, *i.e.*
Component C

Repeat (or perform the tests concurrently) the specific tests given on the following pages for identification of the cation in the nitric acid solution, which was obtained from component C of the mixture. Use 1-ml. samples of the filtrate for each test.

QUALITATIVE TESTS FOR CATIONS

A. Test for Iron(III)

As a test for iron(III) in a known solution, place one drop of 0.1 M $Fe(NO_3)_3$ in a test tube, add one drop of concentrated hydrochloric acid and dilute with 1 ml of distilled water. Then, add one drop of 1 M potassium thiocyanate (KNCS). A red color indicates the presence of iron(III). Now, add a small crystal (about 50 mg.) of sodium fluoride. The red color should disappear. The formation of the red color is a positive test for iron(III) and the disappearance of this color upon the addition of fluoride *confirms* this test. The equations for these reactions are:

$$Fe^{3+} + NCS^- \rightleftharpoons Fe(NCS)^{2+} \qquad\qquad (1)$$
$$\text{red}$$

$$Fe(NCS)^{2+} + 6F^- \rightarrow FeF_6{}^{3-} + NCS^- \qquad\qquad (2)$$
$$\text{colorless}$$

The reaction given by equation (1) forms the basis of a quantitative spectrophotometric determination of iron(III) (see Experiment 23).

To show the sensitivity of this reaction, dilute one drop of 0.1 M $Fe(NO_3)_3$ to 5 ml. with H_2O. Then take one drop of the diluted $Fe(NO_3)_3$ and perform the above test. Perhaps you will want to determine the smallest concentration of iron(III) that will still give a positive test.

B. Test for Cobalt(II)

Cobalt(II) salts dissolve in water to give light pink solutions of $[Co(H_2O)_6]^{2+}$. As a test for Co(II) in a known solution, place one drop of 0.1 M $Co(NO_3)_2$ in a test tube and add 1 ml. of distilled water. Add 5 drops of 1 M NH_4SCN and 1 ml. of acetone to the solution and mix well. The appearance of an intense blue color, due to $[Co(NCS)_4]^{2-}$, confirms the presence of Co^{2+}.

$$Co^{2+} + 4\ NCS^- = [Co(NCS)_4]^{2-} \tag{3}$$

C. Test for Nickel(II)

Nickel ions give light green solutions containing $[Ni(H_2O)_6]^{2+}$. A very specific test for Ni^{2+} involves the use of the organic compound dimethylglyoxime. To 1 drop of 0.1 M $Ni(NO_3)_2$ add 5 ml. of water, 10 drops of a 1% solution of dimethylglyoxime dissolved in alcohol, and 3 drops of concentrated ammonium hydroxide. The formation of a red precipitate is a positive test for nickel(II). If H_2D is used to represent dimethylglyoxime, the red nickel precipitate has the composition of $Ni(HD)_2$. The equation for the reaction is:

$$Ni^{2+} + 2H_2D \rightarrow Ni(HD)_2 + 2H^+ \tag{4}$$
$$\text{red}$$

D. Test for Copper(II)

To test for Cu(II), dilute one drop of 0.1 M $Cu(NO_3)_2$ to 1 ml. with distilled water and then observe the intensity of the blue color. Now, add 3 to 5 drops of concentrated ammonium hydroxide and again observe the color. In the presence of excess ammonia, copper(II) forms the $[Cu(NH_3)_4]^{2+}$ ion, which is more intensely colored than the $[Cu(H_2O)_6]^{2+}$ ion. The reaction is illustrated by Equation 5.

$$[Cu(H_2O)_6]^{2+} + 4NH_3 \rightleftharpoons [Cu(NH_3)_4]^{2+} + 6H_2O \tag{5}$$

A more specific test for Cu(II) in dilute solutions involves formation of the violet ion $CuBr_4{}^{2-}$ (Eq. 6). Dilute one drop of 0.1 M $Cu(NO_3)_2$ to 10 ml. and transfer about 1 ml. of this solution to a test tube. Add 5 ml. of concentrated hydrobromic acid. A light purple color indicates the presence of Cu(II). Very few of the commonly occurring ions interfere with this test for copper(II).

$$[Cu(H_2O)_6]^{2+} + 4Br^- \rightleftharpoons [CuBr_4]^{2-} + 6H_2O \tag{6}$$

→ **E. Test for Magnesium** Positive "C"

Place one drop of 0.1 M $Mg(NO_3)_2$ in a test tube, add 3 ml. of water and 1 ml. of EDTA-NaOH solution. Heat the solution in boiling water for a maximum of one minute. Add 10 drops of 0.1% thiazole yellow and divide the solution, which should be perfectly clear, into two equal parts. To one part of the solution, add 1 ml. of saturated barium nitrate. Add 1 ml. of water to the other one. If Mg(II) is present, the solution to which water is added is yellow or light orange, and the solution to which barium nitrate is added is red or contains a red precipitate.

F. Test for Sodium

Sodium ion is usually detected directly in solution by means of a flame test. If a clean platinum or Nichrome wire is dipped into a solution containing sodium ions and then held in a burner flame, the characteristic yellow flame of sodium is observed. This test for Na^+ is very sensitive but unfortunately some other ions also impart a yellowish color to the flame. Thus, it is very easy to assume the presence of Na^+ when it is actually not present.

Pour 2 to 3 ml. of concentrated nitric acid (HNO_3) into a test tube and dip the wire into the acid. Hold the wire in the hottest part of a flame until the flame shows no color except the red-hot wire. You can hasten this operation by repeating the nitric acid treatment of the wire after each five to ten seconds of heating.

After the wire is clean, dip no more than 1 cm of the wire into a 0.01 M NaCl solution and hold the wire in the hottest part of the gas flame. How many seconds does the yellow coloration persist with the known? Again, be sure the wire is clean and dip the wire into 2 ml. of your filtrate. Obtain a piece of cobalt glass and again observe the sodium flame through this glass. When the flame test shows a *persistent* yellow color that is *completely* absorbed by cobalt glass, the presence of Na^+ is indicated.

After the platinum wire is again cleaned, dip it into 0.01 M $CaCl_2$ and heat it in the flame. Be sure to observe the calcium flame through cobalt glass. Can you distinguish the color of the flame in this case from that obtained from sodium? Would calcium interfere with the detection of sodium by the flame test?

G. Test for Potassium

Try the flame test described in F on 0.01 M KCl. Observe the potassium flame through the cobalt glass. Is the potassium flame test as sensitive as the sodium flame test?

Another method for detecting the potassium ion is to use the large specific precipitation ion tetraphenylborate, $(C_6H_5)_4B^-$. To make the test on a known K^+ solution, dilute 5 drops of 0.1 M KCl to 5 ml. Then add 3 drops of 4 M NaOH and 0.5 ml. of the sodium tetraphenylborate solution. The formation of the white precipitate indicates the presence of K^+ (Eq. 7).

$$K^+ + NaB(C_6H_5)_4 \rightarrow KB(C_6H_5)_4 \downarrow + Na^+ \qquad (7)$$

A white precipitate with sodium tetraphenylborate is also obtained from NH_4^+. However, if the test for potassium is performed in a basic solution, the NH_4^+ is changed to NH_3 which does not precipitate. To convince yourself, try the test for K^+ but substitute five drops of 0.1 M NH_4Cl for the 0.1 M KCl.

H. Test for Ammonium Ion

Ammonium ion in solution is easily detected, as NH_4^+ is converted to NH_3 with a strong base such as NaOH (Eq. 8).

$$NH_4^+ + OH^- \rightarrow NH_3 \uparrow + H_2O \tag{8}$$

To illustrate the test on a known solution, pick two watch glasses of the same size. Place a piece of red litmus paper and 1 drop of water on one watch glass. On the second watch glass, place 1 drop of 0.1 M NH_4Cl, 5 drops of water and 1 to 2 drops of 4 M NaOH. *Immediately,* cover the second watch glass by inverting the one with the moistened litmus paper over it. In a minute or less, the red litmus paper turns blue if NH_4^+ is present. To increase the sensitivity of this test, place the assembled watch glasses on a beaker of hot water.

QUALITATIVE TESTS FOR ANIONS

A. Test for Sulfate Ion

Barium ions precipitate sulfate ions as a white solid even in strongly acid solutions (Eq. 9). To observe the test, add 1 ml. of 3 M HCl to 1 ml. of solution to be tested for sulfate (use 0.1 M Na_2SO_4 for a known). Heat this solution in a water bath for 30 sec. If any insoluble material is present, centrifuge the solution and pour off the clear liquid for use in the test. If the solution is clear after heating in the water bath, do not centrifuge. Add 0.5 ml. of the 10% barium chloride solution and wait approximately 60 sec. The formation of a white precipitate indicates the presence of sulfate.

$$Ba^{2+} + SO_4^{2-} \rightarrow BaSO_4 \downarrow \tag{9}$$
$$\text{white}$$

B. Test for Chloride Ion

The chloride, bromide, and iodide ions all form insoluble compounds with silver ion in acidic aqueous solutions. The formation of a precipitate with silver nitrate is a specific and sensitive test for the three halides. This test will not distinguish between these three halides and, therefore, is only a general test for these three halides.

To illustrate the test on a known chloride solution, dilute 5 drops of 0.1 M NaCl to 5 ml. with water. Add 10 drops of 3 M HNO_3 and then 10 drops of 0.1 M $AgNO_3$. Does a precipitate form? Describe the color and general appearance of this precipitate. Centrifuge the mixture and discard the clear liquid. Add 1 ml. of distilled water to the precipitate, centrifuge and decant (pour off carefully) the clear liquid. Now, add 2 ml. of 2 M ammonia solution to the precipitate and shake. What do you observe? Acidify the clear solution with 3 M HNO_3 and observe the mixture. (AgCl dissolves in 2 M or stronger ammonia to form $Ag(NH_3)_2^+$ and Cl^-. Then, when the solution is acidified, the silver(I) complex, $Ag(NH_3)_2^+$, is decomposed and AgCl reprecipitates). The reactions for formation and dissolving of the AgCl precipitate are Equations 10 and 11, respectively.

$$Ag^+ + Cl^- \rightarrow AgCl \downarrow \qquad\qquad (10)$$
$$\text{(white)}$$

$$AgCl + 2NH_3 \rightarrow Ag(NH_3)_2^+ + Cl^- \qquad\qquad (11)$$

C. Test for Bromide Ion

Dilute 5 drops of 0.1 M KBr to 5 ml. with water. Add 10 drops of 3 M HNO_3 and 10 drops of 0.1 M $AgNO_3$. Is the precipitate formed in this case the same color as that from chloride? Centrifuge the solution and discard the clear liquid. Add 1 ml. of distilled water and shake to wash the precipitate, centrifuge and decant the clear liquid. Treat the washed precipitate with 2 ml. of 2 M ammonia solution and if the precipitate does not dissolve completely, centrifuge the mixture. Decant the clear liquid into a clean test tube and acidify the liquid with 3 M HNO_3. What do you observe? From this experiment, is silver bromide very soluble in 2 M ammonia? Is there enough difference in the solubility of silver chloride and silver bromide in 2 M ammonia so that you could make a good separation of these two substances? How would you perform this separation if both were present? The reaction for the bromide precipitation is:

$$Ag^+ + Br^- \rightarrow AgBr \downarrow \qquad\qquad (12)$$
$$\text{(cream color)}$$

D. Test for Iodide Ion

Repeat all of the steps given above for the bromide ion using 5 drops of 0.1-ml. KI instead of KBr. Is the solubility difference between silver chloride and silver iodide in 2 M ammonia sufficient for you to separate silver chloride from silver iodide if both these substances were present in the mixture? The reaction for formation of the iodide precipitate is:

$$Ag^+ + I^- \rightarrow AgI \downarrow \qquad\qquad (13)$$
$$\text{(yellow)}$$

E. Test for the Carbonate Ion

Place a one-hole rubber stopper and bent delivery tube in a test tube as shown in Figure 3-3. Add about 3 ml. of the test solution to this test

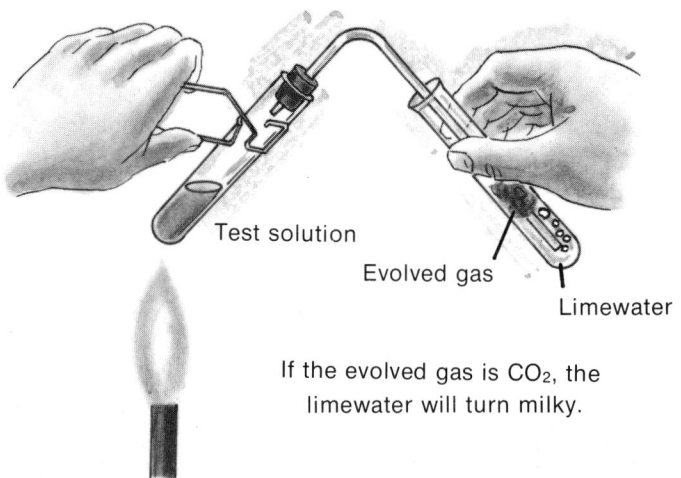

Test solution

Evolved gas

Limewater

If the evolved gas is CO_2, the limewater will turn milky.

Warm over a low flame.

FIGURE 3-3

Apparatus for the CO_2 test.

tube. Now insert the delivery tube into some clear limewater, $Ca(OH)_2$, in another test tube. When ready, remove the stopper just enough to add a little 6 M HCl to the test solution. Immediately close the stopper again, and heat the test solution in the tube gently to boiling to drive any carbon dioxide gas into the limewater. Be careful not to let any of the boiling liquid escape through the delivery tube into the limewater. A white precipitate in the limewater indicates $CO_3{}^{2-}$ or $HCO_3{}^-$ in the test solution. The reactions for this test are:

$$CO_3{}^- + 2H^+ \rightarrow H_2O + CO_2 \uparrow \qquad (14)$$

$$H_2O + CO_2 + Ca^{2+} \rightarrow CaCO_3 \downarrow + 2H^+ \qquad (15)$$
$$\text{(white)}$$

F. Test for Nitrate Ion

Three drops of 0.01 M nitrate test solution are placed in a test tube. Add 3 drops of the diphenylamine-sulfuric acid solution; then add 10 drops of concentrated (18 M) H_2SO_4. The test tube is shaken carefully to mix the reagents. If nitrate ion is present in the test solution, the mixture turns deep blue in color and then slowly changes to a deep violet color.

If your unknown gave a positive test for either bromide or iodide, these ions must be removed before the above test for nitrate is performed. The Br^- and I^- ions can be removed by precipitation with a saturated solution of silver sulfate; the resultant mixture containing the precipitate is centrifuged and 3 drops of the clear liquid are used for the nitrate test.

NOTEBOOK AND REPORT

Accurate and detailed records are especially important in any analysis work. Record the results of the tests on the knowns in your notebook clearly, so you can refer to them for information and comparisons with the results obtained on your unknown. Report the name and the observed melting point of the organic compound and the ions found on the Report Sheet.

NOTES

1. Some compounds decompose before they reach a definite melting or boiling point; in those cases, the decomposition point is reported.
2. As the accuracy of a quantitative elemental analysis depends on the skill of the investigator and the limitations of the equipment and analytical method, commercial carbon analyses are accurate to ± 1% *relative* error. That is, a compound containing exactly 25.00% C would be expected to analyze within the range 24.75 to 25.25% C. Your % C data conforms to these relative error limits.

QUESTIONS

1. If a sample contained both iodide and bromide, could you identify both ions on the basis of the tests in this experiment?

2. Why must iodide and bromide be removed before performing the nitrate test?

3. If a solid organic acid, *i.e.*, a compound of type RCO_2H, were used as the organic component of the mixture, would any difficulty be encountered in separating this component by the procedure given?

4. What is the difference between the freezing point and the melting point of pure glacial acetic acid?

TABLE 3-1 ORGANIC COMPOUND

Name	Empirical Formula	M.P. (°C)	Structural Formula
acetamide	C_2H_5NO	81	$CH_3\overset{\|}{\underset{O}{C}}-NH_2$
N-acetyl-m-toluidine (m-acetatoluidide)	$C_9H_{11}NO$	65.5	$CH_3-C_6H_4-\overset{H}{\underset{}{N}}-\overset{}{\underset{O}{C}}-CH_3$
acetone oxime	C_3H_7NO	60-61	$(CH_3)_2C=N-O-H$
N-acetyl-N-methyl-p-toluidine	$C_{10}H_{13}NO$	80	$CH_3\overset{\|}{\underset{O}{C}}-\overset{}{\underset{CH_3}{N}}-C_6H_4-CH_3$
allyl urea	$C_4H_8N_2O$	85	$CH_2=CHCH_2-\overset{H}{\underset{}{N}}-\overset{}{\underset{O}{C}}-NH_2$
azobenzene	$C_{12}H_{10}N_2$	68	$C_6H_5-N=N-C_6H_5$
benzil	$C_{14}H_{10}O_2$	95	$C_6H_5-\overset{}{\underset{O}{C}}-\overset{}{\underset{O}{C}}-C_6H_5$
benzoyl acetonitrile	C_9H_7NO	80.5	$C_6H_5-\overset{}{\underset{O}{C}}-CH_2-C\equiv N$
N-benzylacetamide	$C_9H_{11}NO$	60-61	$CH_3\overset{}{\underset{O}{C}}-\overset{}{\underset{H}{N}}-CH_2C_6H_5$
biphenyl	$C_{12}H_{10}$	68-70	$C_6H_5-C_6H_5$
p-bromoacetophenone	C_8H_7BrO	50-51	
β-bromonaphthalene	$C_{10}H_7Br$	59	
o-chloroacetanilide	C_8H_8ClNO	88	$Cl-C_6H_4-\overset{H}{\underset{}{N}}-\overset{}{\underset{O}{C}}-CH_3$

Handwritten annotations: Component A (pointing to biphenyl, circled); 93.5% CARBON

TABLE 3-1 (Continued)

Name	Empirical Formula	M.P. ($^\circ$C)	Structural Formula
m-chloroacetanilide	C_8H_8ClNO	72.5	$Cl-C_6H_4-\overset{\overset{\displaystyle H}{\vert}}{N}-\underset{\underset{\displaystyle O}{\parallel}}{C}CH_3$
p-chloroaniline	C_6H_6ClN	70-71	
4-chlorobenzophenone	$C_{12}H_9ClO$	77-78	$Cl-C_6H_4-\underset{\underset{\displaystyle O}{\parallel}}{C}-C_6H_5$
2-chloronaphthalene	$C_{10}H_7Cl$	58-60	
1-chloro-4-nitrobenzene	$C_6H_4ClNO_2$	83-84	
cyclohexanone oxime	$C_6H_{11}NO$	85-87	
cyclooctanone	$C_8H_{14}O$	40-42	
1,10-decanediol	$C_{10}H_{22}O_2$	71-73	$HO(CH_2)_{10}OH$
benzyl phenyl ketone	$C_{14}H_{12}O$	55-57	$C_6H_5-\underset{\underset{\displaystyle O}{\parallel}}{C}-CH_2C_6H_5$
p-dibromobenzene	$C_6H_4Br_2$	83-88	
2,4-dichloroaniline	$C_6H_5Cl_2N$	61-63	
2,5-dichloroaniline	$C_6H_5Cl_2N$	49-50	

TABLE 3-1 (Continued)

Name	Empirical Formula	M.P. (°C)	Structural Formula
p-dichlorobenzene	$C_6H_4Cl_2$	53-54	
o-diethoxybenzene	$C_{10}H_{14}O_2$	41-43	
p-diethoxybenzene	$C_{10}H_{14}O_2$	71-72	
1,1-diethylurea	$C_5H_{12}N_2O$	68-70	
p-dimethoxybenzene	$C_8H_{10}O_2$	55-57	
p–dimethylaminobenzaldehyde	$C_9H_{11}NO$	74-75	
N,N-dimethylbenzamide	$C_9H_{11}NO$	42-44	
diphenylamine	$C_{12}H_{11}N$	52.5-54.0	$(C_6H_5)_2NH$
diphenylcarbonate	$C_{13}H_{10}O_3$	79-81	$(C_6H_5O)_2CO$
1,3-diphenyl-1,3-propanedione	$C_{15}H_{12}O_2$	76-78	

TABLE 3-1 (Continued)

Name	Empirical Formula	M.P.(°C)	Structural Formula
durene	$C_{10}H_{14}$	77-81	
9-fluorenone	$C_{13}H_8O$	83-83.5	
indole	C_8H_7N	52-53	
p-iodoanisole	C_7H_7IO	50-52	
p-iodotoluene	C_7H_7I	34-35	
maleic anhydride	$C_4H_2O_3$	52-54	
4-methylbenzophenone	$C_{14}H_{12}O$	55-57	
2-methyl-1-nitronaphthalene	$C_{11}H_8NO_2$	79-81	

TABLE 3-1 (Continued)

Name	Empirical Formula	M.P.($^\circ$C)	Structural Formula
naphthalene	$C_{10}H_8$	79-80	
2-nitroacetanilide	$C_8H_8NO_3$	92-93	
phenylbenzoate	$C_{13}H_{10}O_2$	68-69	$C_6H_5-\underset{\underset{O}{\parallel}}{C}-O-C_6H_5$
N–phenyl–1–naphthylamine	$C_{16}H_{13}N$	59.5-60.5	
propionamide	C_3H_7NO	80-82	$CH_3CH_2-\underset{\underset{O}{\parallel}}{C}-NH_2$
salicylaldoxime	C_7H_7NO	57-59	
o-terphenyl	$C_{18}H_{14}$	56-58	
m-terphenyl	$C_{18}H_{14}$	85-87	
1,1,3,3-tetramethyl-2-thiourea	$C_5H_{12}N_2S$	77-79	$(CH_3)_2N-\underset{\underset{S}{\parallel}}{C}-N(CH_3)_2$
triphenylmethane	$C_{19}H_{16}$	93-95	$(C_6H_5)_3CH$
triphenylphosphate	$C_{18}H_{15}O_4P$	49-50	$(C_6H_5O)_3PO$

Experiment 3
REPORT SHEET

Name _____   Section No. _____

1. Melting Point of Organic Unknown _____ _____

2. Mixed Melting Points   Compound added   Observed M.P.
                                         to unknown

                                         _____   _____

                                         _____   _____

                                         _____   _____

                                         _____   _____

3. Identity of Organic Compound _____

4. Calculated Elemental Composition of Identified Organic Compound

    _____

5. What compounds (salts) were used for components B and C in the unknown mixture?

Attach sheets reporting:

6. Results of Tests on Known Ion Solutions

7. Results of Tests on Water Soluble Portion of the Mixture, *i.e.,* from Component B.

8. Results of Tests on the Portion that Dissolved in Nitric Acid, *i.e.,* Component C.

CHEMICAL REACTIONS

DETERMINATION OF THE EMPIRICAL FORMULA OF A COMPOUND

OBJECTIVE

To determine the empirical formula of a compound by a synthesis which is performed quantitatively so that the relative masses of the elements in the compound can be measured.

EQUIPMENT

Two porcelain crucibles and lids; clay triangle; analytical balance; high temperature burner; crucible tongs; simple desiccator.

REAGENTS NEEDED

Two pieces of copper wire; powdered sulfur.

INTRODUCTION

Concept of a Mole

Because of the extremely small size of an atom or simple molecule, chemists must use enormous numbers of atoms or molecules to observe a reaction experimentally. For a convenient reference, they have chosen the number of grams of a substance numerically equal to its atomic (or molecular) weight (in a.m.u.) as the standard measure of matter in bulk and have called this a mole. There are as many atoms of sulfur in one mole of sulfur (32.06 g.S) as there are atoms of carbon in one mole of carbon (12.01 g.C). This constant number of atoms contained in a mole of any element (or the number of molecules in a mole of a compound) is Avogadro's number and has the value 6.023×10^{23}.

We shall use the term mole to represent 6.023×10^{23} atoms, ions, or molecules, and to represent 6.023×10^{23} simplest formula units of an electrovalent compound. This number of atoms, ions, molecules or formula units weighs one gram atomic weight, one gram ionic weight, one gram molecular weight and one gram formula weight, respectively. In each case the number of moles corresponding to a specified weight is found by multiplying the weight of the substance by the conversion factor relating moles to weight, as illustrated in the next paragraphs.

Empirical Formula of a Compound

After a chemist has determined by a *qualitative* analysis which elements are present in a compound, he may need to know the empirical (simplest) formula of the compound, i.e., the relative number of moles of each element in the compound. Such information is determined directly from a *quantitative* analysis experiment.

One may calculate the empirical formula from the ratio of the weights of the elements in the compound. The procedure is illustrated by the following example. It was observed that 4.74 g. of sulfur combine with chlorine to produce 10.00 g. of a binary sulfur chloride compound. What is the empirical formula of this sulfur chloride?

$$
\begin{array}{r}
10.00 \text{ g. of sulfur chloride} \\
-\ 4.74 \text{ g. of sulfur in the compound} \\
\hline
5.26 \text{ g. of chlorine in the compound}
\end{array}
$$

$$
\begin{array}{l}
\text{Number of moles of sulfur in} \\
\text{10.00 g. of sulfur chloride}
\end{array} = 4.74\text{g. S} \times \frac{1 \text{ mole S}}{32.06\text{g.S}} = 0.148 \text{ moles S}
$$

(Note that the units and the substance concerned are clearly written in the arithmetical expressions and that the conversion factor relating moles and weight is so written that "grams S" appears in the denominator of the conversion factor so that it cancels "grams S" in the term "4.74 grams S," leaving "moles S" as the unit of the answer. This use of conversion factors is a characteristic of the "factor-label-method" for solving problems in science.)

$$
\begin{array}{l}
\text{Number of moles of chlorine in} \\
\text{10.00 g. of sulfur chloride}
\end{array} = 5.26\text{g. Cl} \times \frac{1 \text{ mole Cl}}{35.45\text{g. Cl}} = 0.148 \text{ moles Cl}
$$

$$
\begin{array}{l}
\text{Ratio of moles of sulfur to moles of} \\
\text{chlorine in the compound}
\end{array} = \frac{0.148 \text{ mole of S}}{0.148 \text{ mole of Cl}} = \frac{1 \text{ mole S}}{1 \text{ mole Cl}}
$$

These data show that in a 10-g. sample of this sulfur chloride there are 0.148 moles of sulfur atoms and 0.148 moles of chlorine atoms, i.e., there is 1 mole of sulfur atoms for every mole of chlorine atoms in the compound. Therefore, the empirical (simplest) formula is SCl.

Copper and sulfur combine under different conditions to form more than one copper sulfide, just as sulfur and chlorine form molecules whose empirical formulas are SCl_2 SCl, S_3Cl_2, and S_2Cl under different reaction conditions. The objective in this experiment is to determine the weight of sulfur that will combine with a given weight of copper metal under the experimental conditions used in our laboratory. This is done by determining the weight gain when a piece of copper is caused to react with sulfur to form a copper sulfide under conditions such that any excess unreacted sulfur will be burned away. The reaction is carried out in a crucible at high temperature.

PROCEDURE

Weighing. Since the primary object of this experiment is to determine quantitatively the relative number of copper to sulfur atoms in the copper sulfide, the determination is strongly dependent upon the care and precision with which one makes the weighings on the analytical balance. Review the procedure and operating instructions in Experiment 1 before starting the present experiment.

Igniting the Crucibles. Prior to weighing the empty crucibles, they should be heated to a dull red to remove any absorbed dust, moisture, and oil film. It is important to note some identification mark for each crucible and to handle the crucible with crucible tongs during the weighings so that oil and moisture are not transferred from your fingers to the crucible. Place a clean, dry covered porcelain crucible on a clay triangle on a ring stand as shown in Figure 4-1, and heat it to redness in the hottest portion of the flame from a Meker or Fisher burner. Allow the crucible to cool somewhat, and then while it is still warm, transfer it to the desiccator to cool to room temperature.

Clean and ignite crucibles.

Allow crucible to cool.

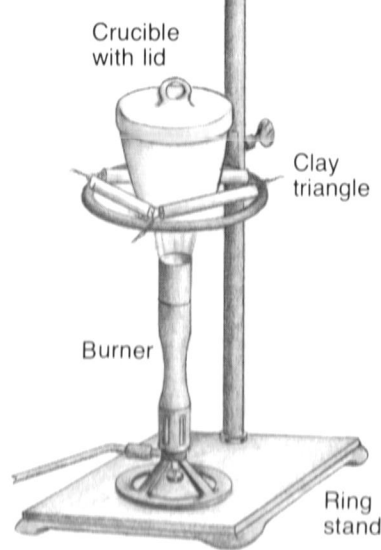

Crucible with lid

Clay triangle

Burner

Ring stand

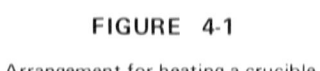

FIGURE 4-1

Arrangement for heating a crucible.

While the empty crucible is cooling, obtain two pieces of clean copper wire, each weighing about 1 g. Use paper or a dry cloth to wrap the wire in order to avoid transferring oil and moisture from your fingers, and turn each piece of wire into a spiral so that it will fit compactly into the bottom of a crucible. With tweezers, place a piece of metal directly on the balance pan, and weigh it to 0.0002g. Record this weight and all subsequent data in your notebook in the form of a data table similar to the one on page 49.

While cooling, weigh metal.

Weigh a cooled crucible without the cover (use tongs; don't touch it with your fingers) to the nearest 0.0002g on an analytical balance, and record the weight in the data table.

Weigh crucible and add metal and sulfur to it.

Add the copper spiral to the crucible and cover it with approximately 1.2 g. of powdered sulfur (weigh the sulfur on a piece of paper to within 0.1 g. with a triple-beam laboratory balance and then add the sulfur to the crucible), place the lid on the crucible, and support the crucible with a clay triangle on a ring stand *in the hood* (Fig. 4-1). Heat the bottom of the

Heat to cause reaction.

crucible slowly for 5 to 6 minutes and then slowly increase the intensity of the flame until the bottom of the crucible becomes a dull red. The excess sulfur may be observed to burn around the edge of the crucible with a blue flame. (What is the chemical equation for the reaction producing this flame?) Continue this intense heating until no more sulfur vapor escapes around the lid, and then heat the sides and the lid of the crucible to vaporize any free sulfur present. Allow the crucible to cool until it stops glowing, remove the lid (with tongs), and inspect the crucible to see whether the excess sulfur has been driven off (as indicated by the absence of the yellow color of sulfur). If the crucible contains no free sulfur, allow it to cool in the desiccator. When the crucible has cooled to room temperature, weigh it without the lid (don't touch it with the fingers).

Cool and weigh crucible and product.

If the inspection shows the presence of free sulfur, this may mean that the flame was not hot enough to vaporize the sulfur and cause complete reaction with the copper. To insure completeness of the reaction, add approximately 0.2 g. of sulfur, replace the lid, reheat the crucible to a red color for several minutes, cool it in the desiccator, and weigh it.

All the copper should have reacted, and the product in the crucible should now be copper sulfide. As a check on your result, add 0.5 g. of powdered sulfur to the crucible containing the copper sulfide and reheat the material as before until the sulfur vapors stop escaping. Allow the crucible to cool, inspect its contents, and reweigh it. Repeat this procedure until the weights in successive weighings are constant to within 0.0004 g.

Add more sulfur and heat. Cool and weigh. Repeat until weight is constant.

This procedure should be repeated exactly, using the second copper wire, but in this experiment use about 2.4 g. of powdered sulfur originally instead of 1.2 g. Begin this duplicate experiment in the spare time you have during the first trial. If the ratio of sulfur to copper atoms in the product does not agree within 7 percent for the two trials, make a third trial and use the average of the two results that agree most closely in calculating the formula of the compound. Place the copper sulfide produced in both trials in a test tube, stopper it with a cork, label it and hand it to your instructor.

Repeat the experiment using double the amount of sulfur.

Possible Alternates

The preceding directions also may be utilized to prepare a lead or iron sulfide by using elemental lead strips or iron wire, respectively.

NOTEBOOK AND REPORT

The data for most of the quantitative experiments you will perform are most clearly recorded in tabular form. A sample data table suitable for this experiment is given on page 49. Copy this in your notebook and use it as a skeleton for filling in the blanks with your experimental data, and turn it in as a part of your report. Be sure to indicate the units of measurement for all data recorded.

The notebook should also contain statements of qualitative observations made during the experiment (Did you smell anything during the heating? Describe the color of the flame from the crucible. Was there a yellow deposit under the lid? Etc. as in Experiments 2 and 3). Note any unclear points or questions about the experiment so that you can determine the answers in class sessions later.

The calculations and the arithmetical steps needed to obtain the empirical formula of the product should appear in your notebook, and a sample calculation must be included in your report. The number of moles of each element, and the ratio, should be calculated and reported to three significant figures.

The report should also contain a brief statement of the purpose of the experiment, a sentence outlining the procedure, complete and balanced equations for reactions which occurred, and answers to assigned questions. Thus, it might take the following form:

Purpose
Procedure
Data Table
Chemical Equations
Sample Calculation
Questions

QUESTIONS–PROBLEMS

1. What evidence indicates that a reaction occurred between the copper and the sulfur?

2. Is the formula you obtained a reasonable one for a compound formed from copper and sulfur? Explain your answer.

3. What happened to the sulfur represented by the difference between Items 11 and 3.d in the data table? Write an equation for the reaction.

4. Is copper sulfide a pure substance or a mixture? Describe briefly how you could verify your answer.

5. Compare your results with those of other members of the class. Considering that your experimental error may be as great as 7 percent, what do you conclude about the product formed from the reaction of copper with sulfur? What is the implication about the stoichiometry of the copper sulfide?

6. For the second trial, you used more sulfur than for the first. How did this affect the composition of the compound that was formed? Explain.

7. Another sulfide of copper, which is unstable above $200°C$, contains 1.98 g. of copper per gram of sulfur. Calculate the empirical formula of this compound.

8. How would the calculated ratio of copper to sulfur have been affected if 5 percent of the copper had reacted with the oxygen instead of sulfur to form copper oxide, Cu_2O?

REFERENCE

1. Addison, W.E.: *Structural Principles in Inorganic Compounds.* John Wiley and Sons, New York, 1961, pp. 147-166.

Experiment 4
REPORT SHEET

Name _____   Section _____

	Crucible Mark	Crucible Mark
1. Weight of copper (to within 0.0002 g.)	_____	_____
2. Weight of crucible (to within 0.0002 g.)	_____	_____
3. Weight of sulfur added		
Initially	_____	_____
After first heating	_____	_____
After second heating	_____	_____
Total weight of sulfur used (to within 0.1 g)	_____	_____
4. Number of moles of copper used as a reactant in the experiment (calculate).	_____	_____
5. Number of moles of sulfur used as a reactant (calculate).	_____	_____
6. Ratio of sulfur atoms to copper atoms in the starting materials	_____	_____
7. Total weight of crucible and product		
First weighing, after heating _____ minutes	_____	_____
Second weighing, after heating an additional _____ minutes	_____	_____
Third weighing, after _____ minutes	_____	_____
Fourth weighing (if necessary)	_____	_____
8. Weight of crucible (from Item 2)	_____	_____
9. Weight of product	_____	_____
10. Weight of copper in the product (copper sulfide)	_____	_____
11. Weight of sulfur in the product	_____	_____
12. Percentage of copper in the product, by weight	_____	_____

13. Number of moles of copper in the weight of product obtained   _____   _____

14. Number of moles of sulfur in the weight of product obtained   _____   _____

15. Ratio (moles of copper) to (moles of sulfur) found in your experiment (to three significant figures)   _____   _____

16. Simplest whole-number ratio of moles of copper to moles of sulfur   _____   _____

17. Empirical formula of product found by your experiment   _____   _____

EXERCISE I

Stoichiometry and Moles

This exercise is designed as a drill on chemical stoichiometry and the concept of a mole to aid in correlating the experimental information obtained in Experiment 4.

1. What is the actual weight in grams of 1 molecule of oxygen, of carbon dioxide, and of sulfur dioxide?

2. What is the weight in grams of 6.02×10^{23} molecules of each gas (oxygen, carbon dioxide, and sulfur dioxide)? What is the terminology commonly used in referring to these weights?

3. For the reaction $3Fe + 2O_2 \rightarrow Fe_3O_4$

    a. How many moles of iron react with 1 mole O_2? _____

    b. How many moles of iron react with 10 moles of O_2? _____

    c. How many moles of oxygen molecules react with 1 mole Fe? ____

    d. How many moles of the product are formed when 6 moles O_2

        react with excess iron? _____

    e. How many grams of iron are required to react with 32 g. O_2? ____

    f. How many grams of iron react with 1 g. O_2? _____

    g. How many grams of the product are formed by the reaction of 3

        moles Fe and 2 moles O_2? _____

4. For the reaction $8KClO_3 + C_{12}H_{22}O_{11} \rightarrow 8KCl + 12CO_2 + 11H_2O$

    a. How many moles of potassium chlorate react with 0.5 mole of

        the sugar $C_{12}H_{22}O_{11}$? _____

    b. How many grams of potassium chlorate react with 0.5 mole of

        the sugar $C_{12}H_{22}O_{11}$? _____

    c. Two moles $KClO_3$ will produce _____ moles CO_2.

    d. Two moles $KClO_3$ will produce _____ moles H_2O.

    e. Two moles $KClO_3$ will produce _____ g. CO_2.

f. Two moles $KClO_3$ will produce _____ g. H_2O.

g. A weight of 122.5 g. $KClO_3$ will produce _____ g. KCl.

h. A weight of 122.5 g. $KClO_3$ will produce _____ moles CO_2.

i. A weight of 122.5 g. $KClO_3$ will produce _____ g. H_2O.

j. How many grams of potassium chlorate are required to produce 1 mole H_2O in the preceding equation?

k. _____ g. $KClO_3$ are required to form 1.0 g. H_2O.

l. _____ g. $KClO_3$ are required to form 10.0 g. KCl.

m. _____ g. $KClO_3$ are required to form 18 moles CO_2.

5. What is the percentage of each element in $C_{12}H_{22}O_{11}$?

6. Calculate the number of moles of carbon, hydrogen, and oxygen in 100 g. $C_{12}H_{22}O_{11}$.

7. How would the mole ratios of carbon: hydrogen, carbon: oxygen, etc. obtained in Question 6 change if one calculated the number of moles of carbon, hydrogen, and oxygen in 62 g. $C_{12}H_{22}O_{11}$?

8. Determine the simplest formula of a compound that has the following composition: 26.52 percent chromium, 24.52 percent sulfur, and 48.96 percent oxygen.

9. A compound has 63.12 percent carbon, 11.91 percent hydrogen, and 24.97 percent fluorine. What is its empirical (simplest) formula?

10. For formation of an ionic compound, how many electrons would be gained or lost by each of the following elements?

Cs _____   F _____   Ba _____

Br _____   Mg _____   Al _____

O _____   K _____   S _____

11. If bismuth atoms easily lose 3 electrons but no more than 3 and if iodine atoms easily gain 1 but not more than 1, what would you predict as the formula for an electrovalent compound of bismuth and iodine? What percent by weight of bismuth would there be in this compound?

12. Calculate the percent by weight of hydrogen in the covalent compound ammonia, NH_3. Would the answer be different if the compound were electrovalent with the formula N^{3-}, $3H^+$?

STOICHIOMETRY AND THE CHEMICAL EQUATION

(Development of an Equation)

OBJECTIVE

To show how the chemist identifies the product of a reaction and develops the chemical equation for the reaction.

EQUIPMENT NEEDED

4-ml culture tubes, dropping tubes, micro spatula, centrifuge.

REAGENTS NEEDED

0.10 molar potassium iodide; 0.10 M lead nitrate; 3% hydrogen peroxide H_2O_2; 5% thioacetamide, CH_3CSNH_2; solid sodium sulfite, Na_2SO_3; 3 M HNO_3; tetrachloroethylene; 0.10 M potassium nitrate.

INTRODUCTION

A chemical equation is a statement in chemical symbols of certain facts about chemical reactions that have been obtained by experiment. When a chemist writes an equation, he is making use of these facts (which may have been discovered by others) and of generalizations derived from the results of experiment. To write the equation for a reaction, the chemist needs to know what reacts, what is formed, and the relationship between the amounts of substances which react to form the product substances. This experiment seeks to illustrate the development of an equation for a simple reaction, that between lead nitrate and potassium iodide.

We know, since both of these reagents are salts, that they must be present in solution only as ions; that is, a solution of lead nitrate contains lead ions and nitrate ions, and a solution of potassium iodide contains potassium ions and iodide ions. When solutions of these two salts are mixed, a yellow precipitate is formed. Since neither of the original solid salts, lead nitrate nor potassium iodide, is yellow, it may be presumed that the yellow solid is formed by the alternate combination of the positive and negative ions into pairs, and is either lead iodide or potassium nitrate. If it is lead iodide, then potassium ions and nitrate ions must be left in the supernatant solution; if it is potassium nitrate, then lead ions and iodide ions must be left in the supernatant solution. Identification of the precipitate might then be made either by examining the solid itself, or by examining the solution left behind. In the latter case, care must be taken not to confuse any excess of ions not needed in forming the precipitate with the "widowed" ions left behind when the precipitate forms.

In this experiment we shall attempt to identify the ions in the precipitate and determine the ratio in which they occur in the precipitate. We shall then have the information needed in order to write the equation for the reaction: we shall know what reacts (lead nitrate and potassium iodide), we shall know what is formed, and we shall know the relative amounts of the two reagents required to form the product.

To carry out the identification, we shall need to have tests for two of the four ions present, one from each solution. Tests for only two are needed, since the presence or absence of the other two can be inferred. Thus if we find lead ion in the precipitate, and we know that the precipitate is either lead iodide or potassium nitrate, we will have identified the precipitate as lead iodide. If we do not find lead ion in the precipitate, then it is not lead iodide and must be potassium nitrate. A test for one of the negative ions is needed to clarify the relative amounts of ions in the precipitate after it has been identified; we shall choose the iodide ion as the easier to test for, rather than the nitrate ion.

A. Tests for Lead Ion and Iodide Ion.

To a 4ml culture tube add 5 drops of lead nitrate solution, 2 drops of 3 M HNO_3 and 20 drops of thioacetamide solution and place the tube in a boiling water bath for 5 minutes. The black precipitate which forms is, in the absence of ions other than those of periodic table Groups I and II, a test for the presence of lead ion. The precipitate is lead sulfide.

Confirm that potassium ion does not give this test by trying it on 5 drops of potassium nitrate solution.

The formation of a black precipitate with thioacetamide thus serves to distinguish material containing lead ion from material containing potassium ion.

To a 4ml culture tube add 5 drops of potassium iodide solution, 5 drops of tetrachloroethylene, 5 drops of 3M nitric acid and 5 drops of 3% hydrogen peroxide. Shake vigorously. A violet color in the layer of tetra-chloroethylene under the water layer indicates the presence of iodide ion. The color is due to the presence of elemental iodine in the solution.

B. Identification of the Precipitate Formed When Lead Nitrate Solution is Added to Potassium Iodide Solution.

Add 5 drops of lead nitrate solution to 5 drops of potassium iodide solution in a 4ml culture tube. Note the yellow precipitate. Insert the culture tube in a centrifuge, balancing the centrifuge by placing another tube containing an equal amount of water exactly opposite, and centrifuge for one minute. Remove the supernatant liquid with a dropping tube, and add a dropping-tube full of distilled water to the culture tube to wash the precipitate. Stir the mixture with a small stirring rod, centrifuge again, remove the wash water, wash again with a fresh quantity of water, centrifuge, and remove the second wash water. The precipitate is now to be tested to identify the ions present.

(1) **Test for Iodide.** Add to the precipitate 5 drops of 3M nitric acid, stir, and warm the mixture by placing the tube in a boiling water bath for a few minutes. Add 2 drops of hydrogen peroxide, stir, and again heat in boiling water. Add 3 drops of water and cool to room temperature. Add 5 drops of tetrachloroethylene, shake and observe the color of the tetrachloroethylene layer. If the bottom tetrachloroethylene layer is violet colored, iodide ion was present in the precipitate; if it remains colorless, no iodide was present.

(2) **Test for Lead Ion.** Centrifuge the solution from (1) above, withdraw the water layer and place it in a clean culture tube. Discard the tetrachloroethylene. Add 5 drops of tetrachloroethylene to the water layer, shake, centrifuge, and again withdraw the water layer to a clean tube. The purpose of this is to remove any color which remains in the water layer so that the lead ion test will be more visible. Discard the tetrachloroethylene. With a spatula, add sufficient solid sodium sulfite to cover the curvature at the bottom of the culture tube and stir to dissolve. The purpose of this addition is to destroy any excess of hydrogen peroxide, which would otherwise react with the thioacetamide to be added next. Make the test for lead ion by adding 20 drops of thioacetamide solution to the water layer, stir, heat in boiling water and centrifuge. A black precipitate of lead sulfide indicates the presence of lead ion; absence of precipitate indicates that the original yellow precipitate does not contain lead ion.

Tests (1) and (2) should enable you to identify the yellow precipitate as either lead iodide or, by elimination, potassium nitrate.

C. Determination of the Ratios in which the Ions Combine to Form the Yellow Precipitate.

(1) **Principle of the Method.** Suppose that two ions, A and B, combine to form a precipitate of formula AB_3. If we should add A and B together in a 1:1 molar ratio, it is evident that precipitation of AB_3 would leave excess A in solution, since it requires 3 moles of B to react with 1 mole of A, or 1 mole of B will react with only $\frac{1}{3}$ mole of A, and $1 - \frac{1}{3}$, or $\frac{2}{3}$ mole of A will remain. A test of the decantate from the precipitate in the 1:1 ratio experiment will thus show the presence of a large amount of A. If in a second experiment we use the molar ratio 1:2, the two moles of B will react with $\frac{2}{3}$ mole of A, and $1 - \frac{2}{3}$, or $\frac{1}{3}$, mole of A will remain. A test on the liquid above the precipitate will still show the presence of A but in smaller quantity than in the first experiment. Not until the experiment is made with a molar ratio 1:3 will the supernatant liquid show no A present.

If the molar ratio is further increased to 1:4, it is now evident that the amount of A becomes the limiting factor, since 1 mole of A can react with only 3 moles of B and 4 − 3, or 1 mole of B will be left over. Tests of the supernatant liquid will then show no A present, but will show the presence of B. Continued increase of the molar ratio to 1:5 and higher will show no A present in the supernatant liquid, but will show increasing amounts of B.

It is thus possible to determine the molar ratio in which A and B combine by examining the supernatant liquid. It is merely necessary to make a number of experiments with different relative amounts of A and B and to find out which component, and approximately how much of it, appears in excess. The ratio at which neither is present in excess is the stoichiometric ratio, corresponding, in our example, to AB_3.

We shall now carry out such experiments for the yellow precipitate, using the molar ratios from 3:1 to 3:9 to cover the range of possible formulas from A_3B to AB_3, namely A_3B, A_2B, AB, AB_2, AB_3.

Measure quantities carefully.

(2) **Procedure.** Since the experiments are to be semi-quantitative it is necessary to measure quantities *exactly*. Use always the same dropping tube for lead nitrate solution and, after cleaning, for potassium iodide solution in preparing the set of culture tubes described below. To get 0.5 drop, squeeze the bulb of the dropping tube gently until a droplet appears, then touch the tip of the dropping tube to the inside of the culture tube.

Prepare five marked culture tubes containing the number of drops of each solution and of water as follows:

Prepare five solutions with different ratios of reagents.

	I	II	III	IV	V
Lead nitrate	9	9	9	9	9
Potassium iodide	3	4.5	9	18	27
Water	28	26.5	22	13	4

Stir the solutions.

Rinse a stirring rod and stir each solution in turn to mix, being careful to rinse the stirring rod between each solution. Centrifuge each tube.

With a rinsed dropping tube containing no droplets of water (shake it to remove the drop which tends to remain in the tip) remove some of the supernatant liquid from Tube I and place 5 drops in each of two clean culture tubes. To one of these 5-drop samples add 2 drops of potassium iodide solution. To the second add 2 drops of lead nitrate solution. (It is not necessary here to use the same dropping tube for both solutions.) From these two results decide which ions were in excess in Tube I. If a precipitate forms in either test, centrifuge the tube so that you can compare the quantity of precipitate with the amount formed in later tests.

Determine which ions are in excess.

Repeat these tests on 5-drop samples of supernatant liquid from each of the other precipitation tubes.

From the results, decide upon the formula of the precipitate.

From the results of your experiments, complete and balance the equation

$$Pb(NO_3)_2 \quad + \quad KI \rightarrow$$

and underline the formula of the precipitate.

QUESTIONS–PROBLEMS

1. Describe what would have been observed in each of the five culture tubes in the set if the lead nitrate solution had been 0.05 molar instead of 0.1 molar, while the potassium iodide solution remained 0.10 molar.

2. What is meant by the statement that "the solution is 0.10 molar"? How many moles of nitrate ions are present in 9 drops of 0.10 M $Pb(NO_3)_2$? (There are 20 drops in 1 ml.)

3. If the charge on a potassium ion is 1 unit of electronic charge and positive, what is the charge on an iodide ion if the formula of potassium iodide is KI? From your results, what is the charge on a lead ion?

4. Calculate, from your results, the percent by weight of the components in the yellow precipitate, using data from a table of atomic weights.

5. Write complete and balanced equations for all reactions which occur in the tests for iodide and for lead ion in Part A, given (a) that when hydrogen peroxide acts as an oxidizing agent in acid solution, hydronium ion is a reagent, and the end-product from hydrogen peroxide is water and (b) that in acid solution thioacetamide produces hydrogen sulfide by the reaction

$$CH_3CSNH_2 + H_3O^+ + H_2O \rightarrow CH_3CO_2H + H_2S + NH_4^+$$

Experiment 5
REPORT SHEET

Name _____ Section No. _____

PART A. Observations on the tests

PART B. Observations on the tests

PART C. Observation on mixing   Observations on adding   Observations on adding
          the original solutions   potassium iodide solution   lead nitrate solution to the
                                          to the supernatant liquid   supernatant liquid

Tube I

Tube II

Tube III

Tube IV

Tube V

Attach sheets with answers to assigned questions.

Experiment 6

SYNTHESES, SEPARATIONS, REACTIONS INVOLVING FERROCENE

OBJECTIVE

To illustrate some important features of chemical synthesis including chromatographic separations. To study some principles of oxidation-reduction reactions.

EQUIPMENT NEEDED

Part A: mortar and pestle, burner, two 6-in test tubes, 250 ml Erlenmeyer flask with stopper, 250 and 600 ml beakers, 3-in evaporating dish, Büchner funnel, suction filter flask, rubber or plastic tubing; plastic bag; rubber bands; M.P. capillaries; 260° thermometer.

Part B: two 6-in test tubes, medicine dropper, four 4-ml test tubes, burner;

Part C: 250 and 125-ml Erlenmeyer flask with stopper, drying tube, 600 and 800 ml beakers, 15 X 1.0-cm glass tube or plastic chromatography column, powder funnel, spatula, m.p. tube;

Part D: four TLC developing jars and four beakers (100-250 ml).

CHEMICALS NEEDED

Part A: $FeSO_4 \cdot 7H_2O$, KOH, diglyme, dimethylsulfoxide, cyclopentadiene, Na_2SO_4(anhy);

Part B: ferrocene, 95% ethanol, 0.2M $Ce(HSO_4)_4$, 0.5M solutions of $SnCl_2$, NaClO, $NaNO_2$, $SbCl_3$, $K_2Cr_2O_7$, saturated H_2S, 3% H_2O_2;

Part C: $CaCl_2$(anhy), $NaHCO_3$, 85% H_3PO_4, acetic anhydride, ice, alumina (activity grade III) ferrocene, methylene chloride, ether;

Part D: Mixture of ferrocene and acetylferrocene, four silica gel TLC plates, benzene, ether, petroleum ether, ethyl acetate, Saran Wrap or equivalent; mixed solvents listed on p. 69.

INTRODUCTION

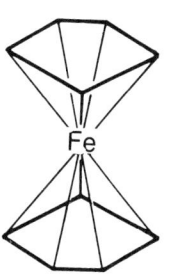

The central compound in this experiment is ferrocene, a "sandwich compound" in which an iron ion lies bonded between two parallel planar C_5H_5 rings. In this unusual structure, each cyclopentadienide ion ($C_5H_5^-$) acts as a six electron donor to an iron(II) ion. By gaining 12 electrons from the two $C_5H_5^-$ ions in addition to its six valence electrons, the iron essentially acquires the 18 valence electrons characteristic of a krypton structure.

61

This electron arrangement affords unusual stability not only to the iron ion at the center of the sandwich, but also to the 2 cyclopentadienide ions bonded to it. Ferrocene is prepared by the reaction of iron(II) ions with cyclopentadiene in the presence of potassium hydroxide. The potassium hydroxide reacts with cyclopentadiene to give cyclopentadienide ions.

$$K^+ + OH^- + C_5 H_6 \rightarrow K^+ + C_5 H_5^- + H_2 O. \tag{1}$$

The iron(II) ions react with the cyclopentadienide ions to give ferrocene.

$$Fe^{2+} + 2C_5 H_5^- \rightarrow (C_5 H_5)_2 Fe \tag{2}$$

The stability of ferrocene has been compared with that of benzene, $C_6 H_6$, and ferrocene undergoes numerous reactions similar to those of benzene. One such reaction is acetylation in which the acetyl group, $CH_3 CO$, is substituted for a hydrogen atom in one of the cyclopentadienide rings.

Another interesting property of ferrocene is its ease of oxidation to the ferrocenium ion, $Fe(C_5 H_5)_2{}^+$, in which the iron can be imagined to be in the III state. Since ferrocene is orange and the ferrocenium ion is blue, reactions of the type,

$$\text{Oxidizing agent} + \text{ferrocene} \rightleftarrows \text{reducing agent} + \text{ferrocenium ion} \tag{3}$$

can be studied so as to provide greater insight into the nature of such reactions and the relative strengths of oxidizing and reducing agents.

Also illustrated in this experiment are the important separation techniques of thin-layer and column chromatography, both of which are used routinely in research in synthetic chemistry.

The four parts of this experiment are arranged so that the student and his instructor may elect to do any or all of them. In Part A, ferrocene is synthesized and purified by sublimation. In Part B, ferrocene is oxidized to the ferrocenium ion and the interconversion of ferrocene and ferrocenium ion by some oxidizing and reducing agents is studied. In Part C, the aromatic nature of ferrocene is illustrated by its reaction with acetic anhydride. The principal product of this reaction, acetylferrocene, is separated from ferrocene using dry column chromatography. In Part D, the effectiveness of several solvents in the thin layer chromatography separation of ferrocene from acetylferrocene is examined.

TECHNIQUES

The more involved techniques in this experiment are *thin layer* and *column chromatography.*

Chromatography is probably the most versatile tool available for carrying out separations involving more than two components. The chromatographic process involves passing the mixture in a fluid such as a gas or a liquid along or through a stationary phase such as a solid in such a way that the several components of this mixture move at different rates across or through the stationary phase. This variation in rate results in a separation of the components of the mixture on or in the stationary phase. Some chromatographic techniques involve carrying out the process until the components are separated on the stationary phase, after which the process is stopped and the stationary phase examined for the positions of the desired components. If the components are colored, a visual examination will reveal their presence. If they are colorless, the solid phase can be sprayed with materials which will react with the components to bring out characteristic colors. Once the components have been identified, they can be separated by separating the stationary phase into sections containing the desired components. Each component can then be dissolved in a suitable solvent, repurified, and characterized.

It is convenient to divide chromatography into two types, *partition* and *adsorption chromatography*. In partition chromatography, the component is distributed or partitioned between the moving phase and a liquid phase coating on the surface of the stationary phase. Since this distribution between the two phases is not likely to be the same for any two components, the several components in the mixture will move across the stationary phase at different rates. Two important applications of partition chromatography are gas-liquid chromatography (glc) and thin layer chromatography (tlc). In adsorption chromatography, the component is directly adsorbed onto the solid phase. The dry column chromatography used in Part C is an adsorption technique.

Thin-layer chromatography involves 1) preparing a thin layer plate made of a smoothly spread layer of some absorbent material on glass or plastic, 2) charging the plate with small amounts of materials to be separated, these materials being dissolved in an appropriate solvent, 3) developing the chromatogram, that is, allowing the liquid phase to move across the stationary phase until a separation has been effected, and 4) dyeing the chromatogram if necessary so the separated components can be observed. The liquid on the surface of the stationary phase in most thin layer chromatography operations is water, even though the plates appear to be perfectly dry. The liquid used in the moving phase usually is a nonpolar solvent or one that is much less polar than water. The components are then partitioned between these solvents of differing polarities.

Column chromatography, which can be used on much larger samples than can tlc, involves, 1) filling a glass or plastic tube with a selected solid adsorbent, 2) charging the column with the mixture to be separated, 3) separating the components by allowing a developing solvent to percolate through the column, carrying the several components with it at different rates depending upon their solubility in the solvent and how strongly they are held on the adsorbent.

The relative rate of movement of a component through or across the stationary phase is characteristic of that component under the fixed conditions of solvent, stationary phase, temperature, and so forth. The relative rate of movement is symbolized by R_f, defined as $R_f = \dfrac{\text{distance a compound moves}}{\text{distance the solvent moves}}$.

The solid support in the stationary phase in thin layer chromatography and the adsorbents used in column chromatography usually are silica gel ($SiO_2 \cdot xH_2O$), sometimes known as silicic acid, or alumina ($Al_2O_3 \cdot xH_2O$). Some commonly used developing solvents are aliphatic hydrocarbons such as pentane and hexane, aromatic hydrocarbons including benzene and toluene, and compounds such as carbon tetrachloride, methylene chloride, chloroform, diethyl ether, acetone, and ethanol. Mixtures of these solvents are also commonly used as developing fluids. The purity of a single developing solvent or the composition of a mixture of such solvents is an important factor in the success of chromatography. Obviously, the greater the developing power of the solvent, the more effective will be the separation. A good developing solvent should effect a satisfactory separation after it has moved approximately 3 inches across the plate or 6 inches down the column.

PROCEDURE

A. Preparation of Ferrocene

Pulverize 1.5 grams of iron(II) sulfate using a mortar and pestle. Dissolve this in approximately 5 ml of dimethylsulfoxide in a 6-inch test tube, heating the tube gently if necessary.[1] [*Caution:* Dimethylsulfoxide is readily absorbed through the skin; therefore, care must be taken to prevent its contact with the skin. If it does come in contact with the skin, wash it off quickly.]

Dissolve FeSO₄ in solvent.

Place 5.0 grams of powdered potassium hydroxide in a 250 ml Erlenmeyer flask. Add 15 ml of diglyme (the dimethyl ether of diethylene glycol) and 1.0 ml of freshly distilled cyclopentadiene.[2] Stopper the flask tightly and vigorously swirl the mixture for about 10 minutes taking care not to splash the solution onto the stopper.

Add other reagents to a 250 ml flask.

Remove the stopper. Add 1 ml of the iron(II) sulfate solution and stopper the flask tightly again. Swirl this mixture for 10 minutes; add another ml of the iron(II) sulfate solution and repeat this procedure until all of the solution has been added. Swirl for 10 more minutes, remove the stopper and quickly pour the mixture into a 250 ml beaker containing about 25 ml of 6M HCl and about 25 grams of crushed ice. Rinse the Erlenmeyer flask with cold 6M HCl, adding the wash liquid to the beaker. Stir the mixture in the beaker for about 10 minutes and filter it using a Büchner funnel and suction filtration. (See Figure 2-1 for the set up of the apparatus.) Wash the precipitate with three 10 ml portions of cold water. Save the filtrate, and allow the precipitate to dry in a 6 in. test tube or in a sealed plastic bag in which is placed a small container of anhydrous sodium sulfate.

Add the FeSO₄ solution, 1 ml at a time, and swirl.

Add to chilled HCl.

Filter and save both parts.

The blue filtrate contains ferrocenium ion which can be reduced to ferrocene and combined with the original yield. Transfer the filtrate to a 250 ml beaker, add 10 ml of 0.5M $SnCl_2$ and stir for 10 minutes. If the solution is still blue, add more tin(II) chloride and repeat the stirring. Filter the yellow precipitate and wash and dry it as before.

Reduce, if necessary.

---

[1]This solution must be used within an hour, otherwise too much iron(II) will be oxidized to iron(III).

[2]The cyclopentadiene must be freshly distilled and maintained at dry ice temperatures to prevent dimerization.

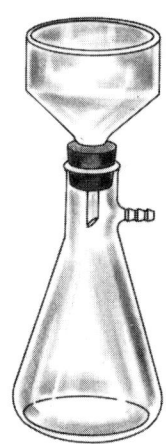

FIGURE 6-1

A Büchner funnel and filter flask.

Purify the crude ferrocene by sublimation as follows: Place the crude product in a 3 inch evaporating dish, mounted on a ring stand; place a carefully cleaned 600 ml beaker containing about 100 ml of ice water on top of the evaporating dish so that the subliming solid can condense onto the bottom of the beaker. Be certain the beaker is securely fastened to the ring stand.

Now gently heat the evaporating dish with a low flame for approximately 10 minutes. Allow it to cool to room temperature and examine the beaker to see if the orange product has sublimed from the residue onto the bottom of the beaker. If not, repeat the sublimation procedure. Finally scrape the orange crystals of ferrocene from the bottom of the beaker onto a sheet of filter paper and transfer them to a previously weighed stoppered bottle. Determine the weight of product, calculate its yield based on the amount of iron(II) sulfate used and determine the melting point of the purified ferrocene.

Purify the product by sublimation.

B. Study of Oxidation-Reduction Reactions Involving Ferrocene[3]

In a 6 inch test tube, dissolve about 50 milligrams of ferrocene in about 10 ml of warm 95% ethanol. Add 10 ml of water and 3 to 4 drops of 6M HCl. Thoroughly shake this mixture and transfer half of it to a second test tube. Stopper this test tube and set it aside momentarily.

Add 8 drops of 0.2M cerium(IV) hydrogen sulfate solution to the ferrocene remaining in the first test tube. Stir and allow the mixture to stand for 5 minutes. If it is not distinctly blue at this point, gently heat it and add several more drops of the cerium(IV) hydrogen sulfate solution, but do not add an excess of this reagent.

Oxidize the ferrocene to ferrocenium ion.

Using a dropping tube (medicine dropper) transfer approximately one-half ml (10 drops) of the blue solution of ferrocenium ions to each of four clean 4 ml test tubes. Clean the dropping tube and transfer one-half ml of the ferrocene solution which was set aside earlier to these same test tubes.

Mix the ferrocene and ferrocenium ion solutions equally.

[3]If you did not prepare and purify ferrocene as in Part A, secure the compound from your instructor or from the designated laboratory area.

Shake each of the test tubes thoroughly and add slowly and with stirring 20 drops of one of the oxidizing or reducing agents listed below to each of the test tubes. Observe the color carefully, recording any changes, and the number of drops needed to bring about a change.

Some oxidizing or reducing agents that can be used here are:

0.5M $SnCl_2$

0.5M $NaClO$

Saturated aqueous H_2S solution

0.5M $NaNO_2$

3% H_2O_2

0.5M $SbCl_3$

0.5M $K_2Cr_2O_7$

The observed color changes can be interpreted by remembering that the formation of blue color means that some ferrocene is being oxidized to ferrocenium ion. Conversely the formation of an orange color indicates that some ferrocenium ion is being reduced to ferrocene. An added reagent that intensifies the blue color is one that is a strong enough oxidizing agent to oxidize ferrocene to ferrocenium ion. Conversely a reagent that causes a color change to orange is one that is a strong enough reducing agent to reduce ferrocenium ion to ferrocene. Using this information, it should be possible to classify the reagents you used as oxidizing or reducing agents and possibly even to compare their relative strengths as oxidizing or reducing agents by noting the number of drops needed to bring about a significant color change. While this method must be used with caution because some strong oxidizing or reducing agents are slower to react than others, it is nevertheless a good indicator as a first approximation.

On the basis of information obtained, it should be possible to identify the element oxidized or reduced in each of the added reagents and to write equations for all of the reactions studied. Write the equations in your laboratory notebook.

C. Acetylation of Ferrocene; Separation by Dry Column Chromatography

Combine the reagents in a flask with constant stirring.

To a 125 ml Erlenmeyer flask add 1 ml of 85% H_3PO_4 dropwise and with constant stirring (swirling) to a mixture of 1.5 grams of ferrocene and 5 ml of acetic anhydride. [*Caution:* Both the phosphoric acid and acetic anhydride are irritants to the skin and soft tissue. Acetic anhydride should be used in well ventilated areas. Should either reagent touch the skin, the affected area should be washed immediately with large quantities of water.] Stopper the flask with a rubber stopper in which is inserted a drying tube containing anhydrous calcium chloride as illustrated in Figure 6-2. Place the flask in an 800 ml beaker of water and heat the water to boiling for 10 minutes. Pour the mixture into a 600 ml beaker containing about 20 grams of ice. When the ice is melted, neutralize the mixture by adding solid sodium bicarbonate and stirring until gas is no longer evolved. This will take about 8 g of sodium bicarbonate. Cool the beaker and its contents on an ice bath for at least 30 minutes to insure precipitation of the ferrocenes. Filter the mixture using a Büchner funnel and suction; wash the residue with water

Neutralize the mixture with $NaHCO_3$.

FIGURE 6-2

Reaction flask for the acetylation of ferrocene.

until the filtrate is a pale orange. Air dry the solid and prepare it for separation of its components using dry column chromatography and/or for studying the effectiveness of several solvents in thin layer chromatographic separations as described in Part D.

Dry-Column Chromatography. Obtain a plastic chromatography column and place a loose plug of glass wool at one end, similar to Figure 6-3. Using a powder funnel, pour about 3 grams of activity III alumina into the tube and on top of the glass wool. Pack the alumina firmly by holding the filled column in a vertical position and tapping the plugged end gently against a folded towel or a clean piece of wood. Gently tap the sides of the tube to compact the alumina throughout (Note 1). Now clamp the column in a vertical position and tap it gently at the top of the alumina packing until   **Pack the column.** this surface layer is level. There should be no cracks or air pockets visible in the alumina column. Keep the column in the vertical position until the separation is complete, otherwise the column material will be disturbed and the chromatogram uneven.

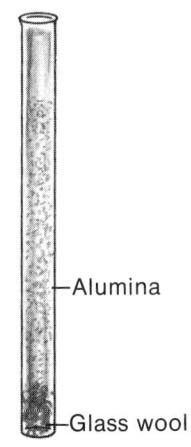

FIGURE 6-3

Column for dry column chromatography.

—Alumina

—Glass wool

Prepare the crude reaction mixture for chromatography by dissolving it in about 5 ml of acetone in an Erlenmeyer flask to which has been added about 1 gram of alumina. Evaporate the solvent by swirling it under a gentle stream of air. When completely dry, add this mixture of alumina and crude product to the top of the column and tap the column gently to form a level surface. Gently add fresh alumina on top of this surface to provide a layer approximately 1/4 inch in depth on top of the sample. Once again tap the top of the column until the surface is level.

Develop the column by carefully pouring methylene chloride, CH_2Cl_2, down the walls of the column until the height of the liquid layer is about 2 inches. As the solvent percolates through the column, add fresh solvent so as to maintain the 2 inch liquid layer on the column. When the solvent has reached a point about *three-fourths inch* from the bottom of the column, remove the excess solvent from the top with a pipet and allow the column to develop until the solvent front has reached about *one-fourth inch* from the bottom. Then place the column on its side to terminate the development. Mark the position of the solvent front and determine the R_f values for all colored bands.

Cut the developed plastic column, with a knife or strong scissors, between each separated component (Note 1). Place each segment in an appropriately labeled beaker and isolate it by washing the mixture with three 25 ml portions of acetone. After each wash, carefully decant the acetone solution into an appropriately sized flask, and evaporate the solvent to obtain the product.[4] [*Caution:* When acetone is used, all flames must be extinguished.] Weigh the products isolated and determine the melting point of the acetyl ferrocene, which is a reddish orange in contrast to the more yellow ferrocene. A third component, the tan colored diacetylferrocene, also may be separated.

NOTE

Note 1. If the plastic chromatography columns are unavailable, obtain a 15 centimeter length of clean dry 10 mm pyrex glass tubing and place a loose plug of glass wool at one end, Figure 6-3. Using a powder funnel, pour about 3 grams of activity III alumina into the tube and on top of the glass wool. Pack the alumina firmly by holding the filled column in a vertical position and tapping the plugged end gently against a folded towel acting as a cushion on the top of the lab bench. Gently tap the sides of the tube to compact the alumina throughout.

At each place where the instructions say "gently tap," if you are using a glass column, place a towel over the top of the glass column and *tap gently* with a piece of wood or some soft object that will not break the glass.

After developing the compound on silica gel in a glass column, insert a plug of glass wool into the top of the column. Place the column on a paper towel so that the alumina can be extruded from the glass tubing onto the towel with a minimum of deterioration. Holding the glass tube firmly place a glass rod against the glass wool plug at the top of the column and carefully extrude the alumina onto the towel. Divide the extruded column into segments, each containing one of the separated components.

---

[4]See Figure 2-2 for a set-up of the apparatus that is used for the solvent evaporation.

D: Effectiveness of Developing Solvents in
Thin-Layer Chromatography

In this section, we shall study the effectiveness of various developing
solvents in separating ferrocene and acetylferrocene using thin-layer chromatography. Obtain four plastic thin-layer plates from your instructor. Handle
the slides by touching only their edges. Do not touch the flat surface since
this will destroy the uniformity of the silica gel layer. Dissolve a very small
portion, approximately 1/4 of a micro-spatula full, of the ferrocene-acetyl-
ferrocene reaction product from Part C in 10 or 20 drops of benzene. Using
a 15 μl disposable pipet apply approximately 10 μl or 2 drops of this extract
at a spot in the center of each of the plates, approximately three-fourths
inch from the lower edge. Prepare four developing tanks by placing enough
developing solvent in the bottom of a small beaker or jar so that the level of
the liquid is no more than 1/2 inch from the bottom. The developing
solvents that can be used for this purpose are: benzene; ether; petroleum
ether; ethyl acetate; and the mixed solvents: 90% benzene, 10% acetone;
90% petroleum ether, 10% ethyl acetate; 70% petroleum ether, 30% ethyl
acetate. From this list, you may select any four developing solvents. When
the solvents have been placed in their respective jars, stopper the jars or
cover them with Saran Wrap[R] and shake them gently to facilitate saturation
of the air with the solvent. Carefully place a plate in each of the jars with the
edge closest to the sample spot toward the bottom. Be certain that the
solvent surface is well below the sample spot, Figure 6-4. When the plate is in
position, cover the container again and allow the chromatogram to develop
until the solvent front has moved approximately 3 to 3-1/2 inches along the
surface of the plate.

When the chromatograms are developed, remove the plates from the
developing tanks, mark the position of the solvent front in each case and
allow the plates to dry. Identify the spots due to ferrocene (yellow), acetyl
ferrocene (orange-red), and diacetyl ferrocene (tan). The latter is present in
small amounts.

Calculate the R_f values for the colored components and compare the R_f
values for each component in the several solvents. Compare the rate of
movement of the components in various solvents with the effectiveness of
the separation.

Account for the differing speeds of movement across the surface in terms
of the molecular structure of the compounds being separated and the
composition of the solvent.

Add the mixture to the tlc plates.

Develop the plates.

Dry the tlc plates. Determine the R_f value.

FIGURE 6-4

Beaker containing a thin layer plate and developing solvent. The beaker must be covered during
development of the chromatogram.

QUESTIONS

1. Write equations using structural formulas for the reaction of
   a. potassium hydroxide with cyclopentadiene
   b. tin(II) chloride with ferrocenium ion
   c. Ce^{4+} ion with ferrocene
   d. ferrocene with an oxidizing agent
   e. ferrocenium ion with a reducing agent

2. Write electron-dot formulas for
   a. cyclopentadienide ion
   b. ferrocene
   c. ferrocenium ion
   d. acetylferrocene

3. Cobaltocene, $(C_5H_5)_2Co$, and nickelocene, $(C_5H_5)_2Ni$, are known. Are they expected to be as stable as ferrocene? Why or why not?

4. On the basis of your observations and those of other students in the laboratory, classify each of the following as a stronger or a weaker oxidizing agent than ferrocenium ion: Ce^{4+}, $Cr_2O_7^{2-}$, Sb^{3+}, H_2O_2, NO_2^-, ClO^-, Sn^{4+}. If possible, prepare a table listing these species in order of decreasing strength as oxidizing agents.

5. Following on 4 above, classify each of the following as a stronger or weaker reducing agent than ferrocene: Sn^{2+}, Cl^-, H_2S, H_2O_2, NO_2^-, Sb, Cr^{3+}. If possible, prepare a table listing these species in order of decreasing strength as reducing agents.

6. Distinguish between partition and adsorption chromatography. Describe briefly in terms of molecular phenomena how components in mixtures are separated in partition, and in adsorption chromatography.

7. List some factors that control the rate of movement of a component through or across the stationary phase in a chromatographic separation. Suggest how each of these factors might be changed to decrease the speed or to improve the effectiveness of the separation.

REFERENCES

1. Jolly, W.L.: *Encounters in Experimental Chemistry*. Harcourt, Brace, Jovanovich, 1972, p. 26.
2. Angelici, R.J.: *Synthesis and Technique in Inorganic Chemistry*. W. B. Saunders Co., 1969, p. 141.
3. Gilbert, J. C. and Monti, S. A.: *J. Chem. Educ., 50:* 369 (1973).
4. Loev, B. and Goodman, M. M.: *Chem. Ind.,* 2026 (1967).

Experiment 6
REPORT SHEET

Name _____ Section No. _____

Part A.

Grams of $FeSO_4 \cdot 7H_2O$ taken   _____

Weight of empty weighing bottle   _____

Weight of bottle and ferrocene   _____

Weight of ferrocene   _____

Yield of ferrocene   _____

Melting point of ferrocene   _____

Equation for preparation of ferrocene:

Part B.

| *Solution* | *Color change observed* | *Element or substance undergoing* | |
		Oxidation	*Reduction*
$SnCl_2$	_____	_____	_____
$NaClO$	_____	_____	_____
H_2S	_____	_____	_____
$NaNO_2$	_____	_____	_____
H_2O_2	_____	_____	_____
$SbCl_3$	_____	_____	_____
$K_2Cr_2O_7$	_____	_____	_____

Equations for reactions observed

Part C.

Measured R_f values: Ferrocene _____ ; Acetylferrocene _____

Weight of substances separated: Ferrocene _____ ; Acetylferrocene _____

Melting points of substances separated: Ferrocene _____ ; Acetylferrocene _____

Part D.

Developing Solvent	Distance moved, cm		R_f Values	
	Ferrocene	*Acetylferrocene*	*Ferrocene*	*Acetylferrocene*
_____	_____	_____	_____	_____
_____	_____	_____	_____	_____
_____	_____	_____	_____	_____
_____	_____	_____	_____	_____

Attach sheets with answers to assigned questions.

SPECTRAL ANALYSIS

SPECTROSCOPY AND COLOR

OBJECTIVE

To provide experience in the experimental and interpretative aspects of emission and absorption spectroscopy and to illustrate important principles of the chemical basis for color.

EQUIPMENT NEEDED

Spectroscope; hydrogen and helium discharge tubes for atomic spectra; high voltage transformer, or induction coil and DC power source for operating the tubes; white light source, such as a flashlight; platinum wires; spectrophotometer; matched cuvettes or cells for spectrophotometer.

REAGENTS NEEDED

Solutions (0.1 M) of sodium chloride, potassium chloride, calcium chloride, strontium chloride, and barium chloride; unknown solutions containing two or more of the preceding chlorides; concentrated hydrochloric acid. Solutions: 0.10 M cobalt(II) nitrate; 0.10 M nickel nitrate; 0.02 M chromium(III) nitrate; 0.001 M iron(III) chloride in 0.1 M HCl; 0.10% titanium(III) chloride, 0.02% aluminon; a solution 0.01 M in copper(II) nitrate, 0.04 M in ammonium hydroxide and 2.0 M in ammonium nitrate.

INTRODUCTION

Color is said to arise from the emission or absorption of photons of visible light by chemical species. Photons are absorbed or emitted as electrons move from one energy level to another in atoms, molecules or ions. Chemical species that have been "excited" by heat, light, electrical or other energy sources may emit photons as electrons return from higher to lower energy states. Electrons in chemical species may absorb photons and move to higher energy levels. Light emission is studied using spectroscopes; absorption is studied using spectrophotometers. In this experiment both types of instruments will be used to study emission and absorption, and to examine some fundamental relationships among energy levels, wavelengths of photons absorbed or emitted, and the observed color of the species or chemical system.

In Part A the arbitrarily numbered scale on the spectroscope is first calibrated in terms of wavelength, using the helium spectrum. The hydrogen spectrum is then examined, and the wavelengths of the various lines are determined and interpreted in terms of electron transitions between energy levels in the hydrogen atom. In Part B the spectrum of helium is examined, and its more complicated structure, as compared to hydrogen, is related to the increased number of energy levels produced by the addition of the second electron. In Part C the spectroscope is used to determine the presence or absence of several cations in a solution. In Part D, the spectrophotometer is used to determine the wavelength(s) absorbed by ions in colored aqueous solutions, and the relation between observed and absorbed color is examined.

A. The Hydrogen Spectrum

The spectrum of an element is the light emitted or absorbed by the atoms of that element. Atoms may absorb light of discrete wavelengths and become excited atoms, which emit light of particular wavelengths. An emission spectrum consists of discrete lines, each of a particular color. Each line corresponds to a single wavelength or to a narrow band of wavelengths of light.

In the spectroscope, light emitted from excited atoms is passed through a prism and separated into its various wavelengths, an effect very similar to the way the sun's white light is separated into its component colors by rain droplets (Fig. 7-1). The atoms may be excited by light absorption, by an electric discharge, by heating to incandescence, or by some similar means. For this experiment, excitation of atoms is accomplished by passing an electric discharge through a gaseous sample of the element. Because each element has its own unique spectrum, a listing of the wavelengths of the lines in its spectrum identifies an element just as fingerprints identify a person.

Scientists think of light as being composed of tiny packets of energy called photons. The energy carried by a photon is related to the wavelength associated with the photon by the Planck equation

$$E = hc/\lambda \tag{1}$$

where E is the energy of the photon in ergs, λ is its wavelength in centimeters, c is the velocity of light in a vacuum, 3.0×10^{10} cm. per second, and h is Planck's constant, 6.62×10^{-27} erg-sec. The energy associated with the photons comprising each line in the spectrum of an element can be calculated from this equation.

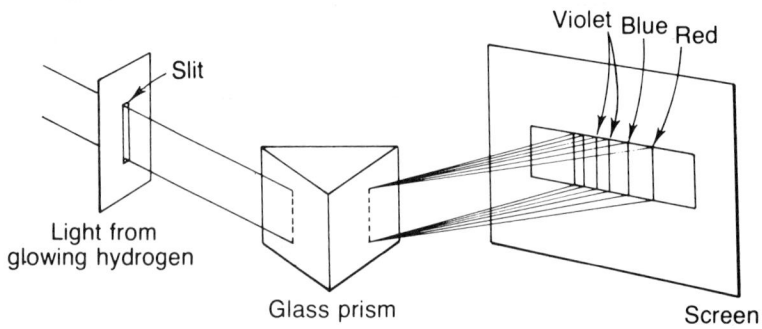

FIGURE 7-1

Dispersion of light by a prism.

Electron Transitions. According to the theory, the energy of photons emitted by excited atoms corresponds to energy differences between energy levels in the atom. For example, an electron falling from the second energy level, E_2, to the first energy level, E_1, in the hydrogen atom emits a line having a wavelength of 1216 Å or 1.216×10^{-5} cm. This corresponds to a photon of energy

$$\frac{hc}{\lambda} = \frac{6.62 \times 10^{-27} \text{ erg-sec.} \times 3 \times 10^{10} \text{ cm./sec.}}{1.216 \times 10^{-5} \text{ cm.}} = 16.3 \times 10^{-12} \text{ ergs}$$

This implies that an electron in the first energy level in the hydrogen atom has 16.3×10^{-12} ergs less energy than an electron in the second energy level in that atom, $E_2 - E_1 = 16.3 \times 10^{-12}$ ergs. Measurement of the wavelengths of lines in atomic spectra thus permits the experimenter to determine *energy level differences* within the atom.

Energy Levels. The equation relating energy levels in the hydrogen atom to the wavelengths of the spectral lines was originally fitted empirically to the experimental data and has the form

$$\frac{1}{\lambda} = R \left(\frac{1}{n_x{}^2} - \frac{1}{n_y{}^2} \right) \qquad (2)$$

where R is the Rydberg constant, having a value of 109,677.72 cm.$^{-1}$, and n_x and n_y are small integers. Bohr showed that a similar equation could be obtained theoretically on the assumption that an electron with 1 unit of negative charge was under the attractive influence of a positive charge Z times as great. Bohr's equation is

$$\frac{1}{\lambda} = RZ^2 \left(\frac{1}{n_x{}^2} - \frac{1}{n_y{}^2} \right) \qquad (3)$$

where n_x and n_y are the principal quantum numbers for the energy levels in the atom between which the electron falls (n_x is the lower and n_y the higher quantum number). If Z is interpreted as the charge on the nucleus, equation (3) reduces to equation (2) for hydrogen since, for the hydrogen atom, the nuclear charge is unity.

For the purposes of this experiment, equation (2) can be rearranged and R incorporated into a single constant to give the relation

$$\frac{1}{n_x{}^2} - \frac{1}{n_y{}^2} = \frac{1}{RZ^2\lambda} = \frac{9.12 \times 10^{-6} \text{ cm.}}{\lambda} \qquad (4)$$

Substituting the wavelengths (in centimeters) of the measured lines in the hydrogen spectrum into equation (4) gives a number that is equal to

$$\frac{1}{n_x{}^2} - \frac{1}{n_y{}^2}$$

From this number it is possible to calculate the exact energy levels between which the electron falls in emitting the photons that give rise to that particular spectral line. For example, if

$$\frac{1}{n_x^{\ 2}} - \frac{1}{n_y^{\ 2}}$$

is found to be 0.75, n_x must be 1 and n_y must be 2, since

$$\frac{1}{1^2} - \frac{1}{2^2} \text{ is } \frac{1}{1} - \frac{1}{4} = \frac{3}{4} \text{ or } .75$$

and the line in question represents the energy emitted when the electron falls from the second quantum level to the first. If

$$\frac{1}{n_x^{\ 2}} - \frac{1}{n_y^{\ 2}}$$

is found to be 0.14, n_x must be 2 and n_y must be 3, since

$$\frac{1}{2^2} - \frac{1}{3^2} = \frac{1}{4} - \frac{1}{9} = 0.14$$

and the electron falls from the third quantum level to the second.

After determining the wavelengths of several lines in the hydrogen spectrum you will be asked to make the calculation just described.

Energy Level Diagram. Once the quantum levels have been identified and the energy level differences established from the measurements, it is possible to record this information in a concise manner in an energy level diagram. Energies of the various levels are plotted vertically – to scale – and the transitions between levels are marked as heavy lines between these levels. The wavelengths of the spectral lines may then be marked on the vertical lines representing the electron transitions, as shown in Figure 7-2. Since only

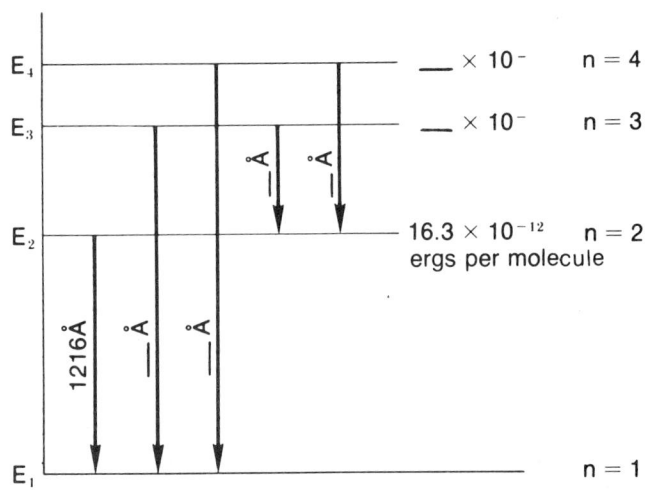

FIGURE 7-2

A portion of the energy level diagram for hydrogen.

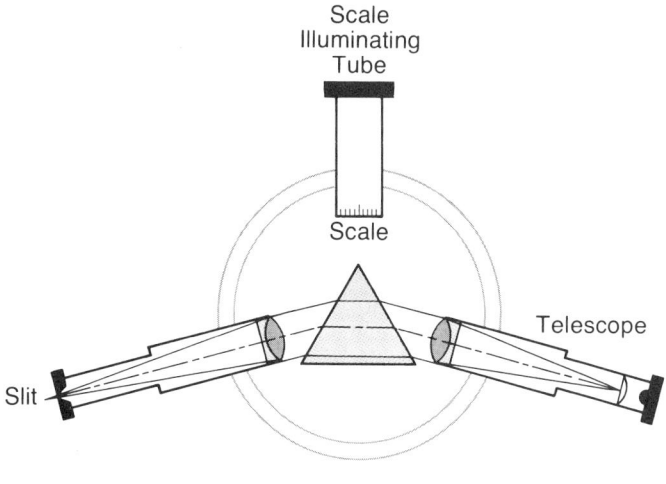

FIGURE   7-3

Spectroscope.

differences in energy may be determined from the wavelengths, the energy or the base level for any series of transitions may be arbitrarily taken as zero.

Spectroscope.   Figure 7-3 is a diagram of the spectroscope used in this experiment. Light from the excited atoms enters the instrument through the narrow slit, falls on the prism where it is separated into various colors, and moves along the telescope to the viewer's eye. A numerical scale is located in the third tube of the instrument. When a white light is placed at the end of this tube the viewer can see both the spectrum of the element and the scale. The scale is *not direct-reading;* i.e., the numbers are arbitrary and do not correspond to any particular measurement value. The purpose of calibration is to translate the numbers on this scale to wavelength values, using lines in the helium spectrum, wavelengths of which have been carefully measured by others.

To operate effectively, the instrument must be adjusted so that the slit is very nearly closed and *directly in front of the source* of the spectrum. If the slit is too wide, the prism will not split the beam effectively.

Your instructor will show you how to align and focus the instrument.

NOTEBOOK AND REPORT

Your notebook for this part of this experiment should contain the calibration data for the spectroscope, the calculations of the n values referred to on page 81, the energy level diagram for the hydrogen lines measured, and the answers to the Questions and Problems that the instructor assigns.

PROCEDURE

Calibration of the Spectroscope.   Carefully align and focus the spectroscope and adjust the width of the slit (on the front of the tube, which points to the light source), using the helium source so that you can see six or seven distinct colored lines.

Observe the spectrum of the helium discharge tube. Notice that there are distinct lines of different colors and that they are at different places on the calibration scale. Record (to the nearest 0.1 scale division) the scale reading for at least six lines in this spectrum. Record the color of the line next to its scale reading. The first strong line should be bright red.

The Hydrogen Spectrum. Substitute the hydrogen discharge tube for the helium source and record the scale reading for the intense red, the intense blue, and the violet line in the spectrum.

CALCULATIONS

The Calibration Curve. Table 7-1 gives the wavelengths, colors, and relative intensities of the helium lines. *You may not be able to see those with relative intensities below about 50, nor those at the highest and lowest wavelengths of the spectrum.* Identify the various lines with your scale readings by the colors listed, and make a plot of wavelength as ordinate against scale reading as abscissa. It should be possible to draw a smooth curve through the plotted points. If your graph does not give a smooth curve, you have made an experimental error or you have not identified some of the lines properly. In the latter case, reassign the lines to scale readings and replot them.

TABLE 7-1 THE HELIUM SPECTRUM

Wavelength (Å)	Color	Relative Intensity (1 to 1000)
7281	red	30
7065	red	70
6678	red	100
5876	yellow	1000
5048	light green	15
5016	light green	100
4922	dark green	50
4713	blue green	40
4471	blue violet	100
4388	violet	30
4121	violet	25
4026	violet	70
3964	violet	50

In the next paragraph, you are instructed to use the calibration curve to determine, by interpolating on the curve at the measured scale reading, the wavelengths of the lines of the hydrogen spectrum. Interpolation is more precise when made from a straight or nearly straight line than from a line of appreciable curvature. In some spectroscopes, a plot of the wavelength of the helium lines against the reciprocal of the scale reading is nearly a straight line. You may wish to make such a plot and use it as the calibration curve.

Wavelengths of the Hydrogen Lines. From the calibration curve determine the wavelengths of the red, blue, and violet lines of hydrogen.

Calculate the energy of the photons in the hydrogen lines (from $E = hc/\lambda$). Use equation (4) to calculate the value of

$$\frac{1}{n_x{}^2} - \frac{1}{n_y{}^2} ;$$

from this value calculate the value of n_x and n_y for each of the lines.

Another line in the visible spectrum of hydrogen appears at 4101Å. Calculate n_x and n_y for this line.

Energy Level Diagram for Hydrogen.   Using the results of the preceding calculation, calculate the energy levels in the visible spectrum of hydrogen, and construct a diagram similar to Figure 7-2, drawn to scale. Assign your value of n_x to the lowest level concerned in the visible spectrum, and give the values of the higher levels in ergs above this level. Indicate the values of n_y for the higher levels. Draw arrows to represent the electron transitions between levels that produce the four lines, and label them with the wavelengths of the lines, as requested in item 4 of the Report Sheet.

B. The Helium Spectrum

You will have observed that the helium spectrum contains many more lines than are observed for hydrogen. This means that there must be many more energy levels in the helium atom, so spaced that electron transitions between them produce photons of such energy that their wavelengths are in the visible region. Basically, there are more energy levels because there are two electrons in helium. Spectral lines are observed when one of these is excited to a higher level and then drops back to a lower level. The energy of both the higher and the lower level is dependent not only upon the quantum states of the electron undergoing the transition, but also upon the quantum states of the second electron.

You will recall that the energy of the electrons in an atom is designated by four quantum numbers, of which the values of n, l and s are important in a discussion of spectra in absence of a magnetic field. The values of l and s are of little importance in determining the energy of the hydrogen atom; the energy levels are determined principally by the value of n. In helium, with its higher nuclear charge and extra electron, this is not true, and the energy level of the atom corresponding, for example, to the quantum numbers of electrons 1 and 2 given by $n_1 = 2$, $n_2 = 2$, $l_1 = 1$, $l_2 = 1$ is different from that given by $n_1 = 2$, $n_2 = 2$, $l_1 = 1$, $l_2 = 0$, and from $n_1 = 2$, $n_2 = 2$, $l_1 = 0$, $l_2 = 0$. Several levels therefore appear in helium, where only one would be observed in hydrogen. The number of levels is further increased by inclusion of the spin quantum numbers, which can be either $+\frac{1}{2}$ or $-\frac{1}{2}$. The more energy levels present, the more possibilities there are for transitions between them, and an increased number of spectral lines can appear.

Spectral Series.   A simplification of the helium spectrum appears because in a high resolution spectroscope some of the lines are observed to be not single lines, but groups of three lines so closely spaced that an ordinary spectroscope does not separate them. These are known as *triplets;* the other helium lines, which are not separated by high resolution, are

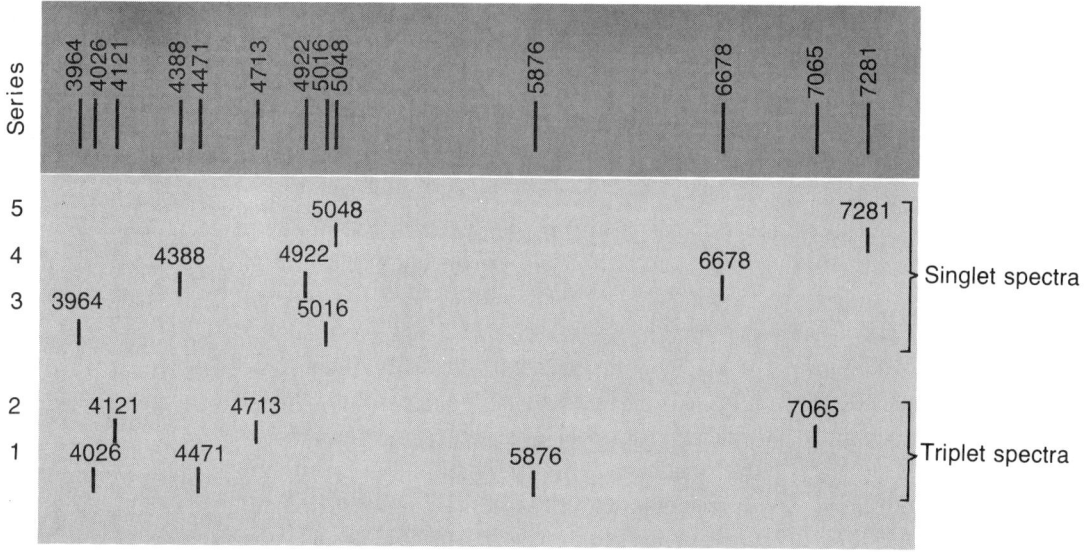

FIGURE 7-4

The helium spectrum.

known as *singlets.* It is easy to separate the singlet spectrum from the triplet spectrum; the lines at 7281, 6678, 5048, 5016, 4922, 4388, and 3964 Å are singlets, and those at 7065, 5876, 4713, 4471, 4121 and 4026 Å are shown by high resolution to be triplets. The singlets and triplets appear in situations in which the electron spins are to be either opposed (↑↓; singlets) or parallel (↑↑; triplets). The helium spectrum is thus separable into a singlet spectrum and a triplet spectrum. (This separation was so obvious that in the early days of spectroscopy it was thought that the two spectra came from two different kinds of helium, known as parahelium and orthohelium; chemical attempts to separate the supposed two kinds, however, were uniformly unsuccessful, and the influence of electron spin was finally recognized.) Consideration of other interactions between the electrons and nucleus and between electrons themselves enables the spectroscopist to separate the spectra into several series, so that the overall spectrum of helium is found to be made up of essentially several overlapping spectra, as shown in Figure 7-4.

Energy Levels. The question now arises as to whether the same considerations that produced equation (3) to represent the hydrogen spectrum will produce a similar equation for the helium spectrum. It does seem as though equation (3) might apply to helium also, since, within each series at least, the line spacings are similar to those found in the hydrogen spectrum. Hence, if the energy level spacings for each series represent changes in the principal quantum number, attempts at calculations of n_x and n_y might be scientifically rewarding. The nuclear charge, Z, for helium, however, is 2 and $Z^2 = 4$, and equation (3) shows that wavelengths for similar changes in principal quantum number should be much smaller for helium than for hydrogen. On the other hand, we cannot expect the full effect of the nuclear charge to be exerted on the electron making the transition (the "first" electron), because there is another electron (the "second" electron) near the nucleus, reducing the charge affecting the first electron.

Let us assume that the shielding of the nucleus is roughly equivalent to 1 unit of charge, and calculate from the data given in Fig. 7-4 a value of the effective nuclear charge on the helium atom. Since other quantum considerations are important in distinguishing the several series in the helium spectrum, we may expect the effective nuclear charge to differ from one series to another. It may not even be constant within a series, because the energy of one electron must of necessity affect the energy of the other; to presume that the two act independently is only an approximation.

CALCULATIONS

Using equation (3) in the form

$$\frac{1}{\lambda} = RZ_{eff}^2 \left(\frac{1}{n_x^2} - \frac{1}{n_y^2} \right) \tag{5}$$

calculate for one of the series in Figure 7-4 that contains three lines the average effective nuclear charge for that series. In each case, presume that n_x is 2, and that n_y is 3, 4 or 5 as λ decreases in the series.

Construct an energy level diagram (drawn to scale) for helium, given that the base levels (corresponding to $n_x = 2$) for the five series of Figure 7-4 are at approximately 3.40×10^{-11}, 3.40×10^{-11}, 3.30×10^{-11}, 3.36×10^{-11}, and 3.36×10^{-11} ergs, respectively, above the ground state of the helium atom. You may wish to separate your diagram into five adjacent vertical parts, one for each series. See item 4 of the Report Sheet.

NOTEBOOK AND REPORT

Your notebook should contain the calculations referred to in the preceding paragraphs, and your report should show the average values of $Z_{effective}$ and the energy level diagram for helium.

C. Qualitative Analysis by Use of the Spectroscope

As discussed in Experiment 2 (in qualitative analysis procedures) chemical and physical properties of substances are used to determine the presence or absence of these substances in a sample of unknown composition. One property used frequently for this purpose is the spectrum.

Often the emission spectrum of an atom or ion is simple enough that an analyst can determine its presence or absence by examining the spectrum of the sample. Sodium, potassium, calcium, strontium, and barium ions have spectra that are simple and easy to recognize. In fact, these ions, when excited, give characteristic colors that can be recognized visually even without a spectroscope — sodium ions appear yellow, potassium ions red-violet, calcium ions orange-red, strontium ions deep red, and barium ions green. If several of these ions are present in a mixture, the eye may not be able to distinguish among them, but a spectroscope can do so.

In this part of the experiment you are asked to excite the ions of each of the five elements listed by using the flame of an ordinary burner, and to observe the light emitted by using first only the naked eye and then the spectroscope. You will then be asked to observe the spectrum of a solution containing two or more of these ions and, on the basis of the observation, to determine which ions are present and which are absent.

PROCEDURE

The ions to be observed can be placed in the flame on a platinum wire (which has been sealed into one end of a piece of glass tubing), but the wire must be cleaned carefully before each new ion is introduced.

To clean the wire, heat it in the hottest portion of a colorless burner flame until the color imparted to the flame by the ions adhering to the wire disappears. If this color does not disappear after the wire becomes red hot, allow it to cool for about 10 seconds and immerse it in a small beaker containing about 2 ml. of concentrated hydrochloric acid. Then return the wire to the flame. Repeat this until the wire no longer imparts a color to the flame. The hydrochloric acid converts any salts adhering to the wire to chlorides, which are volatile and sublime.

The cleaned wire can now be dipped into a solution or powder containing ions having spectra to be observed.

If the flame from the platinum wire is not persistent enough to get a good reading in the spectroscope, a 2 X 2″ piece of asbestos paper can be folded twice to give a ½ X 2″ wick which can be saturated with the solution and held on the flame. The asbestos paper will give a positive test for Na^+.

Place a lighted burner directly in front of the slit of the spectroscope but far enough away so that the flame will not damage the instrument. Examine the spectrum of each of the ions previously discussed by dipping a clean platinum wire into a solution or powder containing the ion and placing the wire in the flame. Observe the color of the flame, and record both the color and the spectroscope reading for each prominent line in the spectrum. It may be necessary to dip the wire into the solution or powder and to return it to the flame several times in order to get all the information needed. Before you proceed to the next spectrum, the wire must be carefully cleaned.

When all the spectra have been viewed and the appropriate data recorded, obtain an unknown solution and, using the same procedure, view its spectrum and record both the color and the spectroscope reading for each prominent line. Use the information you have obtained to determine which ions are present and which are absent in the unknown sample.

NOTEBOOK AND REPORT

The notebook should contain data on the color and spectroscope reading for all prominent lines in the spectra of the five ions and of the unknown solution. The report should contain a description of the spectrum of the unknown and the rationale used in determining the presence or absence of each ion.

D. Spectrophotometry

The measurement of the absorption of radiation by chemical species is known as spectrophotometry. The instruments used are spectrophotometers.

These are constructed so that the sample to be studied can be irradiated with light (or other radiation) of known wavelength and intensity. The wavelength can be varied continuously by the operator (or automatically) and the amount of radiation absorbed or transmitted by the sample determined for each wavelength used. In this way, it is possible to learn which wavelengths of radiation are absorbed by the sample and how effective the species in the sample are in absorbing a particular wavelength. From this information, an absorption spectrum for a species can be obtained and used to identify the species in unknown samples. For example, it is known that the $Co(NH_3)_6{}^{3+}$ ion absorbs wavelengths between 3500 and 4000 Å and between 4500 and 5200 Å. If this ion is present in an unknown mixture and the mixture is placed in a spectrophotometer, one would expect to see absorptions in the 2 wavelength regions indicated.

In many cases, the amount of a substance present in a sample can be determined by spectrophotometry. This is discussed and illustrated in Experiment 23. Diagrams of and instructions for using simple spectrophotometers are given in Appendix II.

In this experiment, we shall use the spectrophotometer to determine the wavelengths absorbed by some simple substances in aqueous solutions and attempt to relate the color of visible light absorbed by the sample with its observed color.

Table 7-2 gives the wavelengths associated with various colors of visible light.

TABLE 7-2 WAVELENGTHS ASSOCIATED WITH VARIOUS COLORS OF VISIBLE LIGHT

Color	Wavelength Range (Angstroms)
Red	7500 – 6100
Orange	6100 – 5900
Yellow	5900 – 5700
Green	5700 – 5000
Blue	5000 – 4500
Violet	4500 – 4000

In this portion of the experiment, you will be asked to determine the absorption spectrum of the colored species in two of the solutions in Group A below and one in Group B.

TABLE 7-3 SOLUTIONS FOR ABSORPTION SPECTRUM DETERMINATION

Group A	Colored species present
Solution	
0.10M $Co(NO_3)_2$	$Co(H_2O)_6{}^{2+}$
0.10% $TiCl_3$	$Ti(H_2O)_6{}^{3+}$
0.01M $Cu(NO_3)_2$, 0.04M NH_3, 2.0M NH_4NO_3	$Cu(NH_3)_4{}^{2+}$
0.02% aluminon	

Group B	Colored species present
0.10 M $Ni(NO_3)_2$	$Ni(H_2O)_6{}^{2+}$
0.02 M $Cr(NO_3)_3$	$Cr(H_2O)_6{}^{3+}$
0.001 M $FeCl_3$ in 0.1 M HCl	$FeCl_4{}^-$

Unknown = CuCl

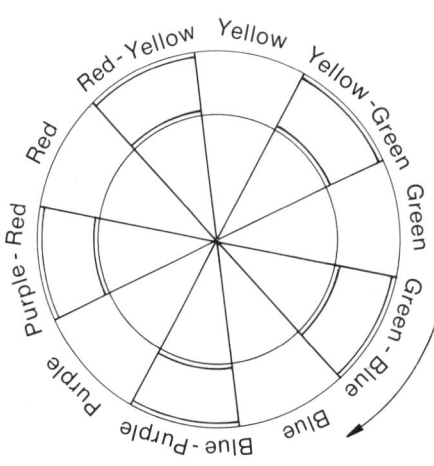

Purple color

FIGURE 7-5

A color wheel showing the complementary relationships among colors.

Figure 7-5 is a color wheel showing the complementary relationship among colors. This might be useful in interpreting the information obtained from the spectrophotometer.

PROCEDURE

Set for zero and 100% transmittance.

The following procedure can be used to obtain the absorption spectrum of your samples. Obtain a pair of matched cuvettes from the instructor. Clean and dry them thoroughly. Fill one with water, the second with the solution to be examined. Switch on the spectrophotometer, allow a 20 minute warm-up period, and turn the wavelength control to 340 millimicrons (340 nanometers or 3400 Å). Set the meter at zero by turning the dark current knob. Place the cuvette containing water in the sample cell holder of the instrument and close the cover. Now set the meter at 100% transmittance by means of the percentage control (light control) knob. Replace the water-containing cuvette with the cuvette containing the solution. Close the sample holder cover and read and record the absorbance from the appropriate scale on the meter.

Obtain absorbance data.

Obtain absorbance as a function of wavelength.

Set the wavelength dial at 360 millimicrons and repeat the above procedure using both the water-filled and solution filled cuvettes. Again, read and record the absorbance value from the meter.

In a similar manner, measure the absorbance of the solution at 20 millimicron intervals from 380 to 650 millimicrons. However, in the regions of maximum absorbance, take readings at 10 millimicron intervals to achieve a smoother curve. Plot the absorbance value as a function of wavelength for all three solutions, identify the colors absorbed by the color-bearing species in the solutions, and relate this to the observed color of the solutions using Figure 7-5.

Plot the spectrum.

Experiment 7C
REPORT SHEET

Name _____   Section No. _____

1. *Data For Known Solutions*

Ion	Observed color	Scale reading	Wavelength (Å) from calibration curve, Part A
Na^+	_____	_____	_____
	_____	_____	_____
	_____	_____	_____
K^+	_____	_____	_____
	_____	_____	_____
	_____	_____	_____
Ca^{2+}	_____	_____	_____
	_____	_____	_____
	_____	_____	_____
Sr^{2+}	_____	_____	_____
	_____	_____	_____
	_____	_____	_____
Ba^{2+}	_____	_____	_____
	_____	_____	_____
	_____	_____	_____

2. *Data For Unknown Solutions*

Observed color	Scale reading	Ion identification
_____	_____	_____
_____	_____	_____
_____	_____	_____
_____	_____	_____
_____	_____	_____
_____	_____	_____

**Experiment 7D
REPORT SHEET**

Name _____ **Section No.** _____

Solution I: Colored species present _____ ; color of solution _____ .

Maximum absorption(s), λ from _____ to _____ ; from _____ to _____ .

Color of maximum absorption   _____   _____

Solution II: Colored species present _____ ; color of solution _____ .

Maximum absorption(s), λ from _____ to _____ ; from _____ to _____ .

Color of maximum absorption   _____   _____

Solution III: Colored species present _____ ; color of solution _____ .

Maximum absorption(s), λ from _____ to _____ ; from _____ to _____ .

Color of maximum absorption

Attach Graphs for Part D

ANALYSIS

GRAVIMETRIC DETERMINATION
OF SULFATE

OBJECTIVE

To study the technique of gravimetric analysis and to determine the percentage of sulfate in a soluble sulfate compound.

EQUIPMENT NEEDED

Two 250-ml. beakers; two crucibles with lids; desiccator; two Meker burners; two plastic funnels; filter support; two clay triangles; two 4-inch rings; two rubber policemen; wash bottle; medicine dropper; tissues; ashless filter paper; crucible tongs.

REAGENTS NEEDED

Unknown sulfate sample (available from instructor); 10 percent $BaCl_2$ solution; 6M HCl; 0.1M $AgNO_3$ test solution; plastic squeeze bottles containing 50-50 ethanol-water mixture and 95% ethanol.

INTRODUCTION

This experiment is an introduction to quantitative analytical chemistry. Because a chemist must often know precisely the composition of the materials or systems he studies, this area constitutes a major segment of the science. For example, he must often know whether a given material contains 10.60 or 10.64 percent sulfate. Consequently, procedures have been developed for determining the amount of an element or ion that might be present in a given sample.

One of the oldest and still one of the most accurate methods for determining composition is gravimetric analysis. This involves: (1) completely precipitating the desired component as a pure compound of precisely known chemical composition, (2) filtering or separating the precipitate onto filter paper that leaves no ash when burned, (3) burning off the filter paper under conditions that will not alter the chemical composition of the precipitate (or will alter it in a precisely known fashion), and (4) weighing the known precipitate. From data obtained by such procedures, it is possible to calculate the percentage of the desired component in the original sample.

In this experiment you will be issued a sample containing a specific amount of sulfate. You will be expected to analyze the sample and to report a value for the percentage of sulfate within a certain range of the correct value. The grade will depend upon how close your answer corresponds to the correct value. If the reported answer is outside accepted limits, you may be expected to repeat the experiment. Therefore, careful work will prove both rewarding and time-saving. It should also provide a deeper understanding of the quantitative basis of chemistry.

The method to be employed involves weighing a sample of the material to be analyzed, dissolving it in water, and adding barium chloride solution. Since barium sulfate is only very slightly soluble (about 1×10^{-5} moles of the salt will dissolve in 1 liter of water at $25°C$), this compound immediately precipitates from the solution. The reaction may be represented by the net ionic equation

$$Ba^{2+}(aq) + SO_4{}^{2-}(aq) \rightleftharpoons BaSO_4(s)$$

After the precipitation has been accomplished it is necessary to filter quantitatively the coagulated precipitate. The filter paper can then be burned off in a porcelain crucible. After ignition is completed (to constant weight), the dry barium sulfate precipitate is weighed, and this weight is used to calculate the percentage of sulfate in the original sample.

TECHNIQUES

Sample Weight from Loss in Weight of Container. Solid materials intended for analysis are usually dried in an oven near $100°C$ and transferred to a small lightweight *weighing bottle,* which is then stored in a desiccator. When a sample is needed, the bottle is weighed, a suitable quantity is transferred to the beaker or flask in which the chemical operation is to be performed, and the bottle is reweighed and returned to the desiccator. The difference in weight before and after the transfer represents the weight of the sample taken, provided no spills occurred.

For a successful operation, the transfer must be quantitatively perfect, since all material removed from the weighing bottle is presumed to have entered the chemical process. The transfer is best accomplished by tilting the weighing bottle over the beaker and tapping it gently with a pencil or spatula to pour out small quantities of the solid. The amount removed may be checked periodically by a rough weighing on the balance. If the material does not absorb water rapidly, the weighing may be made with the weighing bottle uncovered; otherwise the cover must be replaced as soon as material has been removed from the bottle.

It is also necessary that the weighing bottle not pick up additional weight during handling, since this would make the apparent sample weight less than it actually is. The weighing bottle should be carefully cleaned and dried before use, and thereafter should not be touched with the fingers, which would leave perspiration salts and oils on it. It should be handled only when protected from the fingertips by a piece of clean tissue. The cover similarly should not be removed directly with the fingers, but only with tissue.

Filtration. A pure precipitate, once formed, must be separated from the liquid in which it was formed and washed free of any soluble components in this *mother liquor.* This is commonly accomplished by filtration through a

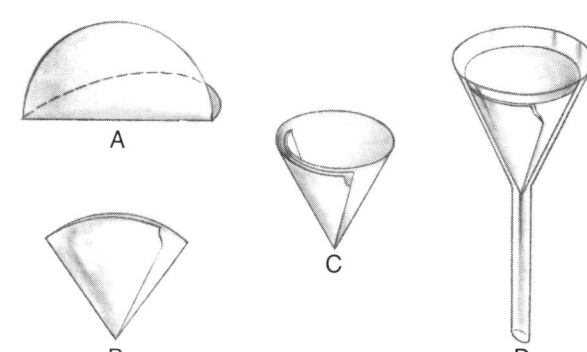

FIGURE 8-1

Illustrations for folding the filter paper.

filter paper, and the washing is carried out during this process also. For quantitative work, ashless filter paper, which has been treated with acids to remove inorganic materials, is employed; the type chosen should have a pore size small enough to retain particles of precipitate but not so small that it increases unnecessarily the time needed for water to flow through the filter paper.

Preparing the Filter. Filter paper is available in several diameters, and the size chosen should be such that the precipitate occupies approximately one-third or less of the filter paper cone. The filter funnel should be of comparable size so that the folded cone comes within 1 or 2 cm. of the top edge of the funnel; it must never extend over this edge.

Fold the filter paper as illustrated in Figure 8-1A. Make the second fold so that the edges do not quite overlap, and tear off a small corner at the outside to give a better fit at the fold (Fig. 8-1B). Place the cone in the funnel and wet it with distilled water, pressing it down gently (Fig. 8-1C and D). A properly fitted paper will let no air in around it; the stem of the funnel should fill completely with water so that the hydrostatic pull of the column of water will aid the filtration. Do not attempt to use an improperly seated filter cone; you will save time by discarding it and folding another one.

Filtering and Washing. Because it is more efficient to wash a precipitate in a beaker than on a filter paper, washing by decantation is usually advisable. Allow the precipitate to settle in the beaker, and then pour off the supernatant liquid through the filter, leaving as much of the precipitate in the beaker as possible. Use a stirring rod to guide the stream of water and to prevent splashing (Fig. 8-2A). Add wash solution to the beaker, stir it, allow the precipitate to settle, and decant the wash solution as before. This washing may be repeated as often as necessary; often a test of the success of the washing may be applied by collecting a small portion of the wash solution that runs through the filter and adding a reagent that will produce an observable sign if the unwanted contaminant is still present in the washings. Sometimes the first portions of mother liquor leaving the filter are cloudy; these should be poured back through the filter after its pores have been clogged by larger particles of precipitate. The chance that refiltration might be necessary indicates the need for collecting the filtrate in a clean beaker, even though the filtrate is later to be discarded. Several washings with small volumes of wash solution are more effective than a single washing with the same total volume of solution.

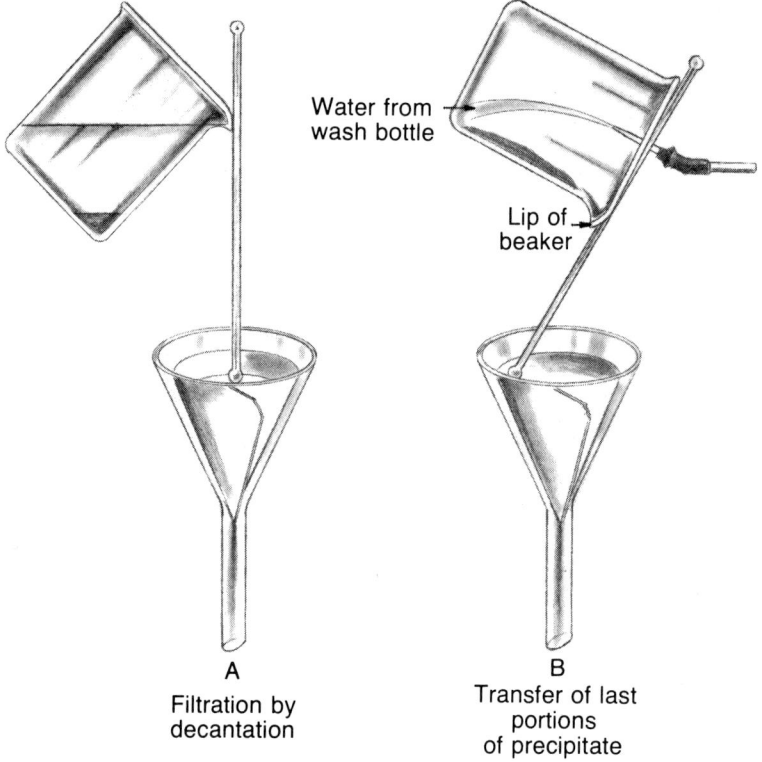

Water from wash bottle

Lip of beaker

A
Filtration by decantation

B
Transfer of last portions of precipitate

FIGURE 8-2

When washing is completed, the precipitate is stirred with a portion of wash solution and poured down the stirring rod without allowing the precipitate to settle. Final residues are transferred by means of a stream from a wash bottle (Fig. 8-2B). The last traces of precipitate adhering to the glass are removed by rubbing the glass with a moist rubber policeman; the policeman is in turn cleaned with a stream from a wash bottle or wiped gently with a scrap of moist filter paper, and the precipitate thus removed is added to the precipitate on the filter. Use a specific policeman for each sample.

Ignition of a Precipitate. The process that ends by heating a crucible to incipient redness preparatory to cooling and weighing is known as *ignition*. It is first carried out with the empty crucible, as in Experiment 4, so that a precise value for the tare weight may be obtained. When a filter paper and precipitate are present, the heating should follow a definite program, as follows:

Drying and Charring the Paper.   Support the covered crucible containing the wet filter paper and precipitate on a clay triangle in a slanted position. Heat it with a small flame impinging about the middle of the underside. Do not hurry the drying by using too hot a flame.

When the paper is dry, lift the cover slightly to allow access of air, and increase the heat to char the paper, moving the flame toward the bottom of the crucible. The paper must not be allowed to burst into flame, which might cause particles of precipitate to be swept out with the escaping gases. If a flame appears, close the cover immediately to smother it. Take care that complete combustion occurs in the burner flame so that reducing gases from it are not swept back onto the precipitate.

Igniting the Precipitate. When the paper is completely charred, remove the cover from the crucible, and increase the heat gradually until the bottom of the crucible is a dull red. It may be necessary to rotate the crucible occasionally to remove soot and tars that may have condensed during the charring. When the dark organic material has completely burned away, set the crucible upright and continue heating it for 10 minutes.

PROCEDURE

Weigh out (to 0.0003 g.), on the basis of loss in weight of the container, about 1 g. each of two samples of the unknown sulfate into marked 250-ml. beakers. Use the sample vial as the weighing bottle and prepare it for weighing by carefully washing the outside with detergent and water, wiping it dry with a tissue, and allowing it to stand in the laboratory atmosphere for 5 or 10 minutes to equilibrate it with the humidity level of the air. If there is a label on the vial it must be removed before the vial is used; transfer any needed identification to the stopper or cap. The weighings should be made to within 0.0003 g. and may be made with the vial uncovered. Record the weights of the vial before and after the samples are removed (Note 1).

Weigh two 1-gram samples.

Add 150 ml. of distilled water and 3 ml. of 6M HCl to each beaker (Note 2) and heat to near boiling, but do not actually boil the samples (Note 3). Counting the drops, add 10 percent $BaCl_2$ solution from a medicine dropper until precipitation appears to be complete. Test completeness of precipitation by allowing the precipitate to settle somewhat and then adding a drop of barium chloride gently so as not to disturb the precipitate. If additional cloudiness appears at the point of entry of the drop, precipitation is not complete (Note 4). After you have determined that precipitation is complete, add an excess of the barium chloride solution equal to one-fifth the number of drops of solution already used. Stir the sample constantly during the addition of barium chloride.

Add H_2O and HCl and heat. Add $BaCl_2$ dropwise.

Keep the solutions nearly at the boiling temperature for approximately 40 minutes (Note 2). Then check again for complete precipitation by adding a drop or two of barium chloride solution. If precipitation is not complete, keep adding 10 percent $BaCl_2$ solution until no further precipitation occurs and then add an excess of one-fifth the total amount. The total volume of $BaCl_2$ solution should vary from 5 to 9 ml., depending on the sample. If it is necessary for you to stop work at this point, cover the beakers with a watch glass or pieces of filter paper and put them away until the next laboratory period.

Check for complete precipitation.

Set up the two plastic funnels and fit each with a piece of 11-cm. ashless filter paper as illustrated in Figure 8-1. Filter and wash the precipitate by decantation, using warm distilled water, until the filtrate shows no chloride ion present, as indicated by silver nitrate test solution. The chloride ion test is best done by placing silver nitrate test solution in a small test tube and allowing a few drops from the funnel to fall into it. A cloudy precipitate of silver chloride shows that chloride ion is still present. Use the policeman and water from the wash bottle to remove the last traces of precipitate from the beaker. Continue washing the precipitate until only a barely detectable turbidity appears in the wash solution. Allow the filter paper to drain in the funnel.

Filter and wash the precipitates.

Test the wash liquid for Cl^-.

Change

Ignite and weigh crucibles.

During the working period, heat the empty, clean crucibles and covers over the full heat of the Meker burner, and place them in the desiccator. As always in cooling crucibles in a desiccator, wait a few moments before covering the desiccator tightly to permit the air in the desiccator to be heated by the hot crucible and to expand safely. Then close the desiccator and allow the crucibles to cool to room temperature. When the crucibles are cool, weigh them without the covers. Repeat this procedure until you get them to constant weight (within 0.0003 g.).

Change

Wash with alcohol.

Wash the precipitate with a 50-50 mixture of water-ethanol and then wash with 95% ethanol and allow the filter paper to drain in the funnel (Note 5). Fold over the top of the paper to encase the precipitate, and carefully transfer the paper to a marked crucible that has been ignited and weighed to constant weight as previously. If you must stop work at this time, cover the crucibles and store them in a desiccator until the next laboratory period.

Change

Dry and char the paper; ignite and weigh the precipitates.

Dry and char the filter paper according to the procedure given on p. 100 and ignite the precipitate (Note 6). Set the crucible in a desiccator, and weigh it when cool. Repeat the ignition, cooling, and weighing until a constant weight is obtained.

Much of the success in this determination depends on the ability of the analyst to obtain reliable weights of crucibles and their contents. Special precautions must be used in handling crucibles so that they are not contaminated by moisture or dust from the air, dirt or paint from the desk top, or oil from the fingers of the analyst. Therefore, crucibles should be handled with crucible tongs and should be kept in desiccators when not in use.

CALCULATIONS

From the weight of barium sulfate precipitated, calculate the weight of sulfate ion in the precipitate, using the ratio of the gram formula weight of sulfate to the gram formula weight of barium sulfate. This amount of sulfate was present in the original weight of the unknown sample. Calculate the percentage of sulfate in each sample trial.

Average the two results and calculate the average deviation in parts per thousand as follows: Find, for each trial, the difference between the measured value and the average; subtract in such a way that the difference is always positive. Add these results and divide by the number of trials. This quotient, divided by the average value and multiplied by 1000, is the average deviation in parts per thousand.

Record the arithmetic of the calculations in your notebook and on the report sheet neatly so that you and the instructor can check them readily for errors.

NOTEBOOK AND REPORT

In this as in most quantitative experiments, the experimental data can best be recorded in tabular form. The notebook record of the experiment will contain a table similar to that given in Experiment 4, with before and after weights of the crucibles, results of the calculations, and other data. A skeleton table may be set up before-hand, so that data will appear in a neat form readily assessed by another person.

In addition to the data and results of calculations, the report should contain a brief statement of the purpose and procedure, with complete and balanced equations for all reactions that occur, and a statement of difficulties encountered or observations that seem noteworthy or not in accord with expectations.

Answers to the questions at the end of the experiment should also be a part of the report.

NOTES

Note 1. Since this is a quantitative analysis, extreme care must be exercised at all times. If the sample is spilled or not *completely* transferred, or if the filter paper ruptures during filtration, discard the sample and start again.

Note 2. Acid must be added to the solution to prevent precipitation of salts other than barium sulfate when the barium ion is added. Hydroxide ions, sulfite ions, and carbonate ions, for example, if present in the sample to be analyzed, would form slightly soluble barium salts in neutral or basic solution, but would not precipitate in acid solution. A second reason for the acid is that it aids digestion of the precipitate. Digestion is the process of forming a precipitate that may be readily separated from the mother liquor by filtration; it is usually carried out, as here, by heating the precipitate in contact with the mother liquor so that the fine particles that may be formed at first recrystallize to form larger, coarse-grained crystals.

Note 3. Barium sulfate solutions have a pronounced tendency to become supersaturated at low temperatures, but not at high temperatures. The near boiling temperatures are necessary to overcome the tendency toward supersaturation.

Note 4. The excess barium ion is added to drive the reaction

$$Ba^{2+}(aq) + SO_4{}^{2-}(aq) \rightleftharpoons BaSO_4(s)$$

as far to the right as possible, thereby removing as many sulfate ions as possible from the solution. If excess barium ion is not added, enough sulfate ion may remain in solution to affect the final result. Because a huge excess of barium ions can also increase the solubility of barium sulfate, the 20 percent excess should not be greatly exceeded. The barium chloride solution is much more dense than the solution in which precipitation occurs. As a result, when precipitation is completed, you may notice a difference in the refractive index of the solution at the point of mixing as drops of barium chloride solution mix with your sulfate solution. Do not confuse this behavior with precipitation.

Note 5. Ethanol will help wash out the water and speed up the drying process. However, heat the crucible containing the sample slowly at first in order to dry the filter paper.

Note 6. Barium sulfate may be reduced to barium sulfide by carbon that is formed from charring of the filter paper. This reduction can be avoided by burning off the paper very slowly with the lid off the crucible to allow sufficient oxygen to remain in the crucible so that the carbon is removed as carbon dioxide.

When ignition is complete the precipitate is anhydrous barium sulfate.

QUESTIONS—PROBLEMS

1. List the principal sources of error in this experiment. In each case, state whether the error tends to make the results high or low.

2. If 0.6045 g. of a soluble sulfate gives 0.4231 g. of $BaSO_4$, what is the percentage of sulfate in the sample?

3. If an error of +1.0 mg. is made in weighing the barium sulfate precipitate and the percentage of sulfate reported is 42.75, calculate the true percentage of sulfate in a 1.00 g. sample.

4. If the unknown sample were to contain two soluble sulfate compounds, how would this affect your calculation of the percentage of sulfate in the sample?

REFERENCE

1. Day, R.A., Jr., and Underwood, A.L.: *Quantitative Analysis, Laboratory Manual.* 2nd edition. Prentice-Hall, Englewood Cliffs, N.J., 1967, Chapter 6.

Experiment 8
REPORT SHEET

Name _____ Section No. _____

	First Crucible Mark ____	Second Crucible Mark ____
1. Weight of crucible		
a) after first heating	_____	_____
b) after second heating	_____	_____
c) after third heating	_____	_____
d) at constant weight	_____	_____
2. Gross weight of sample and vial	_____	_____
3. Weight of vial after removing the 1 g. sample	_____	_____
4. Net weight of sample	_____	_____
5. Total weight of crucible and $BaSO_4$ precipitate		
a) after first ignition	_____	_____
b) after second ignition	_____	_____
c) after third ignition	_____	_____
d) at constant weight	_____	_____
6. Net weight of $BaSO_4$ precipitate	_____	_____
7. Weight of sulfate ($SO_4{}^{2-}$) in the 1 g. sample	_____	_____
8. Percentage sulfate in the sample	_____	_____
9. Average deviation	_____	

10. Attach sheets showing sample calculation and answers to
    assigned questions.

VOLUMETRIC ANALYSIS: PREPARATION AND STANDARDIZATION OF A SODIUM HYDROXIDE SOLUTION

OBJECTIVE

To illustrate acid-base titrations by the standardization of a sodium hydroxide solution against potassium hydrogen phthalate as a primary standard.

EQUIPMENT NEEDED

Buret; two 125-ml. Erlenmeyer flasks; 250-ml. Erlenmeyer flask; 400-ml. beaker; burner; analytical balance.

REAGENTS NEEDED

Potassium hydrogen phthalate, $KHC_8H_4O_4$; 3M NaOH solution (carbonate-free); 1 percent phenolphthalein indicator solution; Alconox or similar cleaning agent.

INTRODUCTION

One of the most important techniques for chemical analysis is *titration* to an *equivalence-point*. To illustrate this procedure, let us examine a specific problem.

Suppose that an investigator wishes to know the exact quantity of acid present in a certain mixture. He can find this value by determining the quantity of a base that must be added to the mixture just to neutralize the acid. The quantity of base needed can be measured by preparing a solution of a known concentration of the base, and measuring the volume of it needed for the neutralization. For example, if the concentration of the base solution is 0.5 mole of base in 1000 ml. of solution, and the investigator finds that exactly 20.00 ml. neutralizes the acid, he knows that (20.00/1000) × 0.5 mole of base has been required. From the chemical equation for the neutralization reaction

$$H_3O^+ + OH^- \longrightarrow 2H_2O$$

he knows that 1 mole of hydroxide ion reacts with 1 mole of hydronium ion; hence $(20.00/1000) \times 0.5 = 0.01$ mole of base reacts with 0.01 mole of acid. The latter figure is the value he set out to find, i.e., the quantity of acid in the mixture.

In order to find out that he needed exactly 20.00 ml. of the base solution he would add the solution from a measuring buret until an *indicator* in the mixture signified that the *end-point* of the titration had been reached; that is, that he had exactly neutralized the acid present. The volume of base solution used would then be read from the buret. In acid-base titrations, the indicator used would commonly be a dyestuff that changes color sharply at the end-point. In the titration just described, this occurs when the hydronium ion originally present is converted to water and a bare excess of hydroxide ion has been introduced. The end-point can be identified in ways other than through a color change, even in acid-base titrations, and an end-point that marks the completion of chemical reactions other than neutralizations can be indicated by a number of methods.

For proper use, a titrating solution must be of known concentration and must contain only a single chemically active reagent. Such solutions are known as standard solutions.

A widely used procedure for preparing and standardizing solutions for titrations is to prepare a solution of approximately the desired concentration and then to titrate it against an accurately weighed quantity of compound of known purity. The compound of known purity used in this manner is called a *primary standard.* Nearly all standard solutions of acids and bases are prepared by this method. For example, sodium hydroxide solutions can be standardized against a known weight of potassium hydrogen phthalate, $KHC_8H_4O_4$.

Potassium hydrogen phthalate is a monobasic acid; it contains 1 mole of neutralizable hydrogen per mole of substance used. This compound can be highly purified (99.97 percent pure material is available), it is not easily oxidized, it can be dried to constant weight, and it has a relatively high molecular weight, permitting accurate weighings to be obtained on samples as small as 0.002 mole. All these are desirable characteristics of a primary standard. One less desirable property of potassium acid phthalate is that it is a weak acid and, therefore, the indicator does not change color as sharply as might be desired. For this reason, it is necessary to carry out the titrations in the absence of carbon dioxide; boiled water must be used to dissolve the acid, and the base to be standardized must be free of carbonate. Phenolphthalein, which is colorless in acid solutions but turns pink at the end-point as base is added, is a suitable indicator in titrations of potassium hydrogen phthalate with sodium hydroxide solutions.

TECHNIQUE

Cleaning Volumetric Glassware. Volumetric equipment, such as pipets, burets, and volumetric flasks, is sufficiently clean if its surface is wetted uniformly by distilled water. Oil films adsorbed on the glassware prevent the glass walls from being wetted uniformly; drainage is then uneven and delivery is not precise. If distilled water drains from the inner walls of volumetric glassware in a uniform film, the equipment is clean; if the film coalesces into streams or droplets, further cleaning is necessary. A general rule is to clean glassware immediately after use, since it is much more difficult to remove chemicals and films that have aged and caked on the surface.

Acids (except hydrofluoric acid) do not attack Pyrex glass. Alkaline solutions etch glass on prolonged contact, and equipment should be rinsed free of such solutions soon after use. Except in unusual cases, all ordinary glassware and volumetric equipment is best cleaned with a warm 1 to 2 percent detergent solution, e.g., Alconox. Mechanical action, such as shaking or scrubbing with a bottle or buret brush, increases cleaning efficiency. After washing the glassware, be sure to rinse away the detergent with tap water and then with distilled water.

If the detergent solution is ineffective, consult your instructor for suggestions on using other cleaning agents, such as a warm 1 to 2 percent solution of trisodium phosphate.

Filling the Buret. Rinse a clean 25- or 50-ml. buret several times with distilled water and then, with the stopcock closed, rinse the buret with about 5 ml. of the reagent solution, twisting and turning it in a nearly horizontal position to cover all surfaces. Fill the buret above the zero mark with the reagent solution and then slowly open the stopcock to allow solution to run into the tip of the buret. Make sure that there are no air bubbles in the tip, because the bubbles will vary in size during the titration and affect the results. Adjust the level of solution in the buret near to the zero mark. Be sure to take your readings across the bottom of the meniscus, *estimating the reading to two decimal places* (Fig. 9-1).

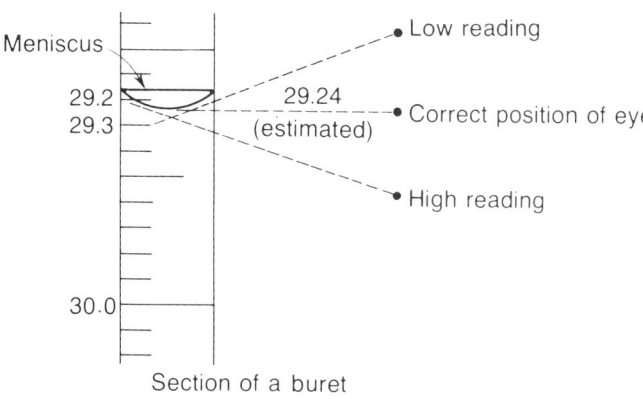

FIGURE 9-1

Estimation of a buret reading to the nearest 0.01 ml. The two etched marks nearest the bottom of the meniscus are for 29.20 and 29.30 ml. Visually divide the distance between the two into 10 equal parts and estimate on which of these imagined divisions the meniscus is. In this figure, the bottom of the meniscus appears to be on the fourth division, and the reading is 29.24 ml.

Reading the Buret. To make the volumetric measurement, one must accurately locate the position of the meniscus, i.e., the surface of the liquid, with respect to the markings on the buret. The apparent reading of the *bottom of the meniscus* is affected considerably by the position of the eye, as shown in Figure 9-1. To eliminate an error due to *parallax,* it is important to establish proper technique in reading the buret. It is easy to bring the eye to the proper level by sighting across the full circle (millimeter graduate marks) nearest the meniscus. In reading, the eye should be kept 12 to 18 inches from the buret. The apparent thickness of the meniscus, which is a

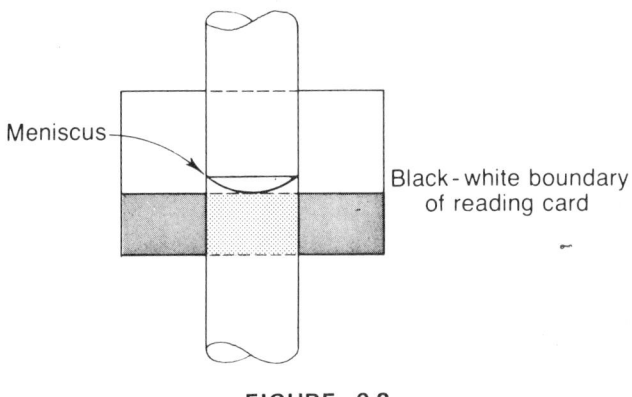

FIGURE 9-2

Proper position for the reading card.

reflecting lens, depends upon the lighting and background, and a reading card is desirable for reproducibility of the background (Fig. 9-2). A reading card may be constructed easily by pasting a strip of dull black paper across a 3- by 5-inch filing card.

NOTEBOOK AND REPORT

Prepare a neat table to show your raw data, i.e., the balance readings and the buret readings before and after titration for each sample. Design a table to contain all necessary data in your notebook before going to the laboratory.

Your report should include a summary of the data, sample calculations, qualitative observations made during the experiment, final conclusions, and answers to the questions at the end. Organize the report according to the outline on page 47.

PROCEDURE

Boil water to expel air.

To prepare carbon dioxide free water, boil about 250 ml. of distilled water in a clean 400-ml. beaker approximately five minutes. Pour about 200 ml. of the boiling water into a flat-bottomed flask, and cool the flask under the water tap, taking care not to shake the water unnecessarily, which would permit carbon dioxide to dissolve in it again. When cool, add to the water 40 ml. of the carbonate-free 3M NaOH solution, stopper the flask to prevent absorption of carbon dioxide, and mix it thoroughly by shaking. Keep this

Cool and add NaOH.

flask stoppered as much as possible hereafter.

The diluted sodium hydroxide solution must now be standardized; i.e., its exact concentration must be determined. For this purpose we titrate against a known weight of potassium hydrogen phthalate.

Clean two 125 ml. Erlenmeyer flasks with detergent and water and rinse them with distilled water.

Weigh out standard acid.

Obtain a dried sample of pure potassium hydrogen phthalate from the instructor and prepare two samples for titration against the sodium hydroxide solution by using the following procedure. Carefully weigh the cleaned glass vial (see Exp. 8) containing the dried phthalate, transfer about

1.2 g. into a 125-ml. Erlenmeyer flask, and reweigh the vial and its contents. Record all weighings. Dissolve the solid in about 40 ml. of recently boiled water. Prepare the second sample in exactly the same way. Be sure to mark the flasks so that you will know which flask contains which sample; note that the flasks have frosted circles for easy marking with an ordinary lead pencil. Add two drops of phenolphthalein indicator to each flask.

Add indicator.

Clean, rinse and fill a buret with the diluted sodium hydroxide solution, and record the initial buret reading.

Fill buret with NaOH.

Place one of the flasks containing the potassium hydrogen phthalate under the buret, and place a sheet of white paper under the flask to provide a suitable background for observing the color change while the base solution is being added.

Titrate the acid by allowing the sodium hydroxide solution to flow slowly into the flask. Control the flow of base solution from the buret with the left hand, and mix the solution in the flask by giving the flask an even, circular motion with the right hand, taking care not to allow liquid to splash from the flask. (Left-handed students will probably find it convenient to reverse this procedure.) The base solution may be added in a rapid stream of drops until it is noted that the pink color begins to linger while the solution is swirling in the flask. At this stage the solution should be added at a slower rate. Near the end-point, i.e., the point at which the entire solution just becomes pink, the base should be added so slowly that the color disappears between the addition of each drop and the next. When it appears that only a few more drops are needed, rinse down the walls of the titration vessel with distilled water before completing the titration (Note 1). Continue to add base until a barely pink color persists throughout the whole solution for at least 20 seconds. (Note 2). If the solution is red, you may have added too much base, and the result of the trial must be regarded with suspicion. Record the final buret reading, estimating the reading to within 0.01 ml.

Titrate with NaOH until barely pink.

Repeat the titration with the second sample of potassium hydrogen phthalate.

Repeat.

CALCULATIONS

The desired quantity here is the *molar concentration* of sodium hydroxide. The molar concentration, or *molarity,* of a solution is the number of moles of reagent present in 1 liter of solution. Realizing that at the end-point

$$\text{moles of potassium hydrogen phthalate} = \text{moles of hydroxide ions used} \qquad (1)$$

$$\text{moles of hydroxide ions used} = \left(\begin{array}{c}\text{molarity of base solution}\end{array}\right) \times \left(\begin{array}{c}\text{liters of base solution used}\end{array}\right) \qquad (2)$$

we can combine the two expressions to get the molarity of the base solution as:

$$\text{molarity of base} = \frac{\text{moles of potassium hydrogen phthalate}}{\text{liters of solution used}} \qquad (3)$$

Using this procedure, calculate M_B, the concentration of the base solution in moles per liter, for each trial. If the two trials do not agree within 2 percent, conduct a third trial, calculate M_B, and average the results of the trials.

NOTES

Note 1. Instead of rinsing the inside walls of the flask with distilled water near the end of the titration, you can achieve the same result by carefully tipping and rotating the flask so that the bulk of the liquid picks up any droplets adhering to the walls.

Note 2. Volume increments smaller than a normal drop may be taken by allowing a small amount of liquid to form on the tip of the buret and then touching the tip to the wall of the flask. This droplet is then combined with the bulk of the solution by rinsing it down with distilled water from a wash bottle or by tipping and rotating the flask.

QUESTIONS—PROBLEMS

1. Explain in concise statements the effect, if any, of each of the following sources of error upon the molarity of the base as determined in the experiment; i.e., would the experimental value for the molarity be too high or too low? Why?
   a. If the Erlenmeyer flask in which the titration was performed contained several milliliters of distilled water from the rinsing at the time the potassium hydrogen phthalate was weighed into it.
   b. If the tip of the buret were not filled with solution before the initial reading was taken.
   c. If a bubble appeared in the tip of the buret during the titration.
   d. If liquid splashed out of the titration flask before the end-point had been reached.
   e. If the buret were not rinsed with the base solution following the rinsing with distilled water.

2. Suppose that, instead of using sodium hydroxide, a base of type formula $B(OH)_2$, such as $Ba(OH)_2$, had been used. What changes in the calculations would then have to be made to determine the molar concentration of the base? Answer this question in words, and illustrate your answer by calculating M_B from the following data:

Moles of potassium hydrogen phthalate taken = 0.040

Ba(OH)$_2$ solution

Initial buret reading	0.02 ml.
Final buret reading	36.70 ml.

3. What is meant by the *normality* of an acid solution? Why is it sometimes more useful to express the concentration of acid and base solutions in normality rather than in terms of molarity?

REFERENCE

1. Skoog, D. A., and West, D. M.: *Fundamentals of Analytical Chemistry.* Holt, Rinehart and Winston, New York, 1969, Chapters 11 and 13.

Experiment 9
REPORT SHEET

Name _____ **Section No.** _____

Data

	Trial 1	Trial 2	Trial 3
Weight of potassium hydrogen phthalate . . ._____	_____	_____
Volume of base . . . . ._____	_____	_____
Calculated molarity of base . . . . ._____ (4 decimal places)	_____	_____
Averaged value of base	_____	
Average deviation	_____	

Sample Calculation

Observations and/or Comments about any of the Titrations

Attach sheets with answers to assigned questions.

DETERMINATION OF THE EQUIVALENT WEIGHT OF AN UNKNOWN ACID

OBJECTIVE

To use a standardized solution of base to determine the equivalent weight of an unknown acid.

EQUIPMENT NEEDED

Buret; two 125-ml. Erlenmeyer flasks; 250-ml. Erlenmeyer flask; 400-ml. beaker; analytical balance.

REAGENTS NEEDED

Standardized sodium hydroxide solution; 1 percent phenolphthalein indicator solution; Alconox or similar cleaning agent; samples of unknown acid (from the instructor).

INTRODUCTION

The *equivalent weight* of a substance in a neutralization reaction is the weight that either contributes or reacts with 1 mole of hydrogen ions in that reaction. Since we now have a sodium hydroxide solution of known concentration, (Expt. 9) i.e., a *standard* solution of a base, we can measure the exact number of moles of base needed to neutralize any sample of an acid and from this result we can calculate the equivalent weight of the acid from the known weight of the acid used. For example, if titration of 0.600 g. of a substance requires 20.00 ml. of a 0.500 molar sodium hydroxide solution, we know that 0.600 g. requires

$$20.00 \text{ ml.} \times \frac{0.500 \text{ mole}}{1000 \text{ ml.}} = 0.01 \text{ mole}$$

of hydroxide ions for neutralization. From this we can calculate the weight of acid that could be neutralized by 1 mole of hydroxide ions, the *equivalent weight*.

In the case cited, the equivalent weight of the acid is 60, i.e.,

$$\frac{0.600 \text{ g.}}{0.01 \text{ mole}} = 60 \text{ g./mole}$$

For this part of the experiment, obtain a sample of an unknown acid from your instructor and determine its equivalent weight.

PROCEDURE

Follow the same procedure used in the preparation and titration of the potassium hydrogen phthalate samples in Experiment 9. If the two results do not agree within 1 percent, repeat the procedure with a third sample. Average the results and report the average to your instructor.

NOTEBOOK AND REPORT

Obtain approval of your results on the unknown acid and include the calculations and results in your report. Calculate the standard deviation of your individual titrations. (See Experiment 35.)

Alternative experiments.

Common commercial antacid tablets containing magnesium hydroxide and aluminum hydroxide may be examined for their neutralizing capacity. Weigh out a measured excess of potassium hydrogen phthalate and add to the antacid tablet in water. Titrate the excess acid with the standardized sodium hydroxide solution after adding phenolphthalein. Alternatively a measured quantity of a standardized solution of an acid may be used to react with the tablet. Your instructor can tell you approximately how much potassium hydrogen phthalate or standard acid is needed to provide an excess. In either case the difference

(moles of acid taken initially) − (moles of acid left for titration)

represents the number of moles neutralized by the antacid pill. Does Brand X neutralize "forty times its weight of stomach acid"?

QUESTIONS—PROBLEMS

1. What is meant by the equivalent weight of an acid? What is the calculated equivalent weight of hydrochloric acid, HCl; of sulfuric acid, H_2SO_4; of oxalic acid, $H_2C_2O_4$; of oleic acid, $C_{17}H_{33}CO_2H$?

2. Explain why the volume of water in which the acid is dissolved need not be known accurately.

3. How would the measure of the equivalent weight of the unknown acid be affected if:

     a. A small amount of the solid were lost in transferring it from vial to flask?

     b. The buret used were contaminated with acid solution?

     c. The unknown acid contained two titratable hydrogen atoms?

4. Can the molecular weight of an acid be calculated from its equivalent weight? Explain your answer.

REFERENCE

1. Skoog, D. A., and West, D. M.: *Fundamentals of Analytical Chemistry*. Holt, Rinehart and Winston, New York, 1969, Chapters 11 and 13.

Experiment 10
REPORT SHEET

Name _____ Section No. _____

Data

	Trial I	*Trial II*	*Trial III*
Weight of container and compound before removing sample	_____	_____	_____
Weight of container and compound after removing sample	_____	_____	_____
Net weight of sample	_____	_____	_____
Buret reading at start of titration	_____	_____	_____
Buret reading at end of titration	_____	_____	_____
Net volume of base used	_____	_____	_____
Molarity of standardized base	_____	_____	_____
Calculated equivalent weight of acid	_____	_____	_____
Averaged value of the equivalent weight		_____	
Standard deviation		_____	

Sample Calculation

Observations and/or Comments about any of the Titrations

Attach sheets with answers to assigned questions.

MOLECULAR BEHAVIOR

VAPOR DENSITY AND THE GAS LAWS

OBJECTIVE

To study some fundamental properties of gases and to determine the molar volume, molecular weight, and the *a* and *b* constants of the van der Waals equation for an unknown volatile compound.

EQUIPMENT NEEDED

250-ml flask modified as shown in Figure 11-1; hot plate; water trough; barometer; thermometer; 1-mm. capillary tubing; Tygon or other inert plastic tubing; platform balance to weigh to 0.001g; Meker burner.

REAGENTS NEEDED

Liquids that boil between 55 and $90°C$ to be used as unknowns; elemental analysis and density of liquids at room temperature must be provided.

INTRODUCTION

From Avogadro's hypothesis, which states that equal volumes of different gases at the same temperature and pressure contain the same number of molecules, many important experimental and mathematical approaches for studying the behavior of gases have been undertaken. In this experiment we shall make use of facets of both these approaches to illustrate their power and scope.

Two inferences can be made directly from Avogadro's hypothesis: (1) The molar volume of a gas should be a constant for all gases at a given temperature and pressure. At standard temperature and pressure ($0°C$ and 1 atm.) this volume is 22.414 liters. (2) The molar gas constant, R, in the perfect gas law $PV = nRT$ is a universal constant. In the perfect gas law expression P, V, n, and T represent the pressure, volume, number of moles, and temperature of the gas, respectively.

Real gases show perfect gas behavior only at low pressures and relatively high temperatures, and the deviations from ideal behavior have led us to important ideas about attractive forces between molecules and about molecular sizes. Students often are confused as to when they can assume that a gas or vapor behaves according to the perfect gas law, and when they must assume that it deviates from ideal behavior. Perhaps the simplest resolution of this dilemma is to assume that although no gas is ideal, deviations from ideal behavior are seldom large enough to change the general pattern of gas behavior. As a consequence, the perfect gas law can be used for real gases with the expectation that the numerical values obtained from calculation are, in many cases, reliable approximations to actual values.

To illustrate this point we shall use the data obtained in this experiment in parallel calculations. In one application — estimation of molecular weight — we use the data in conjunction with the perfect gas law. In a second application we use the same data to estimate the extent of the deviation from ideal behavior.

The experiment involves the determination of the density of the vapor of a volatile liquid. In this experiment the mass of a known volume of volatilized liquid at a known pressure and at the temperature of boiling water is measured. From this the vapor density (mass/volume) is obtained. With this information, the molecular weight of the liquid can be approximated and the deviation of the vapor from perfect gas behavior can be estimated.

If we assume that the vapor approximates perfect gas behavior and if we recall that the number of moles of vapor, n, is equal to the mass of vapor divided by its gram-molecular weight, the perfect gas law becomes

$$PV = nRT = \frac{g}{M} RT$$

where g is the mass of the vapor and M is its gram-molecular or molar weight. Solving this equation for the molar weight we get

$$M = \frac{gRT}{PV}$$

All the data needed to evaluate the molar weight, except the molar gas constant, R, are obtained in measuring the vapor density. In assuming that the vapor behaves ideally, we also are assuming that R is a constant of known value. The value for the molar weight thus obtained is approximate but probably within 10 to 15 percent of the actual value. The error is not entirely attributable to the assumption that the gas is ideal. What other factors might introduce errors in the molar weight thus calculated?

The value for the molar weight should be reliable enough to be used to calculate the molecular formula of the substance if an elemental analysis is provided. For the purposes of this experiment you will be given an unknown liquid and its elemental analysis. Hence you should be able to report not only the approximate molar weight and the molecular formula but also a more exact molar weight.

Now let us set aside considerations of molar weight and use the experimental data to estimate the deviation of the assigned vapor from ideal behavior. A ready measure of this deviation is in the value for the molar gas constant, R. If the gas were ideal, the value for R would be 0.082054 atm. $\times$ liters $\times$ mole^{-1} $\times$ K^{-1}. From our data we can calculate a value for R assuming

$$R = \frac{PV}{nT}$$

We know the values for P, V, and T directly from our data and we also know the mass of the vapor. To determine the number of moles of vapor present we need to know the exact molar weight of the vapor. This is obtainable from the molecular formula, which we have determined previously. Once the value for R is calculated from the data we can compare it with that of an ideal gas, which should give some indication of the non-ideality of our vapor. However, such a comparison may not greatly increase our understanding of non-ideality. Is it possible to use the information we have to estimate the magnitude of the basic causes of non-ideality in gases, i.e., the attractive forces between molecules and the sizes of the molecules? We should at least try to do so.

The van der Waals equation adds the parameters $\underline{a}$ (a measure of attractive forces between molecules) and $\underline{b}$ (a measure of the volume occupied by individual molecules) to the parameters of the perfect gas law to give

$$\left(P + \frac{\underline{a}n^2}{V^2}\right) \times \left(V - n\underline{b}\right) = nRT$$

If the vapor behaves according to this equation we see that we have values available for all parameters except $\underline{a}$ and $\underline{b}$. If we can get the value for either $\underline{a}$ or $\underline{b}$ from another source, we can determine the other from the van der Waals equation. The value for $\underline{b}$ (the volume occupied by 1 mole of molecules when they are packed so close that they can be said to be touching) might be estimated from data on the liquid. If it is assumed that in the liquid state at room temperature the distance between molecules is very small, the volume of 1 mole of the liquid at this temperature can be used as an approximation for $\underline{b}$. To determine the molar volume of liquid we use the density of the liquid and its molar weight. Once a value for $\underline{b}$ has been estimated we can then use this, the accepted value for R, and our measured values for P, V, T, and n to calculate a value for $\underline{a}$. These values for $\underline{a}$ and $\underline{b}$ when compared with corresponding values for other gases should provide greater appreciation for the causes of and the magnitude of deviations of gases and vapors from ideal behavior.

There are many sources of error in this experiment. One of the most serious appears in the calculation of $\underline{a}$ because this value is obtained as a small difference between two large numbers. All the errors and approximations inherent in the values for the other parameters in the equation finally appear in $\underline{a}$. Although this should discourage calculation of $\underline{a}$ in this manner, no scientist would refuse to make such a calculation if there were no other way to estimate the value of an important parameter.

TECHNIQUE

A small volume of a volatile liquid is added to a 250-ml. flask, the orifice of which has been modified as shown in Figure 11-1. The flask is placed in a beaker of water that is heated to boiling. As soon as all the liquid in the flask has vaporized, the flask is removed from the boiling water and allowed to cool. The vaporized liquid remaining in the flask condenses to a liquid. The mass of the liquid is determined by comparing the weight of the empty flask with that of the flask containing the condensed liquid. The volume of the vapor is the volume of the flask; the temperature is the boiling point of

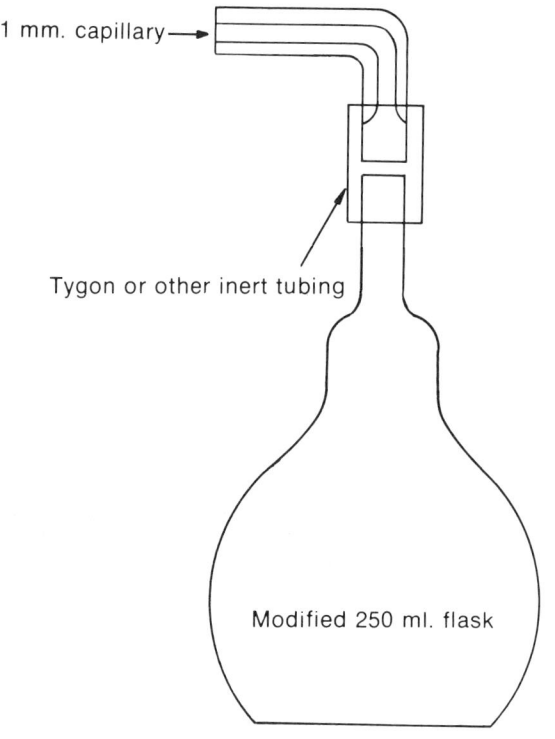

1 mm. capillary ⟶

Tygon or other inert tubing

Modified 250 ml. flask

FIGURE 11-1

Flask for determination of vapor density. Care must be taken to avoid trapping liquid in the junction between the capillary tube and the orifice of the flask.

water, which is not necessarily 100.0°C; the pressure is atmospheric pressure at the time of the experiment.

PROCEDURE

Weigh the flask and add the compound.

Weigh the clean, dry flask (without the capillary tip) to the nearest milligram. Add 5 ml. of the unknown liquid to the flask. Dry the orifice of the flask and replace the capillary tip, being sure to get the junction of the capillary tubing and the end of the flask as snug as possible to avoid trapping liquid at this junction.

Heat to vaporize the compound.

Clamp the flask with a clamp which will serve as a handle and place the flask and its contents in a beaker of mildly boiling water (Note 1). Heat the flask and its contents until the schlieren jet of vaporizing liquid disappears from the capillary tip of the flask (Note 2). Immediately remove the flask from the hot water bath, note the temperature of the bath, and allow the flask to cool by clamping it on a ring stand or by setting it on a *clean* surface. Meanwhile determine the atmospheric pressure from the laboratory barometer. When the flask is cool, wipe it dry, remove the capillary tip, wipe the top of the flask dry, and weigh the flask again (to 0.001g) to obtain (by difference) the weight of vapor in the flask. Determine the vapor density of your sample two more times by adding 5 ml of sample and repeating the heating and cooling steps outlined above.

Weigh the cool flask.

Determine the volume of flask.

Determine the volume of the flask by disconnecting the capillary tip, drying and weighing the dry flask (Note 3), filling the flask level-full with water, and weighing the filled flask on the platform balance. Calculate the volume from the mass of water present and its density at room temperature. Repeat the volume determination to check your first result.

CALCULATIONS

Calculate the molar weight by using the equation $M = \dfrac{gRT}{PV}$ as indicated earlier. Be sure to use a value for R consistent with the units used for pressure and volume. Using the elemental analysis data provided with your unknown, determine its simplest and molecular formulas and its exact molar weight.

Using your experimental data and the exact molar weight, calculate an experimental value for R. Compare this with the literature value and calculate the percent deviation of the experimental from the literature value. Compare your results with those of several other students and comment on these comparisons in your report.

Estimate the value for b in the van der Waals equation for your liquid, using its molar weight and density. The value for the density will be provided by the instructor. Finally, calculate a value for a in the van der Waals equation as described earlier. Compare your values for a and b with values in the literature (if possible) and with those obtained by several other students. Comment in your report on the differences in a and b for different liquids and discuss briefly how the values for a and b are reflected in the bulk properties of the vapor and the liquid.

NOTEBOOK AND REPORT

In addition to the data and calculations already discussed both the notebook and report should include statements on sources of errors. Whenever possible, the magnitude of the error introduced by some procedure should be estimated.

NOTES

Note 1. You should be very careful with flames while working on this experiment. If hot plates are unavailable, use a burner to heat to boiling 500 ml. of water in an 800-ml. beaker secured to a ring stand; then reduce the size of the flame to a size just sufficient to maintain a gentle boiling.

Note 2. If you are unable to see the schleiren vapor stream clearly, obtain a small piece of tissue paper and hold this in front of the capillary to determine when the liquid stops vaporizing rapidly.

Note 3. Dry the flask by removing the capillary tip, rinsing the inside with 5 ml of acetone to remove the organic unknown, rinsing with 5 ml of ether, and drying a few minutes in a 80-110° oven. If an oven is unavailable, insert a long capillary tube to near the bottom of the flask and attach the capillary to an aspirator (*via* tubing) to circulate air in the flask (see Figure 11-2).

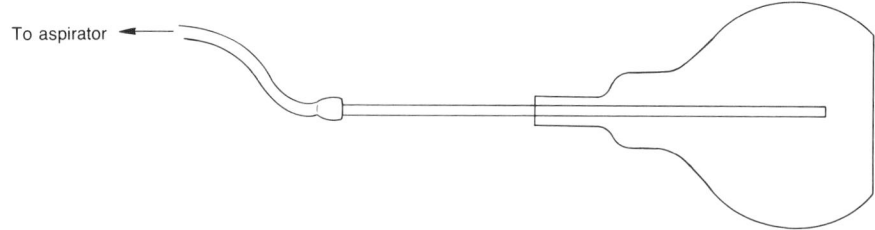

FIGURE 11-2

Drying a flask by sucking air through it.

QUESTIONS—PROBLEMS

1. Why are we justified in using atmospheric pressure as the pressure exerted by the vapor in the flask? What error, if any, is introduced by this?

2. What error is introduced by not removing the flask from the boiling water as soon as the schlieren jet disappears from the tip of the flask? What error is introduced if we remove the flask too soon? You may wish to examine the results of the experiment given in Table 11-1, which records the final weight of the flask and liquid as a function of the time after the schlieren jet disappeared.

TABLE 11-1 WEIGHT OF FLASK AND LIQUID AS A FUNCTION OF TIME

Time After Disappearance of Jet (sec)	Final Weight of Bulb and Unknown (g, ±0.0002 g)	
	cis 1,2-dichloroethylene	*1,1,1-trichloroethane*
5	67.2731	67.5890
30	67.2720	67.5870
60	67.2715	67.5863
90	67.2712	67.5859
120	67.2703	67.5854
180	67.2689	67.5850
240	67.2534	67.5837

Use these data to decide the optimum time to leave the flask in the boiling water. Note that these two compounds have different vapor pressures.

3. Upon what factors measurable or determinable in this experiment does the density of a vapor depend? State this in mathematical terms based on the perfect gas law.

4. Is it reasonable for the $\underline{a}$ in the van der Waals equation to have a negative value? Why or why not?

5. Let us suppose that the only error you make in the entire experiment is that you read the volume of the flask as 10 percent less than it actually is. Estimate the error in each of the following:

   a.   The molar weight.

   b.   The calculated value for R.

   c.   The value for $\underline{b}$.

   d.   The value for $\underline{a}$.

6. Is the value you calculated for R likely to be a constant for the substance you used at all temperatures and pressures at which it is a gas or vapor? Why or why not?

Experiment 11
REPORT SHEET

Name _____ Section No. _____

1. *Sample No.* _____

    Given density of liquid _____

    Given elemental analyses of compound _____

2. *Volume of flask*

    Mass of flask plus water   _____

    Mass of empty, dry flask   _____

    Net weight of water   _____

    Volume of flask   _____

3. *Mass of Unknown Sample in Flask*

    Mass of flask with air and unknown   _____

    Mass of dry flask (*i.e.*, air only)   _____

    Net mass of unknown   _____

    Density of the unknown gas   _____

4. *Molecular Weight of the Gas*   _____

5. *Molecular Formula of the Gas*   _____

6. *Calculated Value of R* _____ *Accepted Value* _____

7. *Calculation of b (use the **liquid** density)*

8. *Calculation of a*

9. *Attach sheet with answers to assigned questions.*

ATTRACTIVE FORCES BETWEEN MOLECULES

(Vapor Pressure and Enthalpy of Vaporization of Liquids)

OBJECTIVE

To measure the vapor pressure of a volatile liquid at several temperatures and to use the values obtained to calculate the heat of vaporization of the liquid.

EQUIPMENT NEEDED

Mercury manometer; 125-ml. filter flask; 125-ml. Erlenmeyer flask; syringe and steel hypodermic needle; thermometer; 800-ml. beaker; 250-ml. beaker; 12 X 100 mm. test tube; melting-point capillary; rubber stopper; neoprene stopper; gas burner; rubber bands; ring stand and clamps; rubber tubing.

REAGENTS NEEDED

Organic liquids as assigned by instructor; ice.

INTRODUCTION

When a liquid vaporizes, the molecules, which are close together in the liquid state, move far apart to the larger separations characteristic of the gaseous state. The energy required for vaporization is thus a measure of the energy required to separate molecules from each other and represents the work done against the attractive forces between the molecules. If the energy required for the vaporization of several liquids is measured, a comparison of the values obtained represents a comparison of the attractive forces between the molecules in each liquid. If the vaporization is carried out at constant pressure, the energy measured is the enthalpy change in the process. (Vaporization at constant pressure also requires that work be done in pushing back the atmosphere, and the corresponding energy is included in the enthalpy of vaporization. This is a small amount of energy compared to the energy needed to separate the molecules, and it does not differ greatly from one liquid to another if the measurements are based upon the vaporization of 1 mole of liquid, since the volumes occupied by 1 mole of different gases are nearly the same at pressures near that of the atmosphere. Hence, a comparison of the molar enthalpies of vaporization for different liquids is a valid comparison of the relative magnitudes of the attractive forces between their molecules.)

Although the enthalpy of vaporization can be determined directly by measuring the quantity of heat energy necessary to vaporize a given quantity of liquid at constant pressure, the experiment is not easy to perform, mainly because heat tends to leak from the apparatus, making the measured heat input greater than the amount actually used for vaporization. It is much easier to measure the *vapor pressure* of the liquid at several temperatures, and to *calculate* the enthalpy of vaporization, which bears a known relation to the vapor pressure and temperature, from those measurements. This is a frequent practice in science; to choose a sequence of measurements that are easy to make with precision in order to arrive at a desired quantity that may not be easy to measure. Although the route to determination of this quantity may seem at first glance to be unnecessarily circuitous, a consideration of the experimental difficulties and likelihood of error in the direct route shows that it is not.

The known relation between vapor pressure and enthalpy of vaporization is obtained from thermodynamic reasoning. It is

$$\log p = -\frac{\Delta H_v}{2.3RT} + C \tag{1}$$

Here p is the vapor pressure, ΔH_v is the molar enthalpy of vaporization, R is the gas constant in calories per degree per mole, T is the Kelvin temperature at which the vapor pressure is measured, and C is a constant the value of which depends upon the liquid being studied. Comparison of this equation with the equation for a straight line

$$y = mx + b \tag{2}$$

shows that if the vapor pressure is measured at several temperatures to give paired values (p_1 at T_1, p_2 at T_2, etc.) and if log p is plotted against $\frac{1}{T}$, a straight line of slope $-\frac{\Delta H_v}{2.3R}$ should be obtained. If such a plot is made and the slope of the line is measured, ΔH_v can be calculated from

$$\Delta H_v = -(\text{slope}) \times 2.3R \tag{3}$$

In this experiment, the vapor pressure is measured by adding a small amount of liquid to an enclosed volume of air attached to a manometer, and measuring the increase in pressure that results when equilibrium is established between liquid and vapor at the temperature of the enclosed air. Since the magnitude of the vapor pressure depends upon the temperature at which it is measured, the temperature at the time of measurement must be known and should remain constant during the measurement. The apparatus to be used is shown in Figure 12-1. The level of mercury in each side of the manometer is read with the flask containing only air. A sample of the liquid to be studied is then added from a syringe by forcing a hypodermic needle through the neoprene insert in the stopper. The liquid quickly comes to

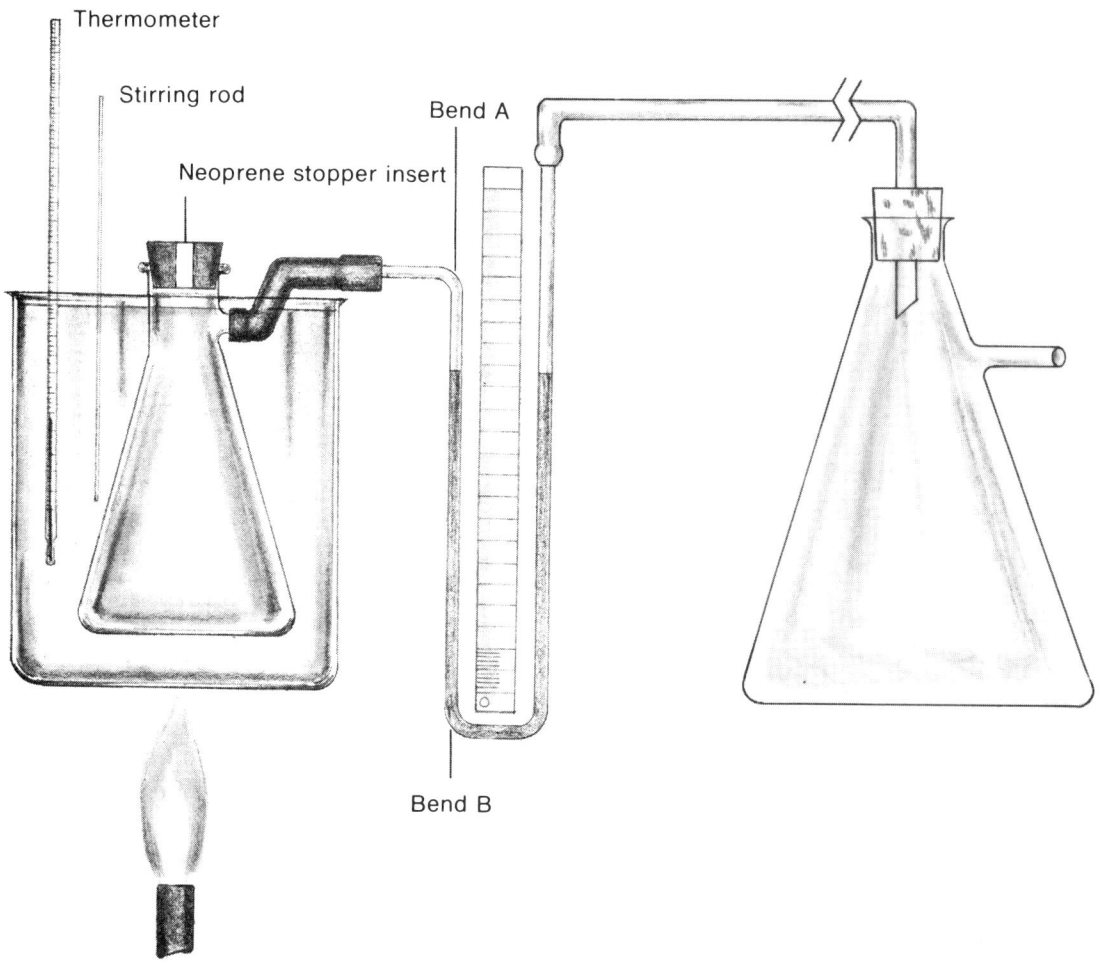

FIGURE   12-1

Apparatus for measurement of vapor pressure.

temperature equilibrium, and the levels of mercury are again read. Except for the small changes represented by the shift in the level of mercury in the left tube of the manometer and the volume of liquid added to the flask, the volume of the system is constant; hence, the pressure of the air remains constant and the whole increase in pressure must be due to vaporization of the liquid. Each gas (air and vapor) acts, according to Dalton's Law, as if it alone were present in the volume occupied by the mixture.

When the apparatus shown in Figure 12-1 is used, the temperature is fixed by the experimenter, and the vapor pressure corresponding to that temperature is determined. The apparatus shown in Figure 12-2 reverses the process. Here the pressure is fixed at atmospheric pressure and the temperature at which the vapor pressure reaches this value, i.e., the boiling point, is measured. A small amount of the liquid under examination is placed in the 12 X 100 mm. test tube fastened to the thermometer, and a capillary melting-point tube is dropped into it with the open end of the capillary tube below the surface of the liquid. As the temperature of the bath is increased, air bubbles emerge from the capillary tube, slowly at first and then in a rapid stream as the liquid boils. The boiling is continued for a few seconds to

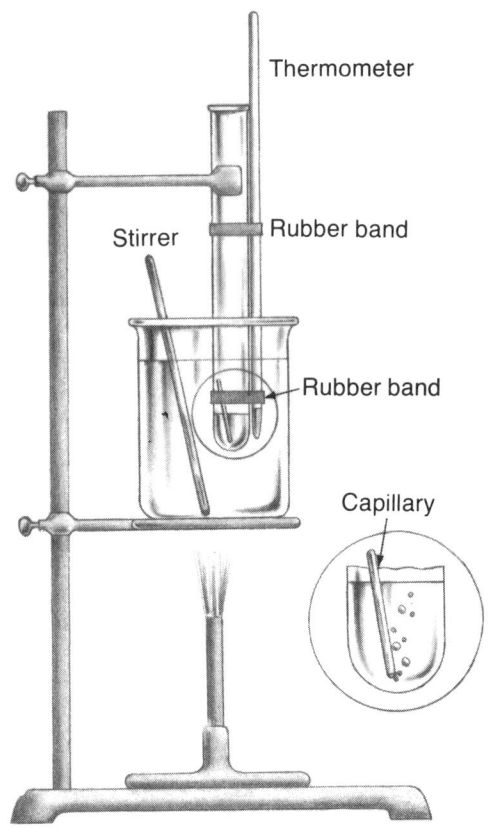

FIGURE 12-2

Determination of the boiling point of a
liquid in the capillary.

remove nearly all the air from the capillary tube and fill it with vapor. The
bath is then allowed to cool slowly, it is stirred constantly, and the open end
of the capillary tube is watched carefully. When the liquid begins to reenter
the capillary tube, the pressure inside is equal to the atmospheric pressure on
the surface of the liquid, and the temperature read on the thermometer is
the boiling point. In this measurement it is presumed that the pressure
exerted by any residual air in the capillary tube is negligible, and the infini-
tesimal effect of the hydrostatic pressure exerted because the capillary
opening is below the surface of the liquid is ignored.

In this experiment each student will first be asked to determine the
vapor pressure of an unknown liquid at several temperatures. From these
measurements he will determine the heat of vaporization of the liquid. Then
the class will be divided into teams of four students. Each team will determine the enthalpies of vaporization of the four compounds in one of the
following sets, with one student assigned to each compound in the set.

Set 1	Set 2	Set 3	Set 4	Set 5
Hexane	Methanol	Heptane	1-Butanol	Benzene
Heptane	Ethanol	3-Methylhexane	2-Butanol	Fluorobenzene
Octane	1-Propanol	2,3-Dimethylpentane	2-Methyl-1-propanol	Chlorobenzene
Nonane	1-Butanol	2,4-Dimethylpentane	2-Methyl-2-propanol	Bromobenzene

NOTEBOOK AND REPORT

Each student should maintain in his notebook the records needed to enable a reader to reconstruct his experiments. He should draw the graph and make the calculations indicated. The report of the second portion of the experiment should be a team effort. It should include the graphs for each of the four compounds examined by the team and the calculated values of the enthalpies of vaporization. The report should include a discussion of the information gained about trends in attractive forces as a result of comparing the enthalpies of vaporization of the four compounds in the set. This discussion might relate, for example, to the effect on the enthalpy of vaporization of changing the molecular weight of compounds of the same chemical nature, of changing the shape, polarity, hydrogen bonding possibilities or other structural features of the compounds without changing the molecular weight, of changing the chemical nature of the compounds, etc.

PROCEDURE

Determination of Vapor Pressure at Room Temperature. Assemble the apparatus as shown in Figure 12-1. Be certain that all connections are tight. Fill the beaker with water at room temperature, and immerse the thermometer to the immersion line.

Assemble V.P. apparatus.

Because the vapor pressures of the liquids to be used are relatively high at room temperature, it is necessary to reduce the pressure in the filter flask before inserting the sample. To do so, insert the needle of the syringe through the neoprene plug in the rubber stopper, and attach rubber tubing from the water aspirator to the needle. Carefully withdraw air from the flask until the level of mercury in the shorter arm of the manometer is about 2 cm. below bend A leading to the flask. Immediately extract the needle from the plug in the stopper. If the apparatus is tight, the level of mercury will remain at this position. If the system has a leak in it, the level of mercury will begin to fall. Any leaks must be found and eliminated by using copper wire to secure the glass-rubber connections.

Evacuate flask carefully with aspirator.

When the system is tight and partially evacuated, record the height of the mercury in both arms of the manometer to the nearest millimeter.

Record Hg column heights.

Draw 1 ml. of the liquid to be studied into the syringe. CAUTION: ALL LIQUIDS USED IN THIS EXPERIMENT ARE INFLAMMABLE. BE SURE TO KEEP OPEN VESSELS CONTAINING THEM, AND ESPECIALLY SYRINGES CONTAINING THEM, AWAY FROM BURNER FLAMES, YOUR NEIGHBORS' AS WELL AS YOUR OWN. Insert the needle through the neoprene plug and inject the sample into the flask; then remove the syringe and the needle.

Add sample without causing a leak.

Allow the system to stand for at least five minutes so that equilibrium between the liquid and vapor can be established at the temperature of the water. Record the height of the mercury in both arms of the manometer to the nearest millimeter. Record the temperature of the water in the beaker; at temperature equilibrium this is the temperature of the air and liquid inside the flask. Calculate and record the increase in pressure produced by adding the liquid; this is the vapor pressure of the liquid at the temperature of the experiment.

Record Hg column heights after attainment of V.P. equilibrium; also record temperature.

Determination of the Vapor Pressure at a Temperature Near 60°. Disconnect the flask from the apparatus and dry it thoroughly. This is most readily accomplished by inserting a glass tube connected to the aspirator and drawing air out of it while waving the bottom of the flask in the general vicinity of a burner. (Do not wave it too close; remember that the contents are inflammable.) Examine the stopper to make sure that it is dry. Dry the hypodermic needle.

Reassemble the apparatus, insert the needle (not attached to the syringe) through the neoprene plug, and heat the water, with stirring, to about 60°. The expanding air in the flask will escape through the open needle. Adjust the burner so that this temperature is maintained as constant as possible. Attach a rubber tube from the aspirator to the needle and reduce the pressure as before, pulling the needle out when the mercury reaches a few centimeters below the bend. Check to see that the system is tight, and read and record the levels of mercury. Maintaining the temperature constant, and with frequent stirring, again insert 1 ml. of the liquid, and read the manometer after five minutes. The temperature at which the readings are made should be the same as the temperature at which the air pressure was determined before inserting the sample of liquid. Record this temperature.

CAUTION: IF MERCURY IS SPILLED, in spite of your precautions, CALL THE INSTRUCTOR AT ONCE FOR HELP IN CLEANING UP. Mercury is a cumulative poison, which may give no notice of its effects for many years after exposure. An undetected droplet in the corner of a drawer can poison other students using your work space for many years to come.

Determination of Vapor Pressure at a Third Temperature. Repeat the whole experiment, including the drying of the flask, at some temperature 25 to 30°C from either of the two temperatures already used so that you have three reliable vapor pressure readings of 10 torr (10 mm. of mercury) or more. For substances of high vapor pressure, the ice point is a convenient and constant temperature, readily obtained by filling the beaker with a stirrable mixture of ice and water. If this temperature would give so low a pressure that the measurements would be unreliable, try a temperature between room temperature and your highest temperature.

Determination of the Temperature at Which the Vapor Pressure is Equal to the Atmospheric Pressure. Set up the apparatus shown in Figure 12-2. The test tube is clamped to the ring stand, and the thermometer is attached to it with rubber bands so that the thermometer bulb is opposite the bottom of the test tube. The capillary tube is about 2 cm. long and is made by scratching a melting-point tube with a file (gently) 2 cm. from the closed end and breaking it off at that point. Put 1 ml. of the liquid for vapor pressure determination into the test tube and heat the water in the beaker, with stirring, looking for the phenomena described in the Introduction. Record the boiling temperature and the barometric pressure.

CALCULATIONS

You now have three or four pairs of vapor pressure-temperature points. Determine the logarithms of the vapor pressures and the reciprocals of the Kelvin temperatures, and plot the data according to equation (1). Draw the best straight line through the points, determine the slope of the line, and calculate the enthalpy of vaporization from equation (3). Use 1.987 calories per degree per mole as the value for R. Record your value of ΔH_v.

QUESTIONS—PROBLEMS

1. Assume that the reading for the difference in levels of mercury at the high temperature in this procedure before liquid was added was 340 mm., with the mercury higher in the right arm of the manometer than in the left and that the difference in levels after the liquid was added was 480 mm. Calculate the pressure of air in the flask and the vapor pressure that your liquid would apparently be exerting under such conditions. Do not forget that the atmosphere is pressing down on the open, right arm of the manometer.

2. Explain in terms of the properties of molecules why some liquids have higher heats of vaporization than do others.

3. Design an apparatus and experiment for measuring the vapor pressure of a solid.

4. Discuss the significance of the error resulting from the volume of the liquid introduced.

5. How does Dalton's Law of partial pressures enter into the measurements in this experiment?

6. The vapor pressure of a solution of a nonvolatile solid in a liquid is less than the vapor pressure of the pure liquid. How can this be explained?

7. When the vapor pressure of a mixture of 1 mole of liquid A and 1 mole of liquid B is measured at some fixed temperature, it is sometimes (though rarely) found to be halfway between the vapor pressures of pure A and pure B at that temperature; for some liquid pairs it is above this value, and for others it is below. The first case is said to represent an ideal solution, and indicates that the attractive forces between molecules of A and B are no different from the attractive forces between molecules of A and A nor between molecules of B and B. What can you suggest about the relative magnitudes of the A-B forces, compared to the A-A and B-B forces, in the other two cases?

REFERENCES

1. Garrett, A. B., Lippincott, W. T., and Verhoek, F. H.: *Chemistry, A Study of Matter.* 2nd Edition. Xerox College Publishing, Lexington, Mass., 1972, Chapter 11.
2. Dunslow, R. R.: *J. Chem. Educ., 5,* 727 (1928).

Experiment 12
REPORT SHEET

Name _____ **Section No.** _____

Unknown No. _____

1. *Data For Room Temperature.* Readings: thermometer _____ barometer _____

Manometer readings	Left arm	Right arm	Difference
Before adding liquid	_____	_____	_____
After adding liquid	_____	_____	_____
Vapor pressure of liquid	_____		

2. *Data For Temperature ~ 60°C.* Readings: thermometer _____ barometer _____

Manometer readings	Left arm	Right arm	Difference
Before adding liquid	_____	_____	_____
After adding liquid	_____	_____	_____
Vapor pressure of liquid	_____		

3. *Data For Third Temperature.* Readings: thermometer _____ barometer _____

Manometer readings	Left arm	Right arm	Difference
Before adding liquid	_____	_____	_____
After adding liquid	_____	_____	_____
Vapor pressure of liquid	_____		

4. *Boiling Temperature of the Liquid.*   Reading: barometer _____

   Boiling temperature _____

5. *Calculated Heat of Vaporization* _____

6. *Attach sheets* showing (a) the graph of the vapor pressure data according to equation (1); (b) the calculation of ΔH_v from this graph; (c) answers to assigned questions; and (d) the data, graphs, calculations, and discussion of the team effort.

SOLUTION BEHAVIOR

SOLUTIONS OF SALTS:
THE VARIATION OF SOLUBILITY
WITH TEMPERATURE
(Purification by Fractional Crystallization)

OBJECTIVE

To study saturation equilibrium as a function of temperature and solvent, and to purify a salt by fractional crystallization from a mixture.

EQUIPMENT NEEDED

Solubility apparatus; two 100-ml. beakers; 400-ml. beaker; filter paper; funnel; spatula; wash bottle; rubber policeman; desiccator; test tube.

REAGENTS NEEDED

Potassium dichromate, $K_2Cr_2O_7$; oxalic acid, $(CO_2H)_2 \cdot 2H_2O$; 3:1 mixture of potassium nitrate and copper nitrate, $KNO_3 + Cu(NO_3)_2 \cdot 3H_2O$; ice; 1M HNO_3; 70:30 mixture of water and dioxane.

INTRODUCTION

This experiment concerns some of the factors that determine the amount of substance that can be dissolved in a given amount of liquid. The substances chosen for investigation are salts, and the liquid chosen is water, because solutions of salts in water are the most common type of solutions observed.

The generalization that systems change from an ordered state to one of less order suggests that there is a natural tendency for all solids to dissolve. The dissolved state is a more disordered arrangement, with the molecules or ions distributed at random through the solution, in contrast to their ordered arrangement in the solid. This tendency to become disordered by dissolving is opposed by the forces holding the molecules or ions to each other in fixed positions in the solid. Work must be done against these forces — energy must be expended — to separate the particles (molecules or ions) of the solid from each other.

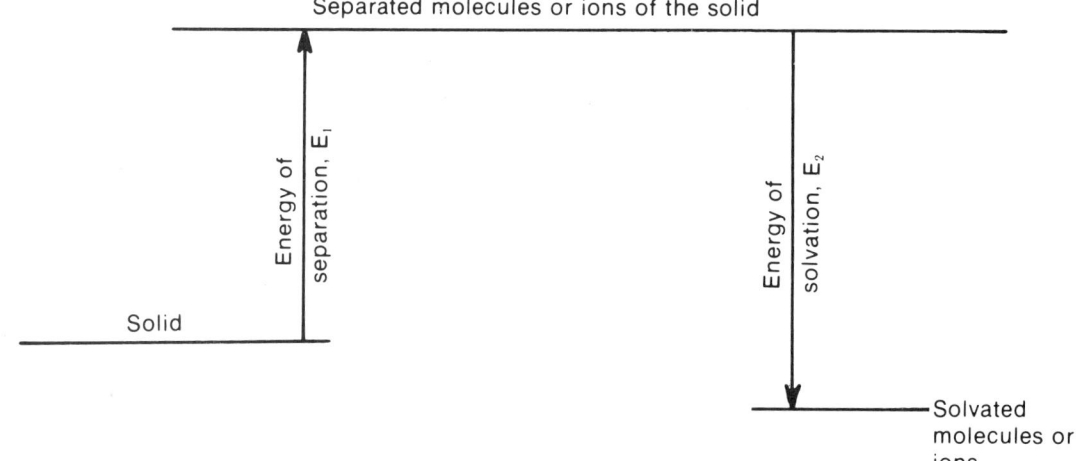

FIGURE 13-1

Energy effects in dissolving solids.

There are also forces of interaction in the dissolved state between the molecules or ions of the solute and the molecules of the solvent. These forces bring the particles of solute and solvent, initially separated before solution occurs, together in the solution; energy is released in this process, which is known as solvation (or in the specific case of water solutions, as hydration).

The net energy change when a solid dissolves in a liquid is the sum of two opposing energy effects: (1) the energy needed to separate the particles of solid from each other and (2) the energy released when molecules or ions of the solid are solvated by molecules of the solvent. We may sketch these effects in an energy diagram (Fig. 13-1). If the energy of separation, E_1, is less than the energy of solvation, E_2, energy will be released in the overall solution process. The opposite case, with E_1 greater than E_2, is more usual for dissolving of salts, and energy (heat) is commonly absorbed when salts dissolve. The solubility represents a compromise between the tendency toward randomness and the requirement of energy absorption.

In accordance with LeChatelier's principle, the change in solubility with temperature depends upon the overall heat change in the solution process. If heat is absorbed

$$\text{Heat + salt + solvent} \rightleftharpoons \text{solution}$$

an increase in temperature will increase the solubility. (Addition of heat causes the position of equilibrium to shift so that heat is used up.) For the few substances for which heat is evolved when a solid salt enters a saturated solution

$$\text{Solute + solvent} \rightleftharpoons \text{solution + heat,}$$

the solubility decreases with increase in temperature, again in accord with LeChatelier's principle.

In this experiment we shall investigate the change of solubility with temperature and solvent and then show how differences in the extent of solubility changes with a change in temperature for different salts can be exploited to provide a method of separating one salt from another in a mixture of the two.

NOTEBOOK AND REPORT

Before going to the laboratory, prepare a data table in your notebook to record all the experimental information needed, e.g., the weights of solute used, the volumes of solvent used, and the temperatures of the solutions. Your report should include a brief statement of the procedure (in your own words), observations about any difficulties experienced in the procedure, calculations, and answers to the questions in the body of the experiment and at the end of it.

A. The Variation of Solubility with Temperature

The variation of solubility with temperature may be determined experimentally by either of two methods; (1) by analyzing solutions saturated at various fixed temperatures to determine the concentration of dissolved solute, or (2) by preparing solutions of various known concentrations and observing the temperature at which each solution becomes saturated. The second method will be employed in this experiment.

PROCEDURE

Weigh accurately 7 to 8 g. of $K_2Cr_2O_7$ or 5 to 5.5 g. of $(CO_2H)_2 \cdot 2H_2O$ on an analytical balance and carefully place it in a large test tube equipped with a rubber stopper, a stirrer, and a thermometer. (Secure this apparatus from the storeroom; do not use a copper stirrer if oxalic acid is used.) From a buret add exactly 10.0 ml of water to the test tube. Put the test tube and its contents into a beaker of boiling water and stir it, as the temperature increases, until all the solid has dissolved, but not much longer.

Weigh out salt.

Add H_2O and heat to dissolve.

After the solid has dissolved, remove the test tube from the hot water and allow it to cool in the air while stirring the contents constantly. Note and record the temperature at which the *first* crystals are observed. If a considerable mass of crystals has been allowed to form before the temperature is noted, the concentration of the solution has changed to such an extent that the observed temperature is not the point of saturation of the original solution. Consequently, you must observe, as nearly as possible, the temperature at which crystallization just begins. If uncertainty exists regarding an observed temperature, the solution may be warmed again and a second set of observations may be made. It is suggested that each point be confirmed at least once, in any case.

Stir while cooling and record temperature at which crystals appear.

After a satisfactory value has been obtained, dilute the original solution by adding exactly 10 ml. of distilled water and observe, in the same manner as before, the temperature at which the new solution is saturated. Repeat this procedure until the saturation temperature of four different solution concentrations has been observed.

Dilute, and repeat.

**Repeat with a
smaller quantity.**

Repeat the foregoing experiment with *accurately weighed portions* of about 2 g. of $K_2Cr_2O_7$ or 1.8 g. of $(CO_2H)_2 \cdot 2H_2O$ and 10.0, 20.0, and 30.0 ml. water as before. In the most dilute solutions, it may be necessary to cool the solutions in ice water to cause the salt to crystallize. While stirring the solution, note carefully the temperature at which crystals first appear.

CALCULATIONS

Calculate the concentration of each solution observed in this experiment, assuming the density of water to be 1.00 g/cc. and expressing the concentration in terms of grams of solute per 100 g. H_2O. Plot the solubility (in grams per 100 g. H_2O) as the ordinate and the temperature as the abscissa on a piece of graph paper, and draw a smooth curve through the points (Graph I).

Many theoretical deductions in chemistry show that the properties of systems are dependent upon the relative number of moles of the two substances present. Transform the solubility in grams per 100 g., as recorded in your data table, into the quantity known as the mole fraction, defined for a solute and solvent system as:

$$\text{Mole fraction of solute} = \frac{\text{moles of solute}}{\text{moles of solute} + \text{moles of solvent}}$$

The mole fraction for a sample of 5.030 g. $K_2Cr_2O_7$ dissolved in 10 g. H_2O would be:

$$\text{Mole fraction of potassium dichromate} = \frac{\dfrac{5.030}{294.2}}{\dfrac{5.030}{294.2} + \dfrac{10.0}{18.0}} = \frac{0.0171}{0.0171 + 0.555} = 0.0299$$

Plot your experimental results on a second graph (Graph II), this time using the temperature as the ordinate and the mole fraction of solute as the abscissa.

Examine the graph. You know that pure water freezes at $0°C$. Can the temperature at which saturation is reached be less than $0°C$? Does it appear that your temperature-mole fraction curve will go through the point (0,0)? How do you explain the curve? Record the answers to these questions in your notebook and report.

B. Effect of Change of Solvent on Solubility

Having examined the solubility of a substance in water, one may now raise the question of whether the solubility and its change with temperature is the same in solvents other than water, or in a liquid which contains water and some other substance. Since one of the criteria for distinguishing between electrovalent and covalent compounds is that electrovalent compounds are soluble in water, but not in non-polar liquids, we readily conclude that addition of a non-polar liquid to water will change the solu-

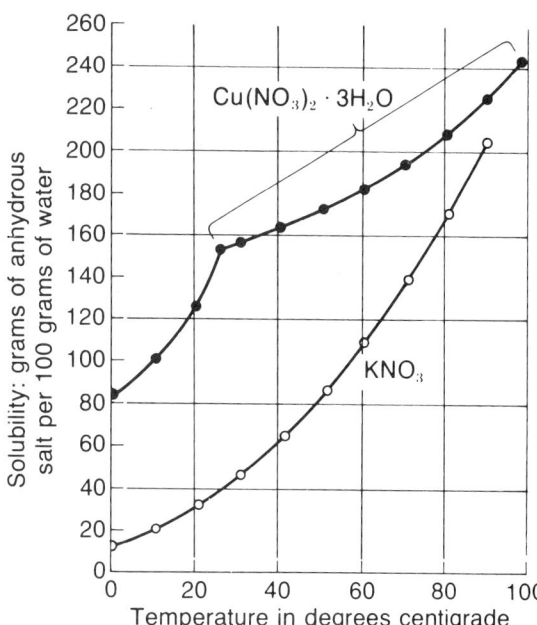

FIGURE 13-2

Solubility of copper nitrate and potassium nitrate as a function of temperature.

bility differently for the two classes of compounds. To test this conclusion, repeat the experiments using a 70:30 (by weight) water:dioxane mixture. Use a precisely weighed 3.5 g. quantity of potassium dichromate or 5.5 g. of oxalic acid. Use 10.0 ml of the mixed solvent initially, and dilute with successive 10.0 ml portions of the same mixture. Record the temperature readings as before.

Taking the density of the water-dioxane mixture as 1.023 g/cc calculate the solubility per 100 g solvent as in part A, and plot the data on Graph I. Draw a smooth curve through the points.

Examine the graph and report the difference in solubility in pure water and in the mixed solvent at 30°C. Suggest a reason for the difference.

C. Separation and Purification by Fractional Crystallization

The graph in Figure 13-2 shows the change in solubility with temperature for the salts copper nitrate, $Cu(NO_3)_2 \cdot 3H_2O$, and potassium nitrate, KNO_3. From the graph, compare the relative solubilities of the two salts at a high temperature with the corresponding quantity at a low temperature. In this experiment you will be given a 3:1 mixture, by weight, of potassium nitrate and copper(II) nitrate 3-hydrate. Assuming that the two salts dissolve independently (which is not strictly true), decide, from examination of the graph, to what temperature you would have to heat 100 g. H_2O in order to dissolve 133.3 g. of such a mixture. Record the value in the data table. If a solution, saturated at that temperature (with which of the salts?) were to be cooled to 30°C, which salt and how much of it would crystallize out? To answer this question, refer to the graph and record the solubility of each salt at 30°C and subtract this value from the weight of that salt in the original weight of mixture taken; record the differences. What does a negative value mean in this instance?

These considerations indicate that the solubility change permits one constituent of the mixture to be separated from the other. Since the crystals that form in the solution of the mixture capture some solution ("occlude the mother liquor" is the chemists' phrase), it is necessary to redissolve the crystals in a new quantity of water and to isolate crystals from this new solution, which, of course, contains a lesser concentration of the second salt, in order to complete the purification. Since the solution containing the redissolved crystals contains much less of the second component than did the first solution, the amount of impurity occluded by the second set of crystals is negligible, and a highly purified material is obtained. In each crystallization, the surfaces of the crystals are covered with the mother liquor when first filtered. This must be washed away if a pure product is to be obtained.

Let us try the separation and purification of the two components of the mixture by this technique of fractional crystallization. In this case, it is easy to follow visually the relative success of the separation since one of the components is colored.

PROCEDURE

Weigh out sample.

Clean and dry a 100 ml. beaker. Take it to a triple-beam balance and weigh into it 20 g. of the potassium nitrate-copper nitrate mixture. Note the two colors of particles in the mixture. Identify which color is due to copper nitrate. Add 12 ml. of distilled water and 3 ml. of 1M HNO_3. Warm the mixture gently, stirring it constantly, until the solids are completely dissolved. Place the beaker in cold water in a 400-ml. beaker. Stir while cooling and add ice occasionally to the cooling bath. When the solution has cooled nearly to the temperature of the cooling bath and crystals have separated, filter the mixture, *collecting and saving the filtrate in a clean beaker for a later part of this experiment.* Use a rubber policeman to be sure that all the solid is transferred to the filter paper. When the filtrate has drained completely from the filter paper, wash the crystals on the paper four times with 2-ml. portions of distilled water that has been cooled in ice, discarding the washings. Allow each washing to drain completely from the filter before adding the next washing. What are the crystals on the filter?

Dissolve in acid solution and cool.

Filter and wash.

Repeat.

Put the funnel into a test tube, and punch a hole in the bottom of the filter paper. Wash the crystals into the test tube with about 5 ml. of distilled water from a wash bottle. Warm the solution gently, stirring it constantly, until the solid is completely dissolved. Then allow the solution to cool by setting the test tube in ice water. Filter and wash the crystals on the filter paper four times with 2-ml. portions of cold distilled water. When the last wash water has drained away, remove the filter paper bearing the crystals, open it, and spread it on a double sheet of filter paper; this will absorb the excess water. Allow the crystals to dry in the desiccator. Calculate the percentage recovered and hand the sample to your instructor in a properly labeled bottle.

Dry the product, weigh and hand in.

Evaporate the filtrate from the first filtration of Part C to about one-third its original volume. Then cool the solution by placing the beaker in ice water. Decant *(save the decantate)* and wash the crystals three times with 2-ml. portions of cold water, removing the wash water by decantation. (See Experiment 8 for decantation technique.) Allow the crystals to dry in the desiccator.

Boil down filtrate for another crop.

Evaporate the decantate to one-half its volume. Decant, dry, and examine the crystals formed. Can you identify them? *Do not allow the water to evaporate completely because some of the crystals will decompose.*

Identify material in mother liquor.

CALCULATIONS

Determine the solubility data requested in the introductory portion of Part C from Figure 13-2, and calculate the amount of the appropriate salt that would crystallize out of solution at 30°C.

QUESTIONS–PROBLEMS

1. According to the solubility curve obtained in this experiment, what is the solubility of the solute you studied in this experiment at 55 and at 75°C?

2. What error in the observed solubility (part A) would be introduced if:
   a. Some water were lost through excessive heating?
   b. Some of the solute were spilled before being transferred to the test tube?
   c. The solute tended to give supersaturated solutions?

3. Suppose you had a saturated water solution of your solute, at 30°C in an amount sufficient to contain 100 g of water, and without change in temperature, added dioxane until the ratio of water to dioxane was 70:30. Would solid crystallize out, and, if so, how much?

4. Suppose it were possible to remove dioxane from a solution containing solute, dioxane and water, so that only solute and water were left. If this was accomplished with a saturated solution of your solute in 70:30 water-dioxane using an amount of solution which contained 100 g of solvent, would solid crystallize out, and, if so, how much?

5. Are the three portions of crystals obtained in Part C pure? What is the basis for your answer? If your answer is that you do not know, describe some methods of discovering whether the crystals are pure. If your answer is yes, describe how you know they are pure.

6. Can any two salts be separated by fractional crystallization? Explain your answer.

7. If you had potassium sulfate and copper nitrate as solutes, what complications might arise in attempting to obtain the pure salts by fractional crystallization?

8. Offer an explanation as to why copper nitrate is much more soluble than potassium nitrate.

9. Can you suggest a method, other than the one used, for separating a solid mixture of potassium nitrate and copper nitrate?

10. What does the abrupt change of slope at 25°C for the copper nitrate solubility curve (Figure 13-2) mean?

REFERENCES

1. Garrett, A. B., Lippincott, W. T., and Verhoek, F. H.: *Chemistry, A Study of Matter.* 2nd Edition; Xerox College Publishing, Lexington, Massachusetts, 1972, Chapter 12.
2. Pimentel, G. C., and Spratley, R. D.: *Understanding Chemical Thermodynamics.* Holden Day, San Francisco, 1969, Section 6-2.
3. Fieser, L. F.: *Organic Experiments.* D. C. Heath, Boston, 1964, Chapter 7.

Experiment 13
REPORT SHEET

Name ———————————————————————— Section No. ———————

Part A

1. Data

2. Sample calculation

3. Discussion and explanation of phenomena near the origin for Graph II.

4. Attach Graphs I and II.

Part B

1. Data

2. Sample calculation

3. Solubility difference in water and water-dioxane at 30°C. Reason for the difference.

Part C

1. Data

2. Calculation

3. Hand in purified crystals in a labeled bottle.

Attach sheets with answers to assigned questions.

FREEZING POINT DEPRESSION

OBJECTIVE

To utilize the fundamental property of the decreased vapor pressure of a solution as a method to determine the molecular weight of a soluble unknown compound.

EQUIPMENT NEEDED

Freezing-point apparatus, including a special thermometer, (from stockroom); buret, 800-ml. beaker.

REAGENTS NEEDED

Ice; water; salt; benzene; unknowns; 0.5 molal solutions of the electrolytes listed in Part B.

INTRODUCTION

In the "mixed melting points" portion of Experiment 3 we observed that the melting point of one pure compound was lowered when it was mixed with a second compound. In a similar manner, the freezing point of a

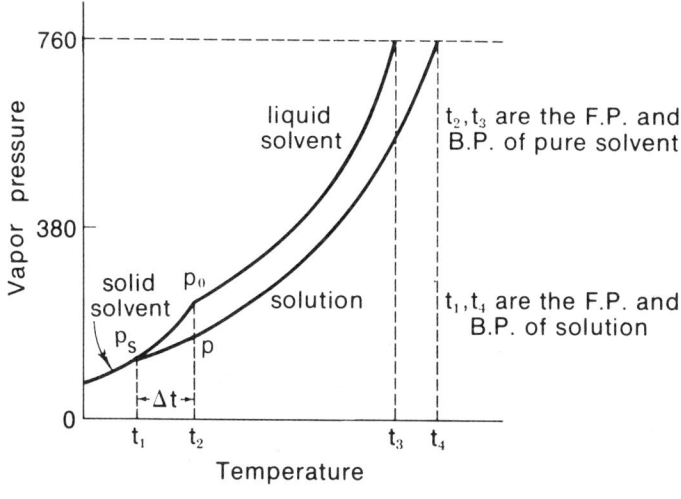

FIGURE 14-1

Diagram showing the freezing point depression and the boiling point elevation of a solution as compared to its pure solvent.

mixture (solution) of solute and solvent is below the freezing point of the pure solvent. This effect may be understood from the following considerations. The addition of a nonvolatile solute to a pure solvent decreases the tendency of the solvent molecules to escape into the gas phase; that is, the presence of the solute molecules decreases the vapor pressure of the solvent. The freezing point of an ideal solution is defined as the temperature at which crystals of the solvent first appear and are in equilibrium with the solution. At the freezing point the vapor pressures of solid solvent and of liquid solvent still in the solution are equal, and since the vapor pressure of the solution is less than that of the pure solvent, the freezing point of the solvent in a solution is below that of the pure solvent (as shown in Fig. 14-1).

The extent of the freezing point depression depends on the concentration of the solution. Raoult found that *the freezing point depression is directly proportional to the molal concentration of the solution.* This law is exact for ideal solutions, i.e., solutions in which the forces between molecules are not altered when molecules of more than one kind are mixed, and it is nearly exact for many covalent molecules in water solution. Table 14-1 gives the data for sucrose solutions as an example.

TABLE 14-1 FREEZING POINT DEPRESSIONS OF SUCROSE SOLUTIONS IN WATER

Molal Concentration m	Freezing Point Depression $\Delta T_f(^\circ C)$	Proportionality Constant $\Delta T_f/m$
0.005	0.0093	1.86
0.010	0.0186	1.86
0.020	0.0372	1.86
0.050	0.0935	1.87
0.100	0.188	1.88
0.200	0.380	1.90

In addition, the depression, ΔT_f, for a fixed concentration is the same for most covalent solutes in the same solvent, but the value for ΔT_f changes from one solvent to another, as shown in Table 14-2. Each solvent exhibits a specific value for the change in temperature per mole of solute per kilogram of solvent.

TABLE 14-2 FREEZING POINT DEPRESSION FOR THE SAME SOLUTES IN VARIOUS SOLVENTS

Solute	Solvent	Density of Solvent at Room Temperature (Grams per Cubic Centimeter)	ΔT_f, Lowering Produced by Adding 1 Mole of Solute to 1000 g. of Solvent
Methyl alcohol	Water	1.00	$1.86^\circ C$
Dextrose (sugar)	Water	1.00	$1.86^\circ C$
Urea	Water	1.00	$1.86^\circ C$
Methyl alcohol	Benzene	0.879	$5.12^\circ C$
Urea	Benzene	0.879	$5.12^\circ C$
Methyl alcohol	Cyclohexane	0.779	$20.00^\circ C$
Urea	Cyclohexane	0.779	$20.00^\circ C$

If the freezing point depression is nearly proportional to the number of solute molecules in solution, determination of the depression is one of the simplest and most accurate means of estimating the apparent molecular weight of a covalent solute. The equation that relates the freezing point depression to the concentration of solution is

$$\Delta T_f = K_f \times m \tag{1}$$

where ΔT_f is the experimental freezing point depression, m is the molality of the solution, and K_f is the known freezing point constant of the solvent. For a 1 molal solution that approximates ideal behavior, K_f is found to be 1.86°C for water, 5.12°C for benzene, 8.1°C for nitrobenzene, and 20.0°C for cyclohexane. In solutions of moderate concentration, equation (1) can be used for practical molecular weight determinations.

A. Molecular Weight by Freezing Point Depression

In this part of the experiment you are to determine the molecular weight of a solute by measuring the freezing point of benzene solutions of the solute and of benzene itself.

PROCEDURE

Obtain the freezing-point apparatus and assemble it in the manner shown in Figure 14-2. The apparatus consists essentially of an 800-ml. beaker, which serves as a cooling bath, covered with a metal lid. Through a hole in the center of the lid passes a large test tube, held in place by a tightly fitting cork. Held inside the test tube is another, smaller test tube containing the freezing mixture, a special thermometer, and a small stirrer. The freezing tube therefore is surrounded by a layer of air through which heat must be transferred, thereby insuring a low and uniform rate of cooling of the solution.

Assemble freezing-point apparatus.

Using the buret, measure to 0.02 ml about 25 ml. of pure dry benzene into the clean dry inner test tube. Put the thermometer in place, and immerse the test tube containing the thermometer and stirrer in the cooling liquid — an ice-water mixture that should contain a few pieces of ice at all times. Use enough cooling liquid to cover the level of the solvent in the test tube. The bath should be stirred frequently to insure a uniform temperature throughout (Note 1).

Obtain an accurately measured quantity of pure benzene.

With the lid removed from the bath, place the test tube containing the pure solvent directly into the ice bath and hold it there until the solvent partially freezes (Note 2). Take the test tube out and dry the outside thoroughly. Hold the tube in your hand and stir it until the solvent just melts. Place the tube in the air jacket, as illustrated in Figure 14-2, and continue to move the stirrers. (Stirring the benzene solution regularly is relatively more important than stirring the ice-water cooling bath.) Obtain temperature readings every 30 to 40 seconds, tapping the thermometer gently with the finger before taking each reading (Notes 3 and 4). Record both the temperature and the time for each reading. The liquid usually supercools

Determine freezing point of pure benzene.

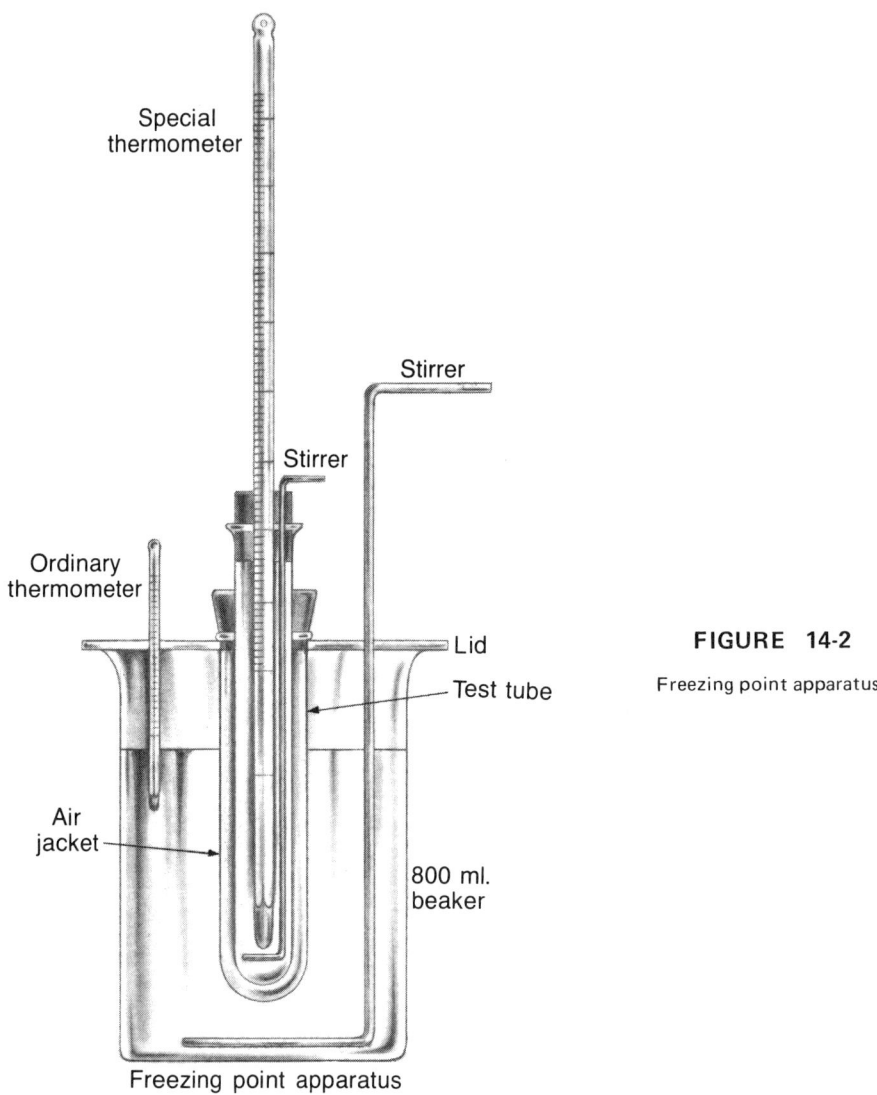

Special
thermometer

Stirrer

Stirrer

Ordinary
thermometer

Lid

Test tube

Air
jacket

800 ml.
beaker

Freezing point apparatus

FIGURE 14-2

Freezing point apparatus.

before it crystallizes. If it does, the temperature increases slightly when the first crystals form and then becomes constant until all the liquid is changed to crystals. This constant temperature is the freezing point. Plot the temperature readings as ordinate against the time as abscissa, and draw a curve through the points. Record the temperature of the horizontal portion as the freezing point.

Remelt solid and repeat f.p. determination.

Remelt the solid with the hand, as before, and redetermine the freezing point of the benzene. If the two values do not agree within 0.10°C, determine the freezing point a third time and average the values.

Weigh unknown accurately and dissolve it in benzene solution.

Weigh (to within 0.0005 g.) about 1 g. of the unknown solute, transfer it into the inner test tube containing the benzene, and dissolve it by stirring.

Equation (1) relates ΔT_f to m, the change in the freezing point to the concentration of the solution. The freezing point is defined as the temperature at which solid and solution are in equilibrium; solid solvent must be present for its determination. But if some solvent has crystallized out, the concentration of the solution is greater than that calculated from the known weight of solute and of benzene added to the test tube. Since the concentration of the solution does not change rapidly with slight changes in the amount of solvent present, we shall, as an approximation, measure the equilibrium temperature when only a small quantity of solvent has solidified, and use that value as the freezing point of the solution of calculated concentration m. It is experimentally easiest to measure this temperature on a warming cycle, as follows.

Remove the air jacket from the ice bath. Place the test tube containing the solution in the ice bath and cool it, operating the stirrer, until a substantial number of crystals are present. Remove the test tube from the ice bath, dry the outside, and place it in the air jacket. Clamp the air jacket at the top with the lower part exposed to the air of the room, operate the stirrer, and read the temperature, tapping the thermometer gently as before. Record as the freezing point the temperature reading when the crystals have almost disappeared. Repeat the measurement until you are satisfied that you have a reproducible value. You may wish to continue the temperature and time readings for a few minutes after the solid has disappeared, and plot the temperature against the time for the readings both before and after the solid disappeared. You should find that the slowly rising curve (with solid present) becomes a curve of greater slope after the solid disappears. If lines are drawn through each straight portion of these curves, the point of intersection of the two lines should be the freezing point (and melting point). Compare the value so obtained with the estimate you made at the disappearance of solid.

Determine freezing-point of solution.

Repeat the measurement as necessary.

CALCULATIONS

Using the density data in Table 14-2, calculate the weight of benzene used as solvent and the proportional weight of solute that would be present if 1000 g. of benzene had been used. From the measured freezing point depression calculate the molality of the solution, using the value 5.12 for K_f of benzene in Equation (1). Then calculate the weight of 1 mole of solute; this is the molecular weight that was to be determined.

NOTEBOOK AND REPORT

Record the freezing point of benzene for each trial, the quantities of benzene and of unknown used, and the temperature-time plots. Give the calculations and the result, and answer assigned questions and problems.

B. Relation of Freezing Point Depression to Chemical Composition of Solute

Many solutes, when dissolved in water (aqueous) solution, give values for the freezing point depression that are larger than those predicted by equation (1) for the concentration of the solution. These are called *abnormal freezing point depressions.* The magnitude of the abnormality is frequently recorded as the value of a quantity i known as the *Van't Hoff i factor,* which may be defined as the ratio of the observed freezing point depression $\Delta T_f'$ to the freezing point depression calculated from equation (1) for the molality of the solution.

$$i = \frac{\Delta T_f'}{\Delta T_f} = \frac{\Delta T_f'}{1.86 \times m} \qquad (2)$$

For common solutes, i factors from 1 (sometimes less than 1) to about 5 have been observed.

The reason for abnormality of freezing point depressions was recognized by Arrhenius as (usually) the result of a dissociation of solutes into ions.

In this part of the experiment, students will operate as teams of three. Each student of each team will determine the freezing point depression for a 0.5 molal aqueous solution of one of the solutes in the set assigned to his team. The results of each member of the team are then to be compared, and any trends in the data are to be noted, discussed, and interpreted in the team report prepared for the instructor.

Set 1	**Set 2**	**Set 3**
Sodium chloride	Potassium nitrate	Lithium sulfate
Magnesium chloride	Calcium nitrate	Magnesium sulfate
Aluminum chloride	Aluminum nitrate	Aluminum sulfate

Set 4	**Set 5**
Cadmium chloride	Hydrochloric acid
Cadmium bromide	Sulfuric acid
Cadmium iodide	Phosphoric acid

PROCEDURE

Determine f.p. of water and aqueous solutions.

Proceed as in Part A to determine the freezing point of distilled water and of 20 ml of your solution. Use a mixture of ice and salt to lower the temperature of the cooling bath below 0°C, so you can obtain the freezing point of water and of the 0.5 molal solution.

CALCULATIONS

Calculate the i value from equation (2).

NOTEBOOK AND REPORT

Your team should prepare a table of i values for the solutions investigated, describe any trends, and interpret them in terms of ionization of the solutes. Some of the values, although abnormal for a covalent substance, may yet be less than expected for a completely ionized substance; your discussion should include comment on the latter aspect of the data as well as on the former.

NOTES

Note 1. Benzene is flammable and somewhat toxic. Thus, extinguish all fires close to your work area and dispose of the benzene solution in the waste solvent can when you have finished this experiment.

Note 2. This is to cool the solvent rapidly to its freezing point.

Note 3. This is to make sure the mercury column is not stuck to the walls of the capillary tube. If it is stuck, a gentle tap will release it so that an accurate reading can be made.

Note 4. If the mercury column in the thermometer becomes separated, immerse the bulb slowly in a slush of dry ice and acetone (or isopropyl alcohol).

QUESTIONS–PROBLEMS

1. What is meant by the term ideal solution?

2. If a liquid cools to a temperature below its freezing point, and then starts to freeze, the temperature increases to the freezing point, even though no heat is added from outside the system. What is the source of the heat that causes the increase in temperature?

3. In Figure 14-1, assume that the lower curve, marked "solution" is for a 0.2 molal solution. Trace this figure in your notebook and add to the tracing a curve to suggest the approximate behavior of a 0.5 molal solution.

4. Cobalt(III) forms a large number of compounds containing other ions and/or molecules which are often called ligands. These molecules or ions retain their identify to some degree in the Co(III) compound. Table 14-Q-1 lists some of these compounds, giving, for each kind of ligand, the ratio of each ligand to the cobalt ion in the compound, as found by analysis. Table 14-Q-2 gives freezing point depression data for solutions of these compounds in water. In this table, the symbol m represents the number of moles of cobalt ions, each with its accompanying ligand groups, dissolved in 1000 g. H_2O.

Calculate the Van't Hoff i factors for each solution in Table 14-Q-2, and using the data of Table 14-Q-1, determine the formula of each compound. Show by an equation how the solid compound ionizes in solution. It will be helpful to consider that the ammonia molecule, NH_3, and the nitrite ion, NO_2^-, remain intact in the Co(III) complex. Also, each ammonia and each nitrite functions as one ligand. Cobalt(III) compounds are uniformly six-coordinate, i.e., they bind to six ligands. Chloride, potassium, and sodium ions exhibit their usual ionic charges in these compounds.

TABLE 14-Q-1 COMPOSITION OF SOME COBALT(III) COMPOUNDS

Compound	Cobalt Ions	Ammonia Molecules	Nitrite Ions	Chloride Ions	Sodium Ions	Potassium Ions
I	1	6	—	3	—	—
II	1	2	4	—	—	1
III	1	3	3	—	—	—
IV	1	4	2	1	—	—
V	1	—	6	—	3	—
VI	1	5	—	3	—	—

TABLE 14-Q-2 FREEZING POINT DEPRESSION FOR SOLUTIONS OF COBALT(III) COMPOUNDS

Compound	m	ΔT_f
I	0.0050	0.031°C
II	0.0052	0.019°C
III	0.0021	0.0039°C
IV	0.0066	0.024°C
V	0.016	0.110°C
VI	0.0046	0.024°C

REFERENCES

1. Daniels, F., and Alberty, R. A.: *Physical Chemistry.* 3rd edition. John Wiley and Sons, New York, 1966, pp. 219-223.
2. Garrett, A. B., Lippincott, W. T., and Verhoek, F. H.: *Chemistry, a Study of Matter.* 2nd edition. Xerox College Publishing, Lexington, Mass., 1972, pp. 258-65 and 268-72.

Experiment 14
REPORT SHEET

Name _____   **Section No.** _____

Unknown No. _____

PART A

1. Volume of benzene used _____.

2. Quantity of solute used _____.

3. Freezing point of pure benzene _____.

_____.

_____.

4. Freezing point of benzene solution of solute _____.

_____.

_____.

5. ΔT_f = _____.

6. Molality of solution of unknown concentration _____.

7. Molecular weight of solute _____.

8. Attach graphs showing temperature-time plots.

PART B

1. Freezing point of distilled water _____.

2. Freezing point of your 0.5m solution _____.

_____.

_____.

3. Identity of salt _____.

4. ΔT_f = _____.

5. What ΔT_f value would you expect if your salt were ionized 100%? _____ . Write the ionization equation and show your logic or method of calculation.

6. i value _____ .

7. i values for the other solutions determined by your team _____ .

8. Attach graphs showing temperature-time plots.

Attach sheets showing answers to the assigned questions.

CHEMICAL EQUILIBRIA

EQUILIBRIA IN CHEMICAL REACTIONS—THE PRINCIPLE OF LE CHATELIER

OBJECTIVE

To study several factors that affect the position of chemical equilibria and the completeness of chemical reactions.

EQUIPMENT NEEDED

Test tubes; 25 × 200 mm. test tube; 10-ml. graduated cylinder; 25- or 50-ml. buret; 50-ml. beaker; 125-ml. Erlenmeyer flask; burner; medicine droppers.

REAGENTS NEEDED

Antimony trichloride, 0.5M in 6.0M HCl; 6M and concentrated HCl; 0.4 M $CoCl_2$; 0.4M $Co(NO_3)_2$; 0.4M $Co(NO_3)_2$ in absolute ethyl alcohol; solid cobalt(II) chloride, $CoCl_2 \cdot 6H_2O$; solid cobalt(II) nitrate, $Co(NO_3)_2 \cdot 6H_2O$; acetone; absolute ethyl alcohol; ice; 1M $Pb(NO_3)_2$; 1M NaCl; 1M Na_2CO_3; 0.1M Na_2S; 6M HNO_3; solution of methyl orange indicator; 0.1 M HCl; 0.1 M CH_3CO_2H; solution of 0.1 M CH_3CO_2H and 0.1 M CH_3CO_2Na; solid CH_3CO_2Na.

INTRODUCTION

Often a chemist finds that he does not isolate 100 percent of the product from a chemical reaction, perhaps because his experimental technique does not allow quantitative isolation of the product or because the reaction actually does not proceed to completion to yield the theoretical quantity of product. In this experiment we are interested in the latter aspect — the manner in which conditions affect the position of equilibrium. In other words, how far does a reaction proceed before reaching equilibrium? For example, does it proceed almost completely to products, proceed only slightly to products, or proceed about 50% to completion? What factors affect this?

Almost every chemical equation may be considered to be reversible. For example, given the reaction

$$A_2 + B_2 \rightleftharpoons 2AB + \text{heat} \qquad (1)$$

171

it may be found by experiment that if A_2 and B_2 are brought together under suitable conditions for reaction, the conversion to product AB is incomplete, even if left for very long reaction times. A condition is reached in which the concentration of the product and the concentrations of the reactants are not changing. This is the condition of *chemical equilibrium*. A system at equilibrium may appear to be undergoing no changes whatsoever, but it actually is a very dynamic situation, undergoing product formation and decomposition at the same rate.

If both the forward and reverse reactions of equation (1) are elementary reactions, we can, according to the law of mass action, represent the rate of formation of the product, AB, by $R_f = k_f [A_2] [B_2]$, where k_f is a proportionality constant and $[A_2]$ and $[B_2]$ are the effective concentrations of the reactants in moles per liter. The rate of the reverse reaction, that of decomposition of the product, is expressed by $R_r = k_r [AB]^2$. Under equilibrium conditions the rate of the forward reaction, R_f, is equal to the rate of the reverse reaction, R_r, and the reactants and products are at their equilibrium concentrations. Therefore,

$$R_f = R_r = k_f [A_2]_{eq} [B_2]_{eq} = k_r [AB]_{eq}^2$$

Since k_f and k_r are both constants at a given temperature, they may be combined into a single constant, K, which is commonly known as the equilibrium constant. The value of K changes only with a change in temperature; *i.e.*, it does not change with different concentrations of A_2, B_2, or AB as long as $R_f = R_r$.

$$K = \frac{k_f}{k_r} = \frac{[AB]_{eq}^2}{[A_2]_{eq} [B_2]_{eq}}$$

A generalization known as Le Chatelier's principle governs systems at equilibrium. According to this principle, *when a change is made in a system at equilibrium, the position of equilibrium shifts in such a direction as to minimize the effect of the change.* The following examples should help to clarify this principle.

Suppose the system represented by equation (1) *had attained equilibrium,* and then you added more of reagent A_2 to the system. What would happen? Le Chatelier's principle suggests that some B_2 would react with some of the excess A_2 to reduce the concentration of the added reagent. This reaction would increase the concentration of AB and reduce the concentration of B_2 and A_2 so that equilibrium would again be established, and the ratio $\dfrac{[AB]^2}{[A_2] [B_2]}$ would be equal to the original K value. In the new equilibrium mixture, $[AB]_{eq}'$ would be greater than the $[AB]_{eq}$ in the original equilibrium mixture; $[A_2]_{eq}'$ would be greater than $[A_2]_{eq}$, and $[B_2]_{eq}'$ would be less than $[B_2]_{eq}$.

By similar reasoning we can see that addition of the compound AB to the system would shift the position of equilibrium to the left until the ratio of the concentrations of the three species again attained the equilibrium constant value, K.

If we add heat to a system that has attained equilibrium, the system shifts in the direction that absorbs heat, *i.e.*, in the direction of the endothermic reaction. In the case of the reaction represented by equation (1), heating the system would shift the position of equilibrium to the left, since this reaction absorbs heat and tends to minimize the effect of the added heat.

A. Effect of Concentration on Equilibrium

Antimony trichloride undergoes hydrolysis in aqueous solutions, forming a heavy white precipitate of antimony oxychloride:

$$SbCl_3 + H_2O \rightleftharpoons SbOCl(s) + 2HCl \qquad \textbf{(2)}$$

Equation (2) shows that equilibrium is established in which hydrochloric acid is a product of the hydrolysis reaction. By increasing the concentration of hydrochloric acid, one would cause some of the solid antimony oxychloride to dissolve, and a new position of equilibrium would be established in which the concentrations of the reactants and products would be different from those under the original conditions.

The solution of antimony trichloride (labeled "for Experiment 15") is prepared by dissolving 0.5 mole of solid antimony trichloride in enough 6M HCl to make 1 liter of solution. At this acid concentration the position of equilibrium is so far to the left that the concentration of antimony oxychloride is negligible and antimony trichloride remains soluble.

PROCEDURE

Dilution of an Acidic Antimony Trichloride Solution with Water

Carefully measure 5.0 ml. of the antimony trichloride (0.5M $SbCl_3$ in 6M HCl) into a 25 × 200 mm. test tube. Add, carefully, 5.0 ml. of distilled water from a graduated cylinder; then shake the test tube to mix the solution thoroughly (or stir the mixture with a clean glass stirring rod). Record your observations in the notebook and calculate the resulting hydrochloric acid concentration in the total volume of 10 ml., assuming that the amount of hydrochloric acid formed via equation (2) is negligible. Why did the solid that formed as water was added disappear on shaking or stirring?

Carefully mix $SbCl_3$ and H_2O.

Now add an additional 3 ml. of distilled water to the solution. Note and explain your observations, taking into account the new hydrochloric acid concentration (note 1).

Continue to add increments of H_2O.

Add 2 ml. more of distilled water to the solution. Record your observations. Recall the total volume of sample at this point and calculate the hydrochloric acid concentration, again assuming that the amount formed according to equation (2) is negligible compared to the amount in the original 6M solution.

Add 5 ml. more of distilled water to the solution. Compare in your record book the result with that when a total of 10 ml. of water had been added.

In calculating the concentrations of hydrochloric acid in each of the four solutions, as a first approximation, you may neglect the amount formed from equation (2). However, does the hydrochloric acid concentration increase or decrease as a result of the hydrolysis reaction represented by equation (2)? As a check on the assumption that you can neglect the amount of hydrochloric acid that is formed *via* equation (2), after you have added 20 ml. of water, make the assumption that 50% of the antimony trichloride in the test tube hydrolyzed according to equation (2). Consider that your original 5 ml. of solution was 0.5M in antimony trichloride, and calculate the amount of hydrochloric acid produced *via* equation (2). Is this value small compared with the amount of hydrochloric acid that is present at every dilution from the 6.0M solution? Was it reasonable to neglect the amount of hydrochloric acid formed by the hydrolysis reaction in the first two or three dilutions?

Addition of Hydrochloric Acid to Antimony Oxychloride (Demonstration of Equilibrium Reversibility)

Add HCl to the suspension.

Add 2 ml. of 6M HCl to the test tube containing the 5 ml. of antimony chloride solution and 20 ml. of water. Mix thoroughly. Record and explain your observation in your notebook. Add 3 ml. of 6M HCl to the solution, and again mix the contents thoroughly. Record your results and explain the observations in terms of equation (2).

Determination of the Equilibrium Constant for Equation (2)

Two additional facets must be considered in calculating an equilibrium constant for the reaction given in equation (2). The first concerns the concentration of water in the expression for the equilibrium constant; the second concerns the concentration of antimony oxychloride in the solid phase.

Since this equilibrium is studied in dilute aqueous solution, water is present in tremendous excess. Moreover, its concentration is virtually unaffected when small amounts of other reagents are added to the equilibrium mixture. Therefore, the concentration of water to be substituted into an expression for the equilibrium constant of equation (2) is a large and constant number. To avoid carrying this large and constant number in the calculation of the equilibrium constant, chemists have agreed, merely as a convenience, to define a new equilibrium constant expression that includes only the concentrations of the reactants and products having concentrations that may change appreciably in the reaction under consideration. In accord with this agreement, the concentration of water is *not* included in the equilibrium constant expression for those reactions performed in aqueous solutions.

A similar situation arises in connection with solid reactants or products. The concentration of a solid in the reaction mixture is the *number of moles of solid per liter of the solid;* it is a measure of the concentration of the solid in its own solid phase and not the concentration of the solid in the reaction solution. Of course, the packing of units in a pure crystalline solid is the same throughout any one crystal and among all crystals of a given solid. Therefore, the concentration of a solid component is a constant; for this reason, concentrations of solids are not included in equilibrium constant expressions.

The equilibrium constant expression used to calculate K for equation (2) includes neither the water nor the antimony oxychloride concentrations, because the concentrations of these materials are not changed by the experimenter (unless he works with very concentrated solutions). It does include the concentration of hydrochloric acid and antimony trichloride because these concentrations change relatively rapidly by adding more water or more of either reagent. Thus, the complete equilibrium constant expression for equation (2)

$$K = \frac{[HCl]^2 \, [SbOCl]}{[SbCl_3] \, [H_2O]}$$

is conventionally written as

$$K' = \frac{[HCl]_{eq}^2}{[SbCl_3]_{eq}}$$

Determine (as accurately as possible) the concentration of hydrochloric acid required to give a permanent precipitation of antimony oxychloride by obtaining another 5-ml. sample of the antimony trichloride solution and adding distilled water carefully from a buret just to the point at which a slight milky color persists after thorough mixing.

Determine the equilibrium constant K'.

From the previous experiments you should be able to estimate approximately how much water is required. When you have added almost enough water to cause precipitation of antimony oxychloride, add the water a drop at a time. Assume that an equilibrium is established between the solid antimony oxychloride and antimony trichloride in solution when the slightly milky color first remains after thoroughly mixing the solution. Record the total amount of water used and then calculate the concentrations of both antimony trichloride and hydrochloric acid in the final volume.

At the point at which the solution has a permanent slightly milky appearance, you may assume that very little antimony oxychloride or hydrochloric acid has been formed, and consequently very little antimony trichloride has reacted. This assumption permits the antimony trichloride and hydrochloric acid concentrations, at the point of antimony oxychloride formation, to be calculated directly from the original concentrations, taking into account the dilution. Record in your notebook all the arithmetic steps necessary to calculate the equilibrium constant K' for equation (2).

B. Effect of Concentration on an Equilibrium
Involving a Geometric Structural Change

Cobalt has oxidation states of II (cobaltous) and III (cobaltic). Unlike the ferrous ion, the cobaltous ion is quite stable to oxidation, and aqueous solutions of cobaltous salts can be exposed indefinitely to air. The pink aqueous solutions of cobaltous salts are characteristic of the complex ion $[Co(H_2O)_6]^{2+}$, in which the cobaltous ion is bonded with six water molecules in a symmetrical arrangement, that of *an octahedron*. The pink cobalt(II) salts may be crystallized from solution and give solids of formula

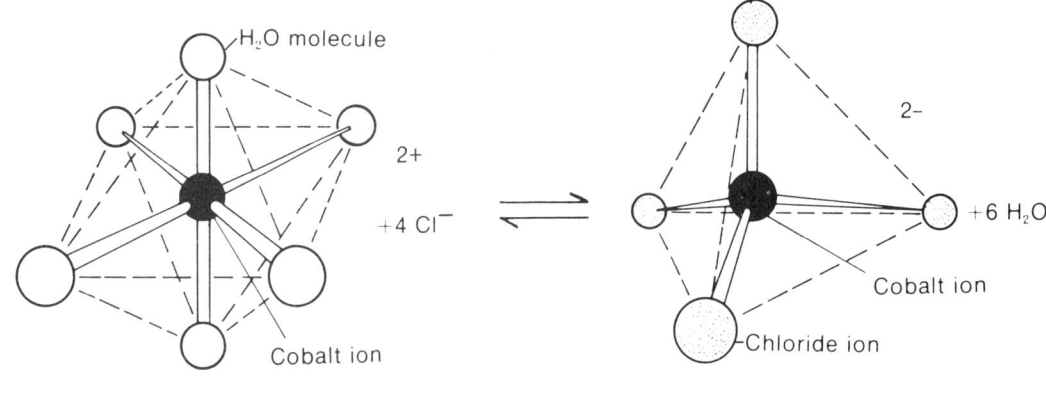

Octahedral structure   Tetrahedral structure

FIGURE   15-1

$CoX_2 \cdot 6H_2O$, where X is the nitrate, acetate, or chloride ion. Both solutions and solids exhibit properties of $[Co(H_2O)_6]^{2+}$ ion.

In this part of the experiment, we shall study the manner in which the concentration of a reagent affects the equilibrium and the way in which this may change the structure and color of the complex ions in solution. Divalent cobalt commonly assumes a coordination number of either 4 or 6 in its complexes. In nonaqueous solvents the chloride ion rapidly replaces water molecules from $[Co(H_2O)_6]^{2+}$ to form the $[CoCl_4]^{2-}$ anion. The concentration of water in dilute aqueous solutions is approximately 55M and, according to the law of mass action, the water molecules effectively compete with chloride for bonding to the cobalt(II) ion. We may express this equilibrium by equation 3 and Figure 15-1 (Note 2):

$$[Co(H_2O)_6]^{2+} + 4Cl^- \rightleftharpoons [CoCl_4]^{2-} + 6H_2O \tag{3}$$

The presence of small concentrations of $[CoCl_4]^{2-}$ is detected easily since this tetrahedral ion is a deep blue color and is approximately 100 times as intensely colored as the pink $[Co(H_2O)_6]^{2+}$ ion. In the following sections we shall investigate the effect of chloride concentration, solvent, and temperature on the equilibrium represented by equation (3). In addition to changing the number of ligands bonded to cobalt, this reaction involves a change in the geometry of the complex ion from an octahedron, $[Co(H_2O)_6]^{2+}$, to a tetrahedron, $[CoCl_4]^{2-}$.

In alcohol and acetone solutions some of the water molecules in $[Co(H_2O)_6]^{2+}$ are replaced by the alcohol or acetone molecules as a result of the very large relative concentrations of these solvent molecules, *i.e.,* the mass law favors bonding an alcohol or acetone molecule rather than water. Also, chloride ion can displace the alcohol or acetone molecules even more readily than water. Consequently, the blue color characteristic of the tetrahedral $CoCl_4^{2-}$ ion appears at much lower concentrations in the organic solvents.

PROCEDURE

Obtain about 0.5 g. of each of the cobalt(II) salts, $CoCl_2 \cdot 6H_2O$ and $Co(NO_3)_2 \cdot 6H_2O$, and record their colors as solids. Add two or three small crystals of each salt to separate dry test tubes and dissolve the crystals in 5 ml. of water. Record your observations. Repeat this procedure, using absolute ethyl alcohol and then acetone. What differences are noted between the chloride and nitrate salt in each of the solvents, water, alcohol, and acetone? Do the colors of each salt in the three solvents produce trends? How are your data explained in terms of the equilibrium represented by equation (3)? For example, which solvent tends to give the deepest blue color with the chloride salt? Consider the huge difference in the number of moles of water present in the aqueous solution compared with the ethyl alcohol and acetone solutions. On this basis do the relative colors make sense if equilibrium (3) is appropriate?

Determine the colors of Co(II) salts in different solvents.

Add 5.0 ml. of 0.4M $Co(NO_3)_2$ solution to a 6-inch test tube. Now add concentrated hydrochloric acid to the solution in eight successive 2.0-ml. increments, and shake the solution thoroughly (or stir) after each addition. Record the color of the solution after each addition and explain the trend observed.

Add increments of HCl.

Now add *three* successive 5.0-ml. increments of distilled water to the cobalt(II) nitrate-hydrochloric acid solution. Record your observations after each dilution.

It should now be interesting for you to repeat the procedure in a mixture of ethyl alcohol and water. To do this, add 5.0 ml. of 0.4M $Co(NO_3)_2$ dissolved in ethyl alcohol to a 6-inch test tube, and then add dropwise concentrated hydrochloric acid to the solution until you obtain a distinct blue color. Compare the concentration of chloride required to produce the blue color in the alcohol solution with that required in water.

CALCULATIONS

Calculate the molarity (concentration, expressed in moles per liter of solution) of concentrated hydrochloric acid. This acid solution contains 37% hydrochloric acid by weight and has a specific gravity of 1.185 g. per milliliter. Use this concentration to calculate the concentration of chloride ion in your cobalt solution after each addition, assuming solution volumes are additive. Also calculate the concentration of chloride ion in the solution after each addition of distilled water. Then calculate the approximate concentration of chloride required to give a visually perceptible blue color in water and in the ethyl alcohol-water mixture.

C. Effect of Temperature on the Position of Equilibrium

In the introduction we stated that the value of the equilibrium constant, K, is specific for a given temperature. This part of the experiment concerns the effect of temperature on the position of equilibrium (3).

Carefully measure 10.0 ml. of 0.4M $CoCl_2$ solution into a small Erlenmeyer flask. Add 6.0 ml. of concentrated hydrochloric acid and swirl the flask to mix thoroughly. The resulting solution should be violet (between the original pink-red and bright blue). If it is not violet, adjust the solution color by carefully adding, a drop at a time, distilled water or concentrated hydrochloric acid, depending upon the color of the solution. Divide the violet solution about equally into three test tubes. Place one test tube in ice, heat one gently to 80 to 90°C., and allow the third one to remain at room temperature. Record the colors of each solution. Determine whether the color changes are reversible.

What can you conclude about the effect of temperature changes on the position of equilibrium for this reaction? Rewrite equation (3) to include the word *heat* on the proper side of the equation.

D. Effect of Solubility on Completeness of Chemical Reactions

Combine Pb²⁺ with Cl⁻.

Repeat with CO₃²⁻.

Mix 10 ml. of 1 M $Pb(NO_3)_2$ and 10 ml. of 1 M NaCl in a 50-ml beaker. Account for your observation by writing a total and a net ionic equation for this chemical reaction in your notebook. Filter and wash the solid lead chloride with 25 ml. of distilled water. Discard the wash solution and add the solid lead chloride plus 25 ml of water to a small Erlenmeyer flask and shake it for three minutes. Allow the precipitate to settle and decant 10 ml of the clear supernatant liquid into a test tube and add 5 ml of 1 M Na_2CO_3. Again explain your observations with words and a balanced equation.

Treat suspension with Na₂S.

Treat suspension with HNO₃.

Shake the test tube to which sodium carbonate has been added and pour half the suspension into a second test tube and add 3 ml. of 0.1 M Na_2S solution. Results? Add 6 or 7 drops of dilute nitric acid to the test tube containing the black solid. Results?

To the remainder of the suspension in the first test tube, add 6 or 7 drops of 6 M HNO_3. What happens to the solid?

For each of the above reactions, consider that some Pb^{2+} ion is present in solution. That is, the solids may be considered to be partially soluble. Formation of each successive precipitate results from the fact that the new solid is less soluble than the preceding one. Thus, each successive reaction proceeds further to completion and removes more of the Pb^{2+} ion from solution. As an example, the lead chloride precipitate is partially soluble and can be represented by equation (4).

$$PbCl_2\,(s) \rightleftharpoons Pb^{2+} + 2Cl^- \qquad \textbf{(4)}$$

You should then be able to write equations for each reaction and to decide which lead compounds are less soluble, i.e., which reactions go further toward completion. Arrange the following compounds in order of decreasing solubility: lead nitrate, lead carbonate, lead sulfide, and lead chloride.

E. Effect of Buffers on Acid-Base Equilibria

INTRODUCTION

Using the Brønsted concept of acids and bases, we have learned to consider acids as proton donors and bases as proton acceptors. We have also learned to think of an acid and its conjugate base (the acid minus its proton) or of a base and its conjugate acid (the base with an added proton). For example:

$$\text{HCN (acid)} \qquad \text{CN}^- \text{ (conjugate base)}$$

$$\text{NH}_4{}^+ \text{ (acid)} \qquad \text{NH}_3 \text{ (conjugate base)}$$

or

$$\text{OH}^- \text{ (base)} \qquad \text{H}_2\text{O (conjugate acid)}$$

$$\text{H}_2\text{O (base)} \qquad \text{H}_3\text{O}^+ \text{ (conjugate acid)}$$

The last two examples show that water is amphoteric; i.e., it may act as an acid or as a base, depending upon the other chemical species in the environment. For example, in the reaction

$$\text{HCl} + \text{H}_2\text{O} \rightleftharpoons \text{H}_3\text{O}^+ + \text{Cl}^-$$

water is acting as a base, receiving a proton from the strong acid hydrogen chloride. However, in the reaction

$$\text{H}_2\text{O} + \text{NH}_2{}^- \rightleftharpoons \text{NH}_3 + \text{OH}^-$$

water is acting as an acid, donating a proton to the strong base, amide ion, $\text{NH}_2{}^-$.

Since all chemical reactions may be regarded as equilibrium processes if the reactants and products remain in the same environment, an acid-base reaction can be viewed in terms of the equilibrium established between the original acid and base and the conjugate base and conjugate acid formed. The equilibrium

$$\text{HCN} + \text{H}_2\text{O} \rightleftharpoons \text{H}_3\text{O}^+ + \text{CN}^-$$

is a typical example of an acid and a base reacting to give a new acid and a new base. The position of equilibrium always favors the weaker acid and weaker base since the stronger acid and stronger base have a greater tendency to react than do the weaker acid and base.

Indicators, which are themselves weak acids or bases, are sometimes used in studying acid-base equilibria. The usefulness of an indicator depends upon the fact that the acid form of the indicator, HInd, has a different color from that of the conjugate base form of the indicator, Ind^-. Thus, for example:

	HInd Form	Ind⁻ Form
Phenolphthalein	Colorless	Pink
Litmus	Pink	Blue
Methyl orange	Pink	Orange

The color of a solution containing an indicator reflects whether HInd or Ind^- is present in higher concentration.

Since most indicators are weak acids (or weak bases), it should be possible to arrange a series of indicators in order of decreasing acid strength by studying the colors of their solutions with some acids of known strength.

If the color of the solution of the indicator in an acid HA is that of the acid form of the indicator, the indicator acid must be weaker than the acid HA; if the color of the solution of the indicator in the acid is that of the base form of the indicator, HInd must be stronger than HA. Perhaps this will be made clear from a consideration of

$$HInd + A^- \rightleftharpoons HA + Ind^-$$

Let us raise the questions: "What statement can be made about the relative strengths of HInd and HA if the position of equilibrium is to the left; to the right?" Recalling that the position of equilibrium favors the weaker acid and base, evidently the color of the solution reveals the relative strengths of the two acids.

In this part of the experiment, you are asked to determine the predominant form of the indicator methyl orange present in solution under several different conditions. Also you are to determine the effect of adding small amounts of acid or base to aqueous solutions of methyl orange in the presence and in the absence of a buffer and to interpret the observations in terms of the equilibria present in the solutions.

A buffer solution of a weak acid is prepared by dissolving the acid and a soluble, completely-ionized salt of the acid in water. For example, acetic acid ionizes according to the following equation:

$$CH_3CO_2H + H_2O \rightleftharpoons H_3O^+ + CH_3CO_2^- \tag{5}$$

and gives a K_{ion} of 1.85×10^{-5} at 25°C. Introduction of added acetate ion would repress the amount of ionization (predicted by Le Chatelier's Principle) compared to the situation with only acetic acid. If a relatively large concentration of acetate ions is present in solution, the acetate ion can readily react with any added proton source to form more CH_3CO_2H and minimize the effect of adding the strong protonic species. Thus, the buffer resists marked changes in hydronium ion concentration (or pH) with introduction of small amounts of strong acids or bases such as HCl or NaOH, respectively.

PROCEDURE

Determine colors of HInd and Ind⁻.

First determine the colors of the acid and base forms of the indicator methyl orange, by placing 1 drop of indicator solution in 2 ml. of 0.1 M HCl and 1 drop in 2 ml. of 0.1 M NaOH. Record the colors of HInd and Ind⁻ for the indicator. Assume the colors of HInd and Ind⁻ are the colors you see when the indicator is placed in 0.1 M H_3O^+ and 0.1M OH⁻, respectively.

Now determine the color of the indicator in 0.1 M acetic acid solution by adding 1 drop of methyl orange solution to 2 ml. of 0.1 M acetic acid in a test tube. Add (one crystal at a time) 4 or 5 small crystals of solid sodium acetate. What do you observe?

Determine form of indicator present in buffer.

Determine the color of the indicator in a solution that is 0.1 M in CH_3CO_2H and 0.1 M in CH_3CO_2Na by adding 2 drops of indicator to 4 ml. of the solution. Now add dropwise 0.1 M HCl to the above solution. What do you observe? Repeat this experiment with 0.1 M NaOH solution using a new solution of methyl orange in the 0.1 M CH_3CO_2H–0.1 M CH_3CO_2Na

solution. How do your observations compare with the above HCl case? Also, how do these experiments compare or contrast with the first determinations of the colors of methyl orange in 0.1 M HCl and 0.1 M NaOH solutions?

Account for the colors of methyl orange in the CH_3CO_2H-CH_3CO_2Na solution (and when a strong acid or base was added) in terms of the equilibrium expressed by equation 5 and the predominant form (HInd or Ind$^-$) of the indicator present in the solutions.

NOTEBOOK AND REPORT

Before going to the laboratory, prepare data tables in your notebook in which to record the volumes of water and acid and concentrations of reagents in Parts A and B, and the temperatures, concentration of chloride, and observations in Part C. Give all necessary data in your report that are useful in obtaining the chloride concentrations in Parts A and B, the equilibrium constant for equation (2), the effect of solvent on equilibrium shown in equation (3), and the effect of temperature on this equilibrium. Include detailed calculations for one example of each type of calculation used in your studies. Give your equations for the tests in Part D and discuss step by step the significance of each precipitate in terms of the appropriate equilibrium. For Part E your notebook should show a careful record of your observations and the relative amounts (*i.e.,* drops) of the HCl or NaOH solutions needed to convert the indicator from one form to the other.

NOTES

Note 1. You will need to calculate the concentration of HCl and $SbCl_3$ several times during this experiment. Recall that $V_aM_a = V_bM_b$. For example, when the initial 5 ml. of antimony trichloride solution, which is 6M in HCl, is diluted to 10 ml. by the addition of 5 ml. of water:

$$5 \text{ ml} \times 6M = 10 \text{ ml} \times M_b$$

Thus, M_b = the concentration of HCl in the diluted solution, *i.e.,* 3M.; on further dilution to 13 ml., M_b becomes 2.3M, etc.

Note 2. The blue tetrahedral cobalt anion is given as $[CoCl_4]^{2-}$ in equation (3) even though some spectrophotometric studies indicate that an appreciable concentration of $[Co(H_2O)Cl_3]^-$ exists in aqueous solutions that are less than 12M in chloride.

QUESTIONS—PROBLEMS

1. In Part A of the experiment, why does a precipitate form when water is added to the antimony trichloride solution and then disappear when the test tube is shaken? How does this behavior relate to differences in local concentrations?

2. Is the position of equilibrium, as shown in equation (3), further to the right in acetone or ethyl alcohol than in water? How is the observation explained?

3. From your observations in this experiment, could you say that water, alcohol, or acetone competes most effectively with the chloride ion for coordination with the cobaltous ion? Why?

4. For a quantitative determination of the heat term involved in the equilibrium of equation (3), what additional data would you need?

5. It has been observed that an indicator gives a different color in a solution of a strong acid from that in a solution of a strong base. Write equations for the reaction of the indicator with H_3O^+ and with OH^-.

6. Explain why the indicator color changes at the equivalence or neutralization point in the titration of a strong acid with a strong base.

7. Explain why buffered solutions should be used for comparing the acid strengths of the indicators or weak acids in general.

8. Write the expression for the ionization constant of acetic acid in terms of the equilibrium concentrations of reactants and products in the reaction of acetic acid and water. In terms of this equation, what is the value of the hydronium ion concentration in a solution 0.1M in acetic acid and 0.1M in the conjugate base?

9. Suppose you were titrating a weak acid, such as acetic acid, into a beaker of sodium hydroxide solution containing methyl orange indicator.
   a. Write the equation for the reaction between HInd and OH^-. (This reaction occurs while OH^- is in excess.) What species gives the color here?
   b. Now write the equation for the reaction of acetic acid with Ind^-. (This presumably should occur when acetic acid is in excess.) From what you know about the relative strength of the acid form of methyl orange compared with acetic acid, where is the position of equilibrium in this reaction? What color will methyl orange be in excess acetic acid?
   c. Would methyl orange be a good indicator in the titration of acetic acid with sodium hydroxide? Explain your answer.

REFERENCE

1. Garrett, A. B., Lippincott, W. T., and Verhoek, F. H.: *Chemistry, A Study of Matter,* 2nd edition, Xerox College Publishing, Lexington, Mass., 1972, Chapters 19 and 24.

Experiment 15
REPORT SHEET

Name _____ Section No. _____

PART A

1. **Dilution of acidic antimony trichloride solution with water**
    a.   What do you see when 5 ml of water is added?

        Resulting hydrochloric acid concentration: _____
        Why did the solid disappear upon shaking?

    b.   After adding 3 ml of water what do you see?

        Hydrochloric acid concentration: _____
    c.   After adding 2 ml of water, what do you see?

        Hydrochloric acid concentration: _____
    d.   After adding 5 ml of water:
        Observations and comparison

    e.   Assuming 50% hydrolysis of $SbCl_3$, how much HCl is produced? _____

    f.   Is the value small by comparison with the amount originally present?

g. Was it reasonable to neglect the amount of HCl formed via equation (2)?

2. **Determination of the value of K′ for the SbCl₃ hydrolysis**

a. Amount of water added _____

b. Concentration of antimony trichloride in the final volume _____

c. Concentration of hydrochloric acid in the final volume _____

d. Calculation of K′

Experiment 15
REPORT SHEET

Name _____ Section No. _____

PART B

Colors

	$CoCl_2 \cdot 6H_2O$	$Co(NO_3)_2 \cdot 6H_2O$
solid		
in acetone		
in ethyl alcohol		
in water		

What differences, similarities, and trends do you notice from the above table?

How do you explain these data in terms of equation (3)?

Upon adding 2 ml increments of hydrochloric acid to the aqueous solution of $Co(NO_3)_2 \cdot 6H_2O$:

a. Complete the table giving the color observed after each added increment.

0 ml _____   8 ml _____

2 ml _____   10 ml _____

4 ml _____   12 ml _____

6 ml _____   14 ml _____

                                         16 ml _____

b. Explain the trend.

c. Concentration of chloride ion at each step in this solution. (Make a table.)

Upon adding 5 ml increments of water to the above solution, what do you see after each dilution? (Make a table.)

Comparison of the chloride ion concentration required to produce the blue color in ethyl alcohol with the concentration required in aqueous solution.

Experiment 15
REPORT SHEET

Name _____ Section No. _____

PART C

What are the colors at the three temperatures?

Are the color changes reversible?

What is the effect of temperature on the position of equilibrium?

Write equation (3) including *heat* on the appropriate side of the arrows. How do you justify this choice?

PART D

Explain what happens when you mix solutions of $Pb(NO_3)_2$ and $NaCl$.

What do you see after adding Na_2CO_3?

What do you see after adding Na_2S?

What do you see after adding nitric acid?

What do you see after adding nitric acid to the second test tube?

What happened to the precipitate?

What are the pertinent equations for each precipitate?

List the compounds in order of decreasing solubility.

Experiment 15
REPORT SHEET

Name _____   Section No. _____

PART E

Data

Interpretations of Your Observations in Terms of Equilibrium (5).

ELECTROCHEMISTRY

ELECTROLYTIC CELLS

OBJECTIVE

To study the relationships between the number of electrons, as represented by a flow of electricity, and the oxidation and reduction reactions that occur at the anode and the cathode, respectively.

EQUIPMENT NEEDED

Electrolytic boxes; sand paper; copper, zinc, or cadmium electrodes; 150-ml. beaker; 250-ml. beaker; clock or watch with second hand; source of direct current; wash bottle; porous cup.

REAGENTS NEEDED

Solutions (1M) of zinc sulfate and copper sulfate; acetone.

INTRODUCTION

Electrochemistry is the area of chemistry that deals with the phenomena associated with the interaction of electricity with matter. Two broad areas of electrochemistry are: (1) that related to electrolytic cells (cells in which electrical energy is used to promote an oxidation-reduction reaction) and (2) that related to galvanic or voltaic cells (cells in which an oxidation-reduction reaction takes place in such a way that electrical energy is generated).

In this and the following experiment, some fundamental laws of the two branches of electrochemistry will be developed. In this experiment an attempt is made to establish a connection between the quantity of electricity passed through the electrolytic cell and the amount of chemical change occurring therein. In the next experiment some factors that determine the voltage of a cell will be investigated and an attempt will be made to relate these factors to the driving force of chemical reactions in general.

Electrolytic Cells. When an electric current is passed through a solution of a salt of a metal, such as copper, lead, zinc, or cadmium, if the weight of metal deposited by a given current passing for a measured period of time is determined, it is possible to discover the laws relating the quantity of electricity passing and the amount of metal deposited. It has been found that these laws apply to all kinds of electrolytic cells and not only to those in which metals are deposited.

The weight of metal plated out may be determined by measuring the increase in weight of a cathode made of that metal. The quantity of electricity that passes through the cell can be measured by multiplying the current (in amperes = coulombs per second) by the time (in seconds) during which the current flows through the cell. The quantity of electricity is thereby expressed in coulombs.

Another way of considering the quantity of electricity passing is to ask the question: How many electrons have passed through the cell? This can be calculated from the number of coulombs passing by recalling that the charge on the electron is 1.6021×10^{-19} coulombs. Thus, the number of electrons passing through the cell is the number of coulombs times this conversion factor.

$$\text{Electrons passing = coulombs passing} \times \frac{1 \text{ electron}}{1.6021 \times 10^{-19} \text{ coulombs}}$$

For convenience, the number of electrons passing through the cell may be expressed as the number of moles of electrons passing. This number of moles is obtained by using Avogadro's number:

$$\text{Moles of electrons passing = electrons passing} \times \frac{1 \text{ mole}}{6.023 \times 10^{23} \text{ electrons}}$$

Electrode Reactions. Chemical reactions occur at both electrodes during electrolysis. At the cathode (the electrode at which reduction occurs), electrons enter the solution and react with substances that can accept electrons, such as metal ions. In the electrolysis of copper sulfate solutions, for example, the reduction reaction at the cathode is

$$Cu^{2+} + 2e^- \longrightarrow Cu \tag{1}$$

At the anode (the electrode at which oxidation occurs), electrons leave the solution. Several reactions are possible and several may occur simultaneously. Three possible anode reactions are:

Discharge of a negative ion, as illustrated by iodide ion

$$2I^- \longrightarrow I_2 + 2e^- \tag{2}$$

Decomposition of water

$$3 H_2O \longrightarrow \frac{1}{2}O_2 + 2H_3O^+ + 2e^- \tag{3}$$

Dissolution of the metal of the anode, as illustrated by the process that occurs at a copper anode

$$Cu \longrightarrow Cu^{2+} + 2e^- \tag{4}$$

The particular reactions that occur at the anode depend upon the chemical species present, their concentrations in the solution, and the voltage of the cell. However, dissolution of the metal anodes is one of the more common anode reactions.

Electron Transfer. Electrons, of course, are not directly observable reactants and products in chemical reactions, and the *overall* reaction in an electrochemical cell is the sum of the *half-reactions* taking place at the anode and cathode, representing a *transfer* of electrons. Thus, the sum of equations (1) and (2) represents the reaction that occurs when a copper(II) iodide solution is electrolyzed

$$Cu^{2+} + 2I^- \longrightarrow Cu + I_2$$

Electrons have been transferred from the iodide ions to the copper ions through the external circuit.

It has been stated that a relation between the quantity of electricity passed and the quantity of material plated out of solution can be obtained by measuring the increase in mass of the cathode as a result of an electrolysis. Since a chemical reaction occurs at both electrodes, it is important to ascertain whether a similar relation also applies to the anode reaction.

Quantitative Electrolysis. In this experiment we shall measure the quantity of material reacting at each electrode by conducting the electrolysis in such a way that the increase in weight of the cathode and the decrease in weight of the anode can serve as measures of the amount of reaction at each electrode.

From the weights of material reacting, the moles of material reacting can be calculated. These values may then be compared with the number of moles of electrons passing through the cell.

The most accurate measurements of this kind are made at very low electric currents over long periods of time. In order to avoid making the experiment too time-consuming, the current suggested will be higher than required for the most accurate results. However, the conditions are suitable for obtaining results to within 10 percent of accepted values.

This experiment may be performed with a variety of cathodes and anodes. The solution should contain a salt of the metal to be plated at the cathode. Examples of typical cells are:

CATHODE	SOLUTION	ANODE
Cu	$CuSO_4$	Cu
Zn	$ZnSO_4$	Cu
Zn	$ZnSO_4$	Cd

NOTEBOOK AND REPORT

You should record in your notebook the cathode, anode, and solution used, the chemical equations for the reactions that occur, the weights of the two electrodes before and after electrolysis, the time and current readings referred to in the following paragraphs, and the calculations and results. Your report should contain a summary of these items and the answers to the questions that the instructor assigns.

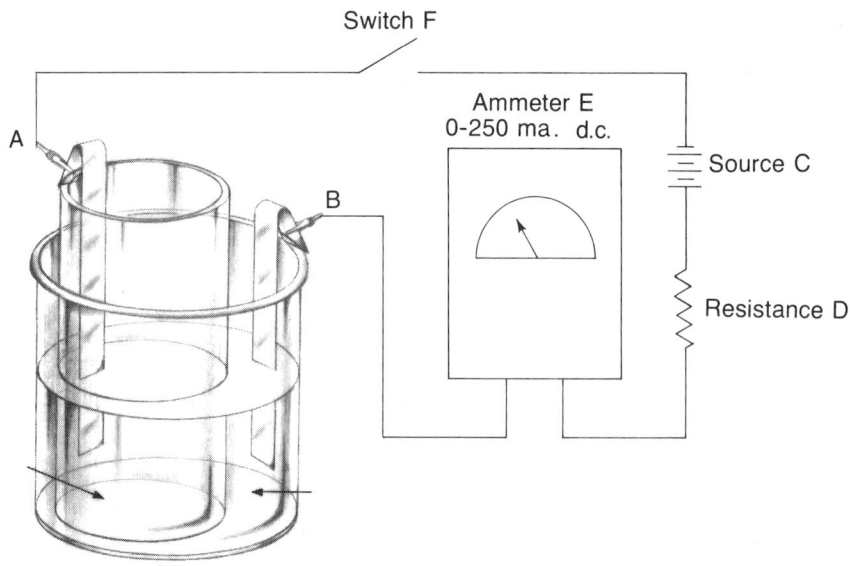

FIGURE   16-1

Diagram of electrolytic apparatus.

PROCEDURE

Select electrodes.

Select the system to be used and polish the two electrodes with sand paper. Handling the polished pieces by the edges, bend them as illustrated at A and B in Figure 16-1. The lengths of the two arms should be such that the electrodes rest on the edge of the beaker or porous cup, with the longer arm extending well down inside the beaker or cup. Weigh each electrode to the nearest milligram, handling it by the edges or with tongs. Record the weights.

Assemble electrolysis apparatus.

Secure an electrolytic box from your instructor, or assemble the apparatus shown in Figure 16-1. The electrolytic boxes are already wired as in the right half of the figure; it is necessary to connect only the DC source at C and the electrodes at A and B. *Be sure the cathode metal is connected to the negative terminal of the source of current.* Have the instructor check your set-up before proceeding.

Add salt solution and start current.

Carefully pour 50 ml. of the appropriate metal salt solution into the beaker containing the porous cup. Dial the resistance in fully, switch on the current, and observe the exact starting time.

Maintain current at steady rate.

Quickly adjust the current to some specific value close to 100 milliamperes, using the resistance, D. At one-minute intervals, record the milliammeter reading, and readjust the current, if necessary, to your selected value. Allow the current to pass through the solution for an exactly measured time of about 20 minutes. At the end of this time, stop the current. Handle the electrodes by the edges and prepare them for weighing as follows:

1. Suspend one of them over the side of a clean 250-ml. beaker to drain.
2. Wash the other electrode *carefully* with distilled water from a wash bottle. Direct the stream of water above the deposit and allow the

water to run over it gently. Do not project the stream directly onto the deposit, as it may be dislodged.

3. Repeat this procedure with the first electrode.
4. Wash both electrodes again with small amounts of acetone (from the plastic wash bottles marked "For Exp. 16"). Acetone evaporates more rapidly than water and helps dry the electrodes.
5. Allow the electrodes to dry for about 10 minutes and weigh them.

CALCULATIONS

For each one-minute interval, add the current flowing as adjusted at the beginning of the interval to that flowing at the end of the interval and divide by two to get the average current during the interval. Add these results for each interval and divide the sum by the number of intervals to get the overall average. Calculate the total number of moles of electrons that passed through the cell, using the equations given in the Introduction.

From the changes in weight of the anode and cathode and the atomic weights of the metals, calculate the number of moles of metal dissolved and deposited.

Find the ratio of moles of metal to moles of electrons, and calculate for each electrode the number of moles of metal that would be dissolved or deposited by the passage of 1 mole of electrons.

Using the chemical equations for the reactions as the theoretical result, calculate the percentage deviation of your result from the theoretical result.

QUESTIONS–PROBLEMS

1. State the error that would result if a portion of the total anode reaction were the decomposition of water to give oxygen, i.e.,

$$3H_2O \longrightarrow \tfrac{1}{2}O_2 + 2H_3O^+ + 2e^-$$

2. What error would result if some of the deposit were to break off the cathode during the electrolysis?

3. What error would result if the ammeter reading were 100 milliamperes when only 90 milliamperes were actually flowing through the cell?

4. How many coulombs constitute 1 mole of electrons? Show your calculations.

5. If the average amount of current flowing in your experiment had flowed for the same length of time through a cell that had a cathode and an anode of metallic silver and an electrolyte of silver nitrate solution, what weight of silver would have deposited on the cathode? What would have been the loss in weight of the anode? Write equations for the reactions occurring at the cathode and anode.

REFERENCE

1. Garrett, A. B., Lippincott, W. T., and Verhoek, F. H.: *Chemistry, A Study of Matter,*
    2nd edition, Xerox College Publishing, Lexington, Mass., 1972, Chapter 21.

Experiment 16
REPORT SHEET

Name _____   **Section No.** _____

A. Cathode metal _____   Electrolyte _____

    1. Weight after electrolysis   _____

    2. Weight before electrolysis _____

    3. Net change in weight   _____

    4. Moles of metal deposited _____

B. Anode metal   _____

    1. Weight after electrolysis   _____

    2. Weight before electrolysis _____

    3. Net change in weight   _____

    4. Moles of metal dissolved   _____

C. Amperes of electricity passed _____

D. Number of electrons passed   _____

E. Ratio of moles of metal deposited per mole of electrons passed

    1. anode _____

    2. cathode _____

F. Equations for the reaction at each electrode.

G. Attach pages with assigned questions.

VOLTAIC CELLS

OBJECTIVE

To determine the cell potentials of several voltaic cells and to examine the effect of concentration changes and complex ion formation on the potential.

EQUIPMENT NEEDED

Porous cup or salt bridge; voltmeter; lead, iron, cadmium, nickel, magnesium, tin, zinc, and copper electrodes; 150-ml. beaker; flexible connecting wires; battery clips; sand paper.

REAGENTS NEEDED

Solutions (1M) of iron(II) sulfate, zinc(II) sulfate, copper(II) sulfate, lead(II) nitrate, cadmium(II) nitrate, magnesium(II) sulfate, tin (II) chloride, and nickel(II) nitrate; solutions (1M) of copper(II) sulfate, cadmium(II) nitrate, and nickel(II) nitrate dissolved in 3M ammonia.

INTRODUCTION

Electrochemical cells that produce electricity are called voltaic or galvanic cells. Sometimes they are called batteries, although this term properly refers to groups of voltaic cells connected together.

In a voltaic cell a chemical reaction is carried out so as to produce an electric current in an external wire that connects the two parts of the cell. The chemical reactions used in voltaic cells are oxidation-reduction processes and involve the transfer of electrons from one part of the cell to the other part. A typical example of such a reaction is:

$$Zn + Cu^{2+} \longrightarrow Zn^{2+} + Cu \tag{1}$$

In this process electrons are transferred from metallic zinc to copper(II) ions with the resulting formation of metallic copper and zinc(II) ions. If a zinc strip is placed in a copper sulfate solution, the transfer of electrons takes place directly at the metal-liquid boundary. This same reaction can be carried out so that the electrons are transferred from the metallic zinc to the copper ions through an external wire connecting the two reactants. This is illustrated in Figure 17-1. Metallic zinc, in the beaker on the left, transfers

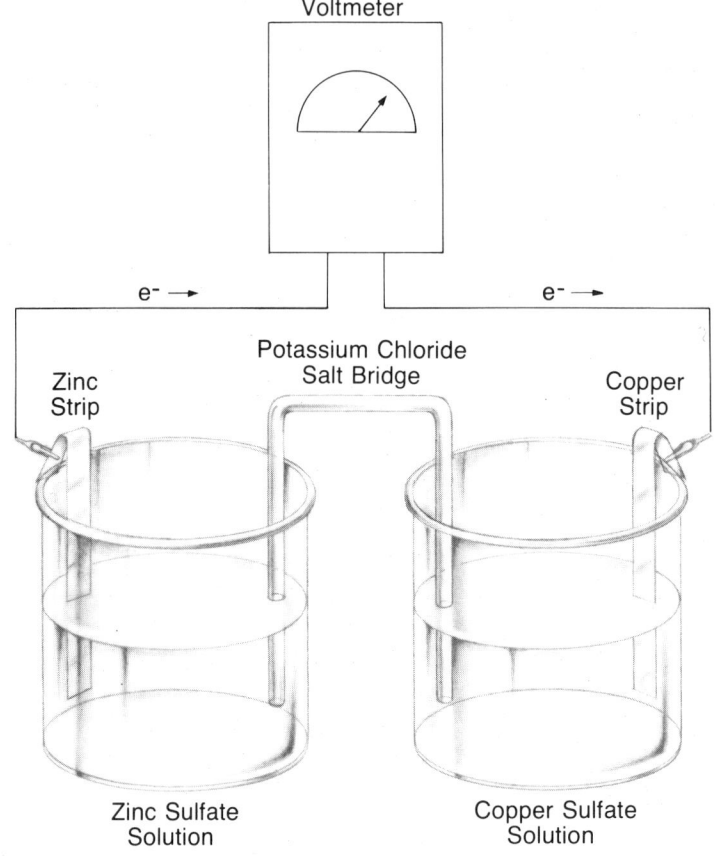

Voltmeter

$e^- \longrightarrow$

$e^- \longrightarrow$

Zinc
Strip

Potassium Chloride
Salt Bridge

Copper
Strip

Zinc Sulfate
Solution

Copper Sulfate
Solution

FIGURE 17-1

A zinc-copper(II) voltaic cell.

electrons to copper(II) ions present in the beaker on the right by means of the metal conductor connecting the two beakers through the voltmeter.

The complete voltaic cell assembly for this reaction is illustrated in Figure 17-1, and includes: (1) a zinc strip immersed in a solution containing zinc ions, (2) a copper strip immersed in a solution containing copper(II) ions, (3) wires connecting the zinc and copper electrodes to a voltmeter, (4) a salt bridge[1] or a porous cup used, as in Experiment 16, to allow ions to migrate from one compartment to the other, and (5) a voltmeter in the external circuit to indicate the cell potential.

The reading on the voltmeter is a measure of the tendency for the electron-transfer process to occur. By measuring the potentials or voltages of cells containing a variety of metals and their ions, at the same concentrations, it is possible to learn just which electron-transfer processes occur readily and which occur with difficulty.

As in electrolysis, cell reactions are often viewed in terms of the processes occurring at the two electrodes. For the reaction in equation (1), the two *half-reactions* are:

$$Zn \longrightarrow Zn^{2+} + 2e^- \text{ (oxidation at the anode)} \qquad (2)$$

$$Cu^{2+} + 2e^- \longrightarrow Cu \text{ (reduction at the cathode)} \qquad (3)$$

---

[1]A salt bridge may be prepared by filling a U-tube with a 1M solution of potassium chloride, plugging the ends tightly with cotton and inverting the tube so that one arm rests in each compartment of the cell. No air bubbles should appear in the bend of the U-tube.

The overall cell reaction is the sum of the two half-reactions. In this case the sum of the reactions in equations (2) and (3) is the reaction in equation (1).

The notion that the voltaic cell reaction is the sum of two half-reactions has two implications:

1. That a variety of cells can be made simply by changing the cathode components (the metal and its ions) and leaving the anode components the same, or by changing the anode components and leaving the cathode components the same. For example, if the beaker containing the copper strip and copper sulfate solution (Fig. 17-1) is replaced by one containing a strip of silver and silver nitrate solution, a new cell would be obtained. Presumably, any metal and a solution of its ions may be used in preparing a cell. Even if the metal reacts with the water of the solution, it may often be dissolved in mercury and the resulting mercury amalgam used as an electrode.

2. That the cell voltage may be regarded as the sum of the contributions from each half-reaction, i.e., from each electrode compartment. If these contributions can be measured for each electrode system, we have a quantitative indication of the tendency of a particular metal to be oxidized or of its ion to be reduced.

In this experiment a number of voltaic cells will be prepared and their voltages measured. You will be asked to devise a scheme for determining the potentials of a number of electrodes. You will also be asked to examine some factors that affect the cell potential.

PROCEDURE

Assemble the voltaic cell as shown in Figure 17-2, using a 150-ml. beaker for the copper metal-copper(II) ion compartment and the porous crucible for the zinc metal-zinc(II) ion compartment. For the moment, leave one of the wires unconnected to its voltmeter terminal. Clean the zinc and copper electrodes with sandpaper before placing them in the appropriate containers. Wash the porous cup thoroughly with water before use. Add 40 ml. of a 1M $CuSO_4$ solution to the beaker and sufficient 1M $ZnSO_4$ solution (about 15 ml.) to the crucible to make the liquid levels the same in both the beaker and the crucible. Attach the last connection to the voltmeter and read and record the initial voltage of the cell.

Assemble voltaic cell.

Read voltage.

Lift the copper electrode until only one corner of it touches the solution and read the voltage. How does the voltage change as the surface area of the electrode exposed to the solution decreases? Having made this observation, disconnect the voltmeter so that the reaction does not continue.

Change exposed surface area of electrode.

For each of the following parts of the experiment, do not leave the voltmeter connected longer than necessary to read it; disconnect it as soon as you have read the voltage.

Determination of the Direction of Electron Flow. It is important for the next set of experiments that you be able to use the voltmeter to determine the *direction* of electron flow. Try to recall whether reaction occurs when a zinc strip is placed in copper sulfate solution or when a copper strip

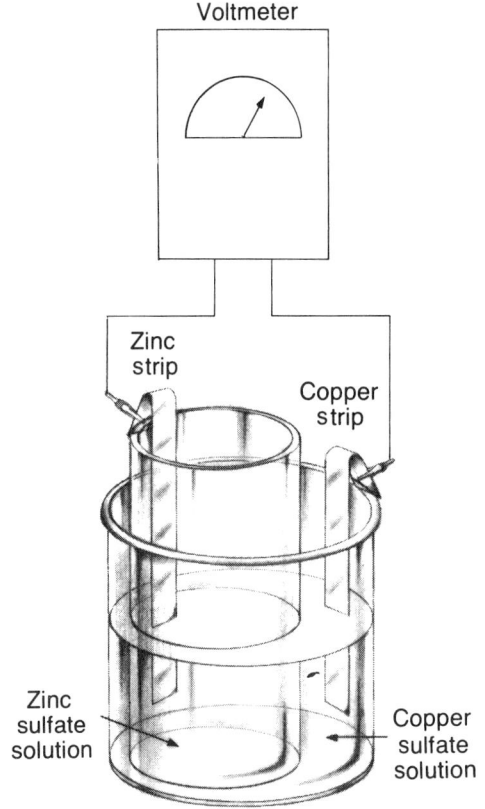

Voltmeter

Zinc
strip

Copper
strip

FIGURE 17-2

A voltaic cell assembly.

Zinc
sulfate
solution

Copper
sulfate
solution

is placed in zinc sulfate solution. In other words, which of the following reactions is spontaneous?

$$Zn + Cu^{2+} \longrightarrow Zn^{2+} + Cu \tag{4}$$

$$Cu + Zn^{2+} \longrightarrow Cu^{2+} + Zn \tag{5}$$

These two correspond to the electrode reactions

$$Zn \longrightarrow Zn^{2+} + 2e^-$$
$$Cu^{2+} + 2e^- \longrightarrow Cu \tag{4'}$$

and

$$Cu \longrightarrow Cu^{2+} + 2e^-$$
$$Zn^{2+} + 2e^- \longrightarrow Zn \tag{5'}$$

Find out whether the voltmeter operates properly when the left terminal is connected to the electrode at which electrons leave the solution (the anode) or the electrode at which electrons enter the solution (the cathode).

Determine direction of electron flow. What happens when you reverse the connections at the two voltmeter terminals? Do not proceed until you can associate the direction of the voltmeter deflection to the electrode at which electrons enter the solution and that at which electrons leave the solution. Describe your experiments and record the results in your notebook.

Measurement of Cell Voltages for Several Cells. Proceed now to replace successively the zinc-zinc(II) sulfate compartment with the following systems:

> Iron-iron(II) sulfate (1M)
> Lead-lead(II) nitrate (1M)
> Cadmium-cadmium(II) nitrate (1M)
> Magnesium-magnesium(II) sulfate (1M)
> Nickel-nickel(II) nitrate (1M)
> Tin-tin(II) chloride (1M)

To do this, first remove the zinc electrode and then quickly remove the crucible from the beaker, thereby preventing the zinc sulfate solution from draining into the beaker. Discard the zinc sulfate solution and carefully wash the crucible. Replace the crucible in the copper sulfate solution and add to the crucible 15 ml. of the next solution, e.g., 1M $FeSO_4$. Place a cleaned electrode of the proper metal in the crucible; attach the wire to it; read and record the cell voltage; disconnect the wire after reading the cell voltage.

Determine voltages of other cells.

Effect of Concentration on Cell Voltage. In each of the cells prepared thus far, the concentration of salt in the solutions has been 1M. The question then arises: What effect does a change in the concentration of the salt solution have on the cell potential? Discover this for yourself. Design your own experiments, using one of the electrode pairs. Dilute one of the solutions by a measured factor of 10; then again by a factor of 10. Dilute the other solution in the same way, making a measurement after each dilution.

Change concentration of ions.

You may wish to try an experiment in which you replace the salt solution used with different salts. For example, what would happen if a sodium chloride solution were used in place of zinc sulfate or copper sulfate?

Write a description of your experiments on the effect of concentration and other salts, tabulate your results, and summarize your conclusions.

Effect of a Complex Ion on Cell Voltage. In the above experiments, you have examined the potentials of several cells at 1M concentration of metal ion and the effect of varying the concentration of the salt solution. We may now be curious whether other factors affect the cell potential. For example, suppose the metal ion is present in some form other than the ion $[M(H_2O)_6]^{2+}$ that exists in aqueous solutions of the Fe^{2+}, Ni^{2+}, and Cu^{2+} salts. The Cd^{2+}, Mg^{2+}, Pb^{2+} and Sn^{2+} ions are tetra-aquo complex ions in water, *i.e.*, $[M(H_2O)_4]^{2+}$. You could change the nature of these ions in aqueous solutions by converting them into other complex ions. For example, addition of concentrated hydrogen chloride solution will convert Sn(II) into $SnCl_3^-$, Pb(II) into $PbCl_3^-$ or $PbCl_4^{2-}$, and Cu(II) into $CuCl_3^-$ or $CuCl_4^{2-}$. Ammonia also forms complex ions, especially with Ni(II), Cu(II), and Co(II). It might be expected that these complex ions would exhibit different cell potentials as compared with the aquo complex ions that were investigated in the 1 M salt solutions above. We shall test this hypothesis by determining the potential of a cell in which the metal salt solution is an ammonia complex.

Determine voltages of different complex ions.

Ammonia solutions (3M in NH_3) of 1M $Cd(NO_3)_2$, $Ni(NO_3)_2$, and $CuSO_4$ are provided. Design your experiment to obtain the voltage of the cell in which the metal ion is an ammonia complex. How much does this voltage differ from the value you obtained originally on the 1M metal solution that contained the aquo complex ion?

Write a description of your cell involving the ammonia complex, and summarize your results and conclusions.

CALCULATION OF ELECTRODE POTENTIALS

It has been suggested that the cell potential is the sum of contributions from the two individual electrode reactions. However, no reliable method for determining the potential of only one half-reaction is available; our measured voltages are only the potential of the overall reaction. Situations of this type in science are often solved by agreeing to a convention by which an arbitrary value is assigned to one quantity and all other values are referred to this "standard." Suppose we adopt the convention that the potential is zero for both of the half-reactions

$$Cu(s) \longrightarrow Cu^{2+} (1M) + 2e^-$$

$$Cu^{2+} (1M) + 2e^- \longrightarrow Cu(s)$$

Then the measured cell voltages for cells having a copper electrode would indicate directly the potential for the half-reaction involving the other metal and its ions. Let us further adopt the convention that the potential for the other half-reaction shall be negative when copper acts as the anode and positive when copper acts as the cathode. The sign of the potential then tells us the direction of electron flow.

For example, the measured voltage of the cell

$$Co \mid CoSO_4 (1M) \parallel CuSO_4 (1M) \mid Cu$$

is 0.63 volt. The connections to the voltmeter show that the direction of electron flow corresponds to the two half-reactions:

$$Co(s) \longrightarrow Co^{2+} (1M) + 2e^- \text{ (cobalt the anode)}$$

$$Cu^{2+} (1M) + 2e^- \longrightarrow Cu(s) \text{ (copper the cathode)}$$

If the second half-reaction is given a potential of zero, the cell voltage 0.63 volt may be taken as the contribution from the cobalt electrode. According to the convention, it is given a positive sign.

$$E_{cell} = E_{Co \rightarrow Co^{2+}} + E_{Cu^{2+} \rightarrow Cu} = 0.63 + 0 = 0.63 \text{ volt}$$

In a similar way, relative half-reactions or electrode potentials may be obtained for a variety of metal-metal ion systems.

NOTEBOOK AND REPORT

Using the raw data recorded in your notebook, organize the data and report the results as outlined on the Report Sheet.

Table A should include, for each electrode pair, the anode reaction, the cathode reaction, the overall reaction, and the observed voltage of the cell when 1M concentrations are used. Since you have discovered that the concentration of the solution affects the value of the voltage, the concentrations should be included in the equations. For example, write

$$Cu^{2+}(1M) + 2e^- \longrightarrow Cu(s)$$

where (s) indicates solid copper.

QUESTIONS–PROBLEMS

1. Discuss the validity of assigning values to the electrode potentials, based on the assumption that the copper-copper ion half-reaction has zero potential.

2. What is the standard half-reaction on which the accepted electrode potentials are based, and what value is assigned to it? Can you suggest reasons this might be a better choice than the copper-copper ion reaction?

3. The voltage of the cell

$$Pt \mid Pt^{2+}(1M) \parallel Cu^{2+}(1M) \mid Cu$$

is 0.88 volt and the cell reaction is

$$Cu(s) + Pt^{2+}(1M) \longrightarrow Pt(s) + Cu^{2+}(1M)$$

Write the half-reactions and add, to your Tables B and C, the oxidation half-reaction for platinum and its potential.

4. The whole of this experiment has been concerned with metal-metal ion systems. Is it possible to use nonmetals and their ions as one or both electrodes? Why or why not?

5. The dissolution of a metal in ionic form may be looked upon as the result of three steps as follows:
   a. Separation of the atoms of the solid into separated gaseous atoms

$$Me(s) \longrightarrow Me(g)$$

   b. Ionization of the separated gaseous atoms

$$Me(g) \longrightarrow Me^{2+}(g) + 2e^-$$

   c. Solution of the gaseous ions to a given concentration in water

$$Me^{2+}(g) + nH_2O \longrightarrow Me^{2+}(1M)$$

The energies involved for these steps are

<table>
<tr><th>FOR ZINC</th><th>FOR COPPER</th></tr>
<tr><td>a. 31.2 kcal. per mole</td><td>a. 81.1 kcal. per mole</td></tr>
<tr><td>b. 630.8 kcal. per mole</td><td>b. 646.0 kcal. per mole</td></tr>
<tr><td>c. −491 kcal. per mole</td><td>c. −507 kcal. per mole</td></tr>
</table>

Calculate the total energies per mole of metal in the processes

$$Zn(s) \longrightarrow Zn^{2+} (1M) + 2e^-$$

$$Cu(s) \longrightarrow Cu^{2+} (1M) + 2e^-$$

and suggest which step is responsible for metallic zinc being more readily oxidized to ions in solution than is metallic copper.

6. Using your data, calculate the free energy change that would occur if 1 mole of zinc was oxidized to zinc ion in the cell

$$Zn(s) \mid ZnSO_4 \ (1M) \ \| \ CuSO_4 \ (1M) \mid Cu(s)$$

under conditions that would maintain the 1M concentrations throughout the oxidation reaction.

REFERENCE

Garrett, A. B., Lippincott, W. T., and Verhoek, F. H.: *Chemistry, A Study of Matter*, 2nd edition. Xerox College Publishing, Lexington, Mass., 1972, Chapter 21.

Experiment 17
REPORT SHEET

Name _____   Section No. _____

A. *Cell Voltages*

	Anode Metal	Cathode Metal	Voltage
1.	_____	_____	_____
2.	_____	_____	_____
3.	_____	_____	_____
4.	_____	_____	_____
5.	_____	_____	_____
6.	_____	_____	_____

Voltaic Cell Reactions

	Anode Reaction	Cathode Reaction	Overall Reaction
1.	_____	_____	_____
2.	_____	_____	_____
3.	_____	_____	_____
4.	_____	_____	_____
5.	_____	_____	_____
6.	_____	_____	_____

B. *Electrode Potentials Based on $E_{Cu, Cu^{2+}} = 0$*

This section should include each different half-reaction listed in part A. Using the cell voltages recorded in A above, and assigning a "standard" value of 0.00 v. to the copper electrode, list the potential for each half-reaction. If the cobalt-cobalt(II) ion potential is included, three entries in your Table B will be

HALF-REACTION	**POTENTIAL (volts)**
$Cu(s) \longrightarrow Cu^{2+} (1M) + 2e^-$	0
$Cu^{2+} (1M) + 2e^- \longrightarrow Cu(s)$	0
$Co(s) \longrightarrow Co^{2+} (1M) + 2e^-$	+0.63

C. *Comparison of Your Electrode Potential Values with Accepted Values*

Using your data from B above, prepare a table with headings as below in which you rewrite all your half-reactions, as necessary, as oxidation half-reactions, and list them in order of decreasing value, algebraically, with the most positive at the top of the list and the most negative at the bottom of the list. Note that the potential of a reduction half-reaction appearing in B will have the opposite sign in Table C, since writing it as an oxidation half-reaction corresponds to an electron flow in the opposite direction in the measuring cell.

Now look up, in a textbook or handbook, the accepted potentials of the oxidation half-reactions written, and record these in a third column of your Table. Calculate for each half-reaction the difference between your value and the accepted value and record this difference in a fourth column.

TABLE C		
Oxidation Half-Reactions	**Literature Value, (volts)**	**Difference between Lit. and Expl. Potential, (volts)**

Examine the magnitudes of the differences in the potentials in Table C. Are they the same or different? Do they show a trend from top to bottom? Interpret the results of your examination in a brief discussion.

D. *Effect of Concentration on the Cell Potential*

Make a neat table here to summarize the results and the experiment you performed to vary the concentration of the metal salts. Summarize briefly the trend.

E. *Effect of the Ammonia Complex Ion on the Cell Potential*

　　Make a neat table here to summarize the results and the experiments you performed for this section. Summarize briefly the effect and/or trend observed.

F. *Attach Sheets with Answers to Assigned Questions.*

VOLUMETRIC DETERMINATION OF IRON IN A SOLUBLE IRON SALT—AN OXIDATION-REDUCTION TITRATION

OBJECTIVE

To learn about oxidation-reduction indicators, and to carry out an analysis of an unknown sample for its iron content.

EQUIPMENT NEEDED

Two 250-ml. Erlenmeyer flasks; 250-ml. volumetric flask; 250-ml. beaker; 50-ml. graduate; stirring rod; buret.

REAGENTS NEEDED

Solid reagent-grade potassium dichromate; 6M HCl; 85 percent H_3PO_4; tin(II) chloride solution; saturated mercury(II) chloride solution; diphenylamine sulfonate solution; unknown solid containing iron(II) ion.

INTRODUCTION

Iron may exist in solution in either the +II or the +III oxidation state. The quantity of dissolved iron can be determined if it can first be reduced to Fe(II), and then oxidized to Fe(III) by titrating with a standard solution of an oxidizing agent until the *equivalence point* is reached, that is, until an amount of oxidizing agent exactly sufficient to react with the amount of Fe(II) present has been added. If, for example, we used Ce(IV) as the oxidizing agent

$$Fe^{2+} + Ce^{4+} \longrightarrow Fe^{3+} + Ce^{3+} \tag{1}$$

the equivalence point would be reached when exactly 1 mole of Ce^{4+} had been added for each mole of Fe^{2+} originally present. This is an example of a general procedure known as an oxidation-reduction titration.

Obviously, we need some means of identifying the equivalence point when we reach it during the titration. One method for identifying the equivalence point is to add to the solution an oxidation-reduction indicator, which changes color at the equivalence point in a manner analogous to the color change of an acid-base indicator in the titration of acids and bases (Exp. 9 and 10).

An important feature of such an indicator is that it exists in oxidized and reduced forms, each of which has a different color.

$$\text{Ind} \rightleftharpoons \text{Ind}^+ + e^- \qquad (2)$$
$$\text{Color A} \qquad \text{Color B}$$

The reaction of equation (2) is similar to the reaction

$$Fe^{2+} \longrightarrow Fe^{3+} + e^- \qquad (3)$$

which is accomplished by addition of an oxidizing agent such as Ce(IV). To be an effective indicator, however, reaction (2) must not occur until reaction (3) (or reaction (1)) has gone effectively to completion. If the opposite were true and (2) was completed before (3), much unchanged Fe(II) might remain in the solution after the color had changed, and the results would be incorrect. The indicator used for a particular oxidation-reduction reaction, therefore, must be carefully chosen to be no more easily (and only slightly more difficultly) oxidized than the species whose concentration is to be determined, Fe^{2+}, in this case.

Comparison of (2) with the indicator reaction for acid-base titration

$$\text{H Ind} \rightleftharpoons \text{Ind}^- + H^+_{aq} \qquad (4)$$
$$\text{Color A} \qquad \text{Color B}$$

suggests that the "electron concentration" plays the role analogous to the hydrogen (or hydronium) ion concentration in the acid-base titration, and that we need to choose an indicator for which reactions (2) and (3) have similar degrees of completion for the same "electron concentrations." Of course, we cannot measure an electron concentration, but we can measure the voltage of a cell in which an oxidation-reduction (electron-transfer) reaction is taking place, and this possibility gives us a tool for comparing the extents of reaction in reactions (2) and (3).

The cell which we might consider is one such as that in Figure 18-1 showing a reference electrode connected through a salt bridge to a reaction solution. The potential difference between the hydrogen electrode and the inert metal conductor in the reaction solution is measured on the voltmeter or potentiometer, V. Soon after the titration begins, the reaction solution will contain Fe^{2+}, Fe^{3+}, Ce^{4+}, Ce^{3+}, Ind, and Ind^+. The reactions among these are rapid, so that equilibrium exists in the reaction solution at all times. Thus the cell reaction may be considered to be

$$Fe^{3+} + \tfrac{1}{2}H_2 \longrightarrow Fe^{2+} + H^+aq \qquad (5)$$

or

$$Ind^+ + \tfrac{1}{2}H_2 \longrightarrow Ind + H^+aq \qquad (6)$$

or

$$Ce^{4+} + \tfrac{1}{2}H_2 \longrightarrow Ce^{3+} + H^+aq \qquad (7)$$

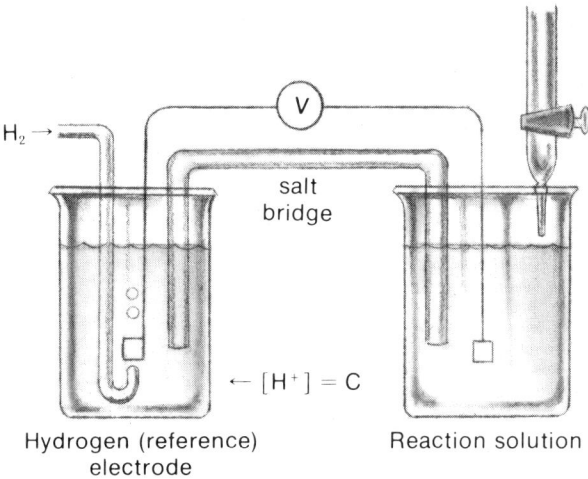

FIGURE   18-1

Apparatus for oxidation-reduction titration.

Because of the equilibrium it makes no difference which of these we use since equilibrium relationships such as

$$\frac{[Fe^{3+}]\ [Ce^{3+}]}{[Fe^{2+}]\ [Ce^{4+}]} = K \tag{8}$$

(from reaction (1)) and

$$\frac{[Ind^{+}]\ [Ce^{3+}]}{[Ind\ ]\ [Ce^{4+}]} = K' \tag{9}$$

always hold.

Considering the visual appearance of the indicator, we find that the eye can follow the color change of $\frac{[Ind^{+}]}{[Ind\ ]}$ in the range from 10 to 0.1, that is

$$\frac{10}{1} \geq \frac{[Ind^{+}]}{[Ind\ ]} \geq \frac{1}{10}$$

Our choice of indicator, then, must be with one which has a concentration ratio in this range when the ratio $\frac{[Ce^{3+}]}{[Ce^{4+}]}$ is that which will exist when the ratio $\frac{[Fe^{3+}]}{[Fe^{2+}]}$ corresponds to the value at the equivalence point of the titration.

It must be remembered that the ratio $\frac{[Fe^{3+}]}{[Fe^{2+}]}$ *has* a finite value at the equivalence point. At the equivalence point, if a moles of Fe(II) were originally present, there will have been added exactly a moles of Ce(IV); *complete* reaction of these would have formed a moles of Fe(III) and a moles of Ce(III). But reaction (1) does not go to completion, so the actual product concentration does not correspond to a moles because of the equilibrium established in reaction (1). If y is the amount of Fe(II) still present at equilibrium at the equivalence point, the concentrations are

$$[Fe^{2+}] = \frac{y}{V}$$

$$[Fe^{3+}] = \frac{a-y}{V}$$

$$[Ce^{4+}] = \frac{y}{V}$$

$$[Ce^{3+}] = \frac{a-y}{V}$$

where V is the volume of the solution at the equivalence point. Setting these into the equilibrium expression (8).

$$\frac{(a-y)^2}{y^2} = K$$

The ratio needed for equation (9) is

$$\frac{[Ce^{3+}]}{[Ce^{4+}]} = \frac{a-y}{y} = \sqrt{K} \tag{10}$$

For our indicator, therefore, we need a substance which has different colors in the oxidized and reduced forms, and which will have a concentration ratio $\frac{[Ind^+]}{[Ind]}$ in the range from 10 to 0.1 when the ratio $\frac{[Ce^{3+}]}{[Ce^{4+}]}$ has the value $\sqrt{K}$. Several substances having the property of changing color on oxidation are then examined, and that one chosen for which the ratio falls within the required range. How this may be done by examining voltages is explained in detail in Appendix VI. The most useful indicator would be one in which the ratio would be $\frac{1}{1}$ at the equivalence point, but it is not always possible to find one that fits that condition.

Oxidation-Reduction Titrations Using Dichromate Ion as the Oxidizing Agent.

Although the previous discussion has concerned the simple reaction of oxidation by ceric ion, corresponding to the reduction reaction (cf. Eq. (7))

$$Ce^{4+} + e^- \rightarrow Ce^{3+} \tag{11}$$

the use of dichromate ion is in some ways more convenient. Dichromate ion, as potassium dichromate, is readily obtained pure. It is inexpensive. Standard solutions may be prepared directly by weighing an appropriate quantity of the pure solid substance. No special precautions must be observed in dissolving the solid; the solutions are stable. From the theoretical view the use of dichromate ion is not so satisfactory, because although the electronic partial equation for the reduction of dichromate ion has the form

$$Cr_2O_7{}^{2-} + 14H^+_{aq} + 6e^- \longrightarrow 2Cr^{3+} + 7H_2O \tag{12}$$

the voltage of a cell in which this reaction is presumably taking place does not show the dependence on $[H^+]$ or on $[Cr^{3+}]$ predicted by this equation. It appears that the reduction occurs in several steps in several reactions before finally reaching Cr^{3+}. In spite of this difficulty, the formal potential (Appendix VI) for the dichromate reduction has been measured and combined with that for ferrous ion oxidation to give a potential at the equivalence point close to 0.9 volt. The indicator diphenylamine sulfonic acid, which changes from a colorless to a purple compound of such intense color that it overcomes the green of Cr^{3+} at about 0.85 volt, is suitable for the titration. The addition of phosphoric acid, which forms a more stable complex with Fe(III) than with Fe(II) and thus changes the formal potential for $Fe^{2+} \longrightarrow Fe^{3+}$, also aids in adjusting the voltage at the equivalence point to a useful value. In this experiment, iron(II) will be titrated with dichromate ion, in acid solution, using diphenylamine sulfonic acid as the oxidation-reduction indicator.

Analysis for Iron by Oxidation-Reduction Titration Using Dichromate. A weighed quantity of the substance containing iron is brought into solution.

For an effective titration all iron ions present must be brought to the II state; this is accomplished by adding tin(II) ion in the presence of chloride ion

$$Sn^{2+} + 2Fe^{3+} \longrightarrow 2Fe^{2+} + Sn^{4+} \tag{13}$$

The presence of chloride ions in this reaction is desirable because the speed of the reaction between chloride ion complexes of Fe(III) and Sn(II) is greater than between the simple or hydrated ions.

Any Sn(II) left in the solution would reduce the dichromate titrant and vitiate the results; hence, any excess must be removed. This is accomplished by adding mercuric chloride solution

$$Sn^{2+} + 2Hg^{2+} + 2Cl^- \longrightarrow Sn^{4+} + Hg_2Cl_2 \tag{14}$$

The mercurous chloride precipitates and is not affected by the addition of dichromate ion.

Finally, the titration reaction is

$$6Fe^{2+} + Cr_2O_7{}^{2-} + 14H^+_{aq} \longrightarrow 6Fe^{3+} + 2Cr^{3+} + 7H_2O \tag{15}$$

The first few drops of excess dichromate oxidize the diphenylamine sulfonic acid indicator to its intense purple color.

In equation (15) each mole of potassium dichromate contains 2 moles of Cr(VI), which are reduced to Cr(III). One-sixth of a mole of potassium dichromate gains 1 mole of electrons in the analytical reaction and 1 mole of Fe^{2+} loses 1 mole of electrons.

NOTEBOOK AND REPORT

In your notebook you should record the weights and volumes of reagents used, the calculations for the analysis of the unknown, and the answers to the problems at the end of the experiment. Do not forget to record the code number of the unknown, if given.

PROCEDURE

Weigh out K$_2$Cr$_2$O$_7$.

Dissolve and transfer to volumetric flask.

Mix thoroughly dilute to the mark, and mix.

Preparation of Standard Potassium Dichromate Solution. Weigh accurately on a weighed piece of clean paper about 1.2 g. of pure dry potassium dichromate. Transfer all of it, brushing the paper, to a 250-ml. beaker, add about 50-ml. H$_2$O, and stir to dissolve the dichromate. Transfer the solution to a clean, though not necessarily dry, 250-ml. volumetric flask without losing a drop, and rinse the beaker with three additional 50-ml. portions of water. Add the rinsings to the volumetric flask without spilling any. Shake the liquid in the volumetric flask to mix it thoroughly and add water little by little to bring the level of liquid up to the mark. Mix the contents for not less than five minutes by shaking and inverting the flask to insure homogeneity. Calculate the molarity of the solution from the weight of salt taken, and record this on the label.

Weigh samples.

Fill buret.

Dissolve in HCl.

Reduce with SnCl$_2$.

Add HgCl$_2$.

Analysis of the Unknown. Weigh accurately two samples (about 1.2 g. each) of the iron unknown into marked 250-ml. Erlenmeyer flasks. Using proper technique, fill a buret with the standard potassium dichromate solution. This should be done before the reduction because each sample must be titrated very shortly after reduction has been completed. Dissolve each of the weighed samples by adding 30 ml. of 6M HCl followed by 70 ml. of H$_2$O to each flask. Heat each flask nearly to boiling; then carefully reduce the iron(III) by the slow drop by drop addition of the stannous chloride solution. The iron(III) is detected by the yellow color of the iron(III)-chloride complexes. If only a small amount of iron(III) is present, the solution is only a faint yellow, in which case only 1 or 2 drops of the stannous chloride solution may be sufficient. The stannous chloride is added until the yellow of the solution is discharged. Add 1 or 2 drops in excess but no more. Cool the flask under the cold water tap, being careful not to contaminate the contents. Measure 15 ml. of the mercuric chloride solution into a graduated cylinder. Then, while swirling the solution in the flask, rapidly add the mercuric chloride. *DANGER: MERCURIC CHLORIDE IS POISONOUS. NEVER PIPET THIS ORALLY!*

If the procedure is properly performed and only a slight excess of stannous chloride is used, a silky white precipitate of mercurous chloride is produced. Formation of a gray or black precipitate, which contains finely divided metallic mercury, indicates that too much stannous chloride was added. Because the free mercury reacts with the dichromate, such solutions must be discarded.

Add H$_3$PO$_4$ and indicator, and titrate.

After the addition of the mercuric chloride solution, wait one or two minutes for the reaction to be complete. If the mercurous chloride precipitate is white, add 5 ml. of 85 percent H$_3$PO$_4$ and 5 drops of diphenylamine sulfonic acid solution, and titrate immediately with the potassium dichromate solution. Too much indicator is undesirable because a dimer having a less intense color is formed.

Since an appreciable amount of the dichromate reacts with the indicator to produce the color change, it is necessary to make a correction to the volume of titrant used. This correction is about 0.04 ml. of dichromate solution per drop of indicator. Repeat the procedure until two concordant results are obtained.

CALCULATIONS

On the basis of your data, calculate the percentage of iron in your sample. Results for duplicate samples should be within 0.1 percent. However, you should not analyze more than four samples before reporting your results.

QUESTIONS—PROBLEMS

1. (a) Write the equation for the reaction of dichromate ion with Sn(II) (b) If 0.0005 moles excess of Sn^{2+} remained in your solution after reducing the Fe(III), how much of your dichromate solution would be required to titrate it?

2. What does the fact that mercurous chloride is insoluble have to do with the fact that it is not affected by the addition of dichromate ion?

3. Suppose you had titrated your first two samples with 0.01 molar Ce^{4+}. How much ceric ion solution would have been required?

4. Suppose you were to analyze a prescription pill containing iron to combat anemia and this pill contained an organic binder oxidizable by dichromate ion, but no metallic substance other than iron. How would you proceed to make an accurate analysis?

REFERENCE

1. Day, R. A., and Underwood, A. L.: *Quantitative Analysis.* 2nd edition. Prentice-Hall, Englewood Cliffs, N. J. 1967, Chapter 5.

**Experiment 18
REPORT SHEET**

Name _____ **Section No.** _____

Data   Unknown No. _____

	Trial 1	Trial 2	Trial 3
Weight of unknown iron sample (g.)	_____	_____	_____
Volume of standardized $K_2Cr_2O_7$ solution (ml.)	_____	_____	_____
Molarity of $Cr_2O_7^{2-}$ solution	_____	_____	_____
Moles of iron in sample	_____	_____	_____
Percent iron in sample	_____	_____	_____
Averaged percent iron in sample		_____	

Sample Calculations

Attach sheets with observations and answers to assigned questions.

CHEMICAL DYNAMICS

CHEMICAL KINETICS: THE RATE OF DECOMPOSITION OF TRINITROBENZOIC ACID

OBJECTIVE

To measure the rate and to determine the rate law of a chemical reaction.

EQUIPMENT NEEDED

Buret; five 25- $\times$ 200-mm. test tubes; 600-ml. beaker; 400-ml. beaker; burner.

REAGENTS NEEDED

Standardized 0.0500M NaOH solution; nearly saturated trinitrobenzoic acid solution; 1 percent phenolphthalein solution; 95 percent ethanol; boiling chips; ice.

INTRODUCTION

Chemical kinetics is the study of the rates of chemical reactions. The rate of a reaction in solution may be measured by measuring the change in concentration of a reacting substance per unit of time. According to the law of mass action, the rate at any instant depends upon the concentration of the reacting substance at the instant at which the rate is measured.

It is not easy to measure the rate at which the concentration of a substance is changing at a particular instant. The best we can usually manage is to measure the concentration, c_1, at some time, t_1, and to measure it again at a later time, t_2. If we call the concentration at the later time c_2 then $c_1 - c_2$ is the change in concentration that has occurred in the time $t_2 - t_1$, and the rate is approximately given by the quotient $(c_1 - c_2)/(t_2 - t_1)$. This is not exactly the rate, however, because this quotient applies to the time *interval* $t_2 - t_1$ and is not the value appropriate to a particular *instant* of time, such as t_1. To obtain the value *at* t_1 rather than for the *interval* $t_2 - t_1$, we can imagine that $t_2 - t_1$, and hence the concentration difference $c_2 - c_1$ appropriate to the time interval, becomes smaller and smaller. The limiting value of the quotient for $t_2 - t_1 = 0$ is the desired quantity, i.e., the rate of reaction at t_1. Students who have included calculus in their mathematical training will recognize this as the definition of the derivative of concentration with respect to time:

$$\text{Rate} = \lim_{t_2 \to t_1}\left(\frac{c_1 - c_2}{t_2 - t_1}\right) = \lim_{t_2 \to t_1}\left(-\frac{c_2 - c_1}{t_2 - t_2}\right) = \lim_{\Delta t \to 0}\left(-\frac{\Delta c}{\Delta t}\right) = -\frac{dc}{dt} \quad (1)$$

The geometric interpretation of finding the value of $(c_1 - c_2)/(t_2 - t_1)$ at the limit where $t_2 = t_1$ (or $t_2 - t_1 = 0$) may be made clear if we show how the concentration changes with time in Figure 19-1. The chemical reaction uses up the reacting material; hence, its concentration steadily decreases with time, as shown by the curved line in the figure. The line AB is drawn from B, where the measured coordinates are (t_2, c_2) through P (t_1, c_1); the approximate measure of the rate is given by the slope of this line, $\frac{c_2 - c_1}{t_2 - t_1}$. Now, if we take another point, D, with coordinates (t_2', c_2'), a line, CD, drawn from it through P will have the slope $\frac{c_2' - c_1}{t_2' - t_1}$. Moving from B to D corresponds to making the interval $t_2 - t_1$ smaller, and if we continue to make the interval smaller, the points corresponding to B and D move up the curve closer and closer to P. At the same time, the slopes of the lines corresponding to BP, DP, etc., become steeper and steeper, until, at the limit where t_2 and t_1 are the same, they have the slope corresponding to the line EF, which is *tangent* to the curve at P. The slope of this tangent thus represents the instantaneous rate of reaction at the instant at which the concentration has the value c_1 corresponding to the point of tangency at P. One way to determine the rate of a reaction, therefore, is to determine by experiment the shape of a *concentration-time curve,* such as that shown in Figure 19-1, and then to determine the slope of a tangent to that curve at a particular concentration.

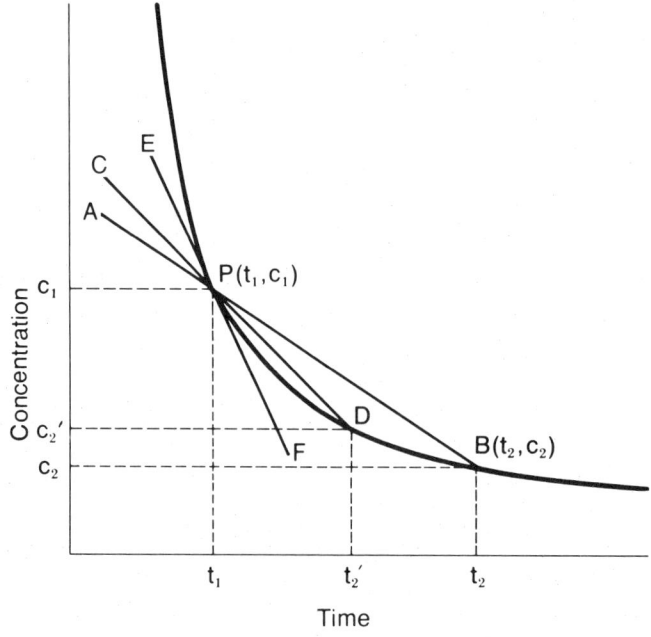

FIGURE 19-1

Finding the slope at t_1.

The relationship between the rate of reaction and the concentration of reactants can often be cast in the form

$$\text{Rate} = k\,c^n \qquad (2)$$

where n is a constant giving the *reaction order,* and k is a constant known as the *rate constant.* If the rates of reaction at several concentrations are determined, as measured by the slopes of tangents T_1, T_2, T_3, etc., at concentrations c_1, c_2, c_3, etc., and the rates are plotted against c^n, a straight line through the origin is obtained. The slope of this line gives the rate constant, k. The value of n may be determined by trial and error to see which value gives the straight line. For example, if a plot of the rate against c gives a curved line, but a plot of rate against c^2 gives a straight line, it may be concluded that n = 2, and that the reaction is of the second order. If n were 1, we would call it a *first-order reaction;* if n were 3/2, a *three-halves-order reaction,* etc. Whatever the reaction order, the rate must be zero at zero concentration (no reaction if no reacting substance is present), and all curves previously mentioned must pass through the origin.

Most reactions with which you have been dealing have been rapid reactions between ionic substances, and they are too fast to measure without elaborate equipment. A reaction that occurs slowly enough to measure with simple equipment is that of the decomposition of trinitrobenzoic acid, $C_6H_2(NO_2)_3CO_2H$, which loses carbon dioxide in water solution to form trinitrobenzene, $C_6H_3(NO_2)_3$.

$$C_6H_2(NO_2)_3CO_2H \longrightarrow C_6H_3(NO_2)_3 + CO_2 \qquad (3)$$

The rate of this reaction depends only upon the concentration of the single substance trinitrobenzoic acid. Also, the reaction is of the first order and the expression for the rate according to the law of mass action is

$$\text{Rate} = k_1\,[C_6H_2(NO_2)_3CO_2H] \qquad (4)$$

The progress of the reaction may be followed by titrating, with a standard base solution, the amount of acid left after the reaction has proceeded for a measured length of time. If this is done for several samples from the same original solution, a graph of concentration plotted against time, like Figure 19-1, can be made.

Since the reaction rate is different at different temperatures, meaningful rate measurements can be made only by carrying out the experiments at a fixed and constant temperature. Constancy of temperature is maintained in the present experiment by taking advantage of the fact that water boils at the constant temperature determined by the prevailing pressure of the atmosphere.

NOTEBOOK

The data might be recorded in your notebook in a table like the following.

Test Tube	Ml. Solution Added	Time Placed in H_2O	Time Removed from H_2O	Buret Readings	
				Initial	Final
___	___	___	___	___	___
___	___	___	___	___	___
___	___	___	___	___	___
___	___	___	___	___	___
___	___	___	___	___	___

Molarity of NaOH _____

Temperature of Boiling Water _____

The data should be treated as indicated in the calculations and the results tabulated and plotted on graphs as suggested. The arithmetic of the calculations should appear in the notebook. The final result to be reported to the instructor should include these graphs and the numerical value of the rate constant, in appropriate units, for comparison with the results of other students. If assigned, certain questions and problems require further treatment of the data.

PROCEDURE

Clean and dry five 25 × 200 mm test tubes. Add to each, from the buret marked "trinitrobenzoic acid for Experiment 19," an accurately measured quantity (estimated to the nearest 0.02 ml.), amounting to approximately 20 ml., of trinitrobenzoic acid solution. Your later calculations will be less time-consuming if exactly the same volume is placed in each test tube, but this is not necessary.

Obtain accurate quantities of the reagent.

Prepare a large beaker of water, add a couple of boiling chips, and bring it to a rapid boil. The water should continue to boil throughout the remainder of the experiment. Prepare another beaker with a mixture of ice and water. Insert one of the test tubes into the boiling water, noting the time, and swirl it around so that its contents reach the temperature of boiling water as quickly as possible. Remove the test tube after a measured time interval, plunge it into the ice-water mixture, and swirl it around to cool it quickly. Repeat this with the four other test tubes. More than one test tube may be in the boiling water at the same time, but the water should be boiling vigorously when each is added, in order to maintain a uniform time of heating from room temperature to the temperature of boiling water. The length of time each test tube remains in the boiling water must be accurately measured. One test tube should be left in the boiling water for only three minutes, and the others for about 7, 15, 25, and 40 minutes, respectively.

Prepare hot and cold water baths.

Heat reagent.

While the test tubes are heating for the longer times, the cooled tubes may be titrated by using the 0.05M standardized sodium hydroxide solution. Before using the buret for this solution, be sure to clean and rinse it with distilled water and then with the sodium hydroxide solution. The titration may be made directly in the large test tubes after adding 3 drops of Titrate solutions. phenolphthalein. If the white precipitate of trinitrobenzene interferes with the observation of the end-point, add 5 to 10 ml. of alcohol to the test tube to dissolve it.

Record the temperature of the boiling water bath.

CALCULATIONS

From the volumes of solution used and the known concentration of the sodium hydroxide solution, calculate the concentration of the acid remaining in each test tube at the time it was removed from the boiling water. Plot the data, using the time in the water as abscissa and the concentration as ordinate, and lightly draw a smooth curve through the points with a sharp pencil (Graph 1).

Mark, on the curve, four conveniently spaced points covering the total time, and at each of these draw a light tangent to the curve, using a sharp pencil. To draw a tangent is not easy; use your best judgment so that the curve bends away from the straight line by the same amount on each side of the point of tangency. Read the coordinates of two points on each tangent and determine the slope of each, using the formula that the slope is the difference in ordinates divided by the difference in abscissae. Read the ordinate for the concentration at the point of tangency, and record all results.

Plot the reaction rate as ordinate against the concentration at the point of tangency as abscissa (Graph 2), and draw a straight line through the origin that comes as close as possible to each of the experimental points. Determine the slope of the line and record it as the rate constant, k, of the reaction. What are the units of the rate constant?

QUESTIONS—PROBLEMS

1. What would be the appearance of the expression for the law of mass action if the rate-determining step in the reaction had required two molecules of trinitrobenzoic acid instead of one, as in the following equation?

$$2C_6H_2(NO_2)_3CO_2H \longrightarrow 2C_6H_3(NO_2)_3 + 2CO_2$$

2. In general, does the rate of reaction increase or decrease with increase in temperature?

3. If the rate of a reaction doubled with each 10°C increase in temperature, how many minutes would be required at 30°C for a reaction to proceed as far toward completion as it went in 40 minutes at 100°C? What bearing does the answer to this question have on the stability of the stock solution of trinitrobenzoic acid used to fill the buret referred to on p. 230?

4. Recognizing that trinitrobenzoic acid is a strong acid and, hence, almost completely present in water solution in the ionized form, discuss the possibility that the *mechanism* of the reaction is not given by equation (3)

but by a process in which the *rate-determining step* is a decomposition of the trinitrobenzoate anion, followed by fast reactions to give the final products.

5. If one takes logarithms of both sides of equation (2), one obtains

$$\log (\text{rate}) = \log k + n \log c$$

Show how n could be determined by examining the slope of a line obtained by plotting log (rate) against log c.

Make such a graph (Graph 3) from your data, and determine the best value of n.

6. If equations (1) and (4) are combined, letting c represent the concentration of trinitrobenzoic acid, one obtains

$$-\frac{dc}{dt} = k_1 c$$

Integration of this equation gives

$$2.303 \log c = -k_1 t + \text{constant of integration}$$

Tabulate log c for each of your measured points and plot log c against t (Graph 4). Draw the best straight line through the points and calculate k_1 from the slope of this line

$$k_1 = 2.303 \times (\text{slope})$$

Compare the value of k_1 with that determined from Graph 2.

REFERENCE

1. Garrett, A. B., Lippincott, W. T., and Verhoek, F. H.: *Chemistry, A Study of Matter.* 2nd edition. Xerox College Publishing, Lexington, Mass., 1972, Chapter 22.

Experiment 19
REPORT SHEET

Name _____ Section No. _____

Data

Concentration of trinitrobenzoic
    acid in each test tube   Heating time

1. _____   _____

2. _____   _____

3. _____   _____

4. _____   _____

5. _____   _____

Molarity of standardized base

Temperature of the hot water bath

Sample Calculations

Rate Law for the Reaction

Calculation of the Rate Constant of the Reaction

Attach Graphs and Answers to Assigned Questions

THERMODYNAMICS

CALORIMETRY

OBJECTIVE

To measure enthalpies of neutralization and formation, and to use these values to calculate enthalpy changes in other reactions.

EQUIPMENT NEEDED

250-ml. electrolytic beaker, 15- × 120-mm. test tube; 1-ml. pipet graduated to 0.01 ml.; No. 12 rubber stopper; 5-mm. glass tubing; glass rod; rubber tubing; 1500-ml. beaker or plastic container; 5 ml., and 10 ml. pipets; micropipet; 3 styrofoam coffee cups; –15° to + 15°C thermometer graduated in tenths of degree.

REAGENTS NEEDED

Magnesium turnings; 6 M HCl; standardized solutions of 3.0M HCl and 3.0M NaOH; crushed ice; two copper cylinders about 1.2 cm in diameter and 3.1 cm long.

INTRODUCTION

The change in enthalpy accompanying a chemical or physical change can often be measured in a calorimeter — an instrument that measures the heat changes in a system. The calorimeter usually is used to measure the heat liberated in exothermic processes. Several types of calorimeters are in common use. In most calorimeters the amount of heat evolved is calculated by multiplying the heat capacity of the calorimeter and its contents by the observed increase in temperature. The heat capacity is the amount of heat necessary to increase the temperature of the calorimeter and its contents 1°. A calorimeter of this type will be used in Part II of this experiment. In Part I of this experiment we shall use a simple ice calorimeter in which the heat liberated is used to melt ice, and the temperature of the calorimeter remains constant at 0°C.

The enthalpy changes associated with two reactions will be measured in this experiment. These enthalpy changes will be used to determine the enthalpy change for a third reaction, which is not measured directly.

In the first determination, the enthalpy of formation of a metal ion is obtained by observing the heat liberated in the reaction:*

$$M(s) + 2H_3O^+ \rightleftharpoons M^{2+}(aq) + H_2 + 2H_2O \quad \Delta H_{(1)} \tag{1}$$

Any of several metals, such as zinc, magnesium, iron, or cadmium, added to a hydrochloric acid solution will satisfy this equation.

In the second determination, the enthalpy of neutralization is obtained by adding a solution of a base such as sodium hydroxide to a hydrochloric acid solution. The reaction is:*

$$H_3O^+ + OH^- \rightleftharpoons 2H_2O \quad \Delta H_{(2)} \tag{2}$$

Utilizing Hess' law of heat summation and the ΔH values obtained by these determinations, we can determine the enthalpy change for a reaction having an equation that is the sum of or difference between equations (1) and (2). For example, multiplying equation (2) by two and subtracting it from equation (1) gives:

$$M(s) + 2H_2O \rightleftharpoons M^{2+}(aq) + 2OH^-(aq) + H_2 \tag{3}$$

The ΔH for this reaction is $\Delta H_{(1)} - 2\Delta H_{(2)}$.

Similarly, the enthalpy change in the reaction

$$Mg^{2+}(aq) + 2OH^-(aq) \longrightarrow Mg(OH)_2(s)$$

can be calculated by making use of the measured enthalpy changes and tabulated values of the enthalpies of formation of $Mg(OH)_2(s)$ and $H_2O(\ell)$, namely, -221.0 kcal and -68.3 kcal, respectively.

Part I. The Ice Calorimeter

In the ice calorimeter the heat evolved is used to melt some of the ice in a mixture of ice and water surrounding the reaction mixture. The amount of ice melted is determined by the change in volume of the ice-water mixture because as the ice melts, forming the more dense water, the volume of the ice-water mixture decreases. This change in volume is readily measured if the ice-water mixture is enclosed so that some of the water is forced up into a capillary tube or small pipet. Then as the ice melts the water level in the capillary tube or pipet drops. This drop in level is a very sensitive measure of the decrease in volume due to the melting of ice and from this the heat evolved in the reaction being studied can be calculated.

The calorimeter used here has been described by Mahan (ref. 3) and is shown diagrammatically in Figure 20-1. This calorimeter is best used to determine the enthalpy changes in fast exothermic reactions. The change in volume of the ice-water mixture is determined by measuring the change in volume at regular time intervals during the period immediately before and

---

*See the Discussion section for a more precise explanation of enthalpy changes accompanying reactions (1) and (2) and their relation to the data obtained in this experiment.

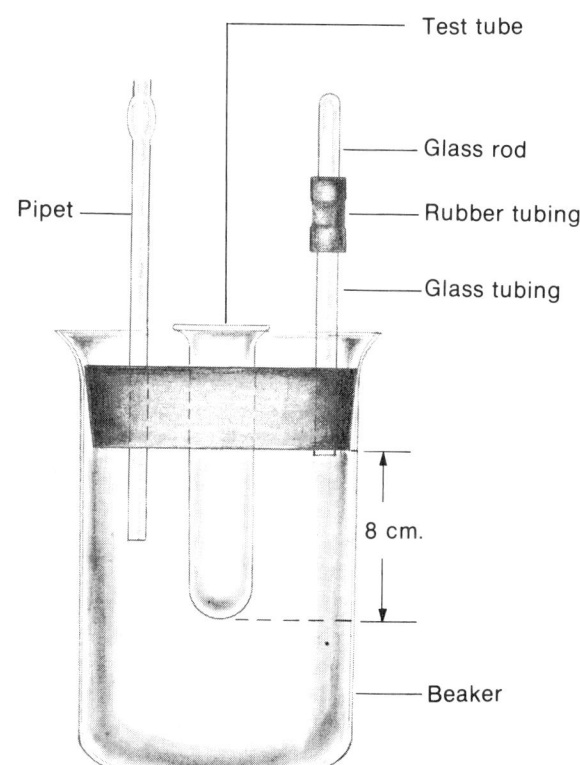

FIGURE 20-1

Diagram of an ice calorimeter. The test tube should be centered in the beaker with its bottom well above the bottom of the beaker. The glass tube should not extend below the rubber stopper. All seals must be airtight. The glass rod must be firmly in the rubber tubing but capable of being slid up and down inside the tubing to act as a plunger.

following the mixing of reactants. A graph of ΔV vs. time is plotted and the overall volume change due to the reaction is obtained from the graph as shown in Figure 20-2.

Pipetting. A pipet is a device for adding a fixed and accurately known volume of liquid to a reaction vessel. It is filled by means of a suction bulb to a point a centimeter or two above the etched graduation mark. The bulb is quickly replaced with a forefinger, and then the liquid height is adjusted until the bottom of the meniscus is exactly at the mark by cautiously admitting air around the finger. If too much liquid runs out so that you overshoot the mark, draw up more liquid and try again. Be careful not to draw liquid into the rubber bulb; this is likely to occur if the bottom end of the pipet is removed from the liquid during the suction.

FIGURE 20-2

Changes in volume with time for a reaction in the ice calorimeter. ΔV_0 is the overall change in volume, which is proportional to the heat liberated.

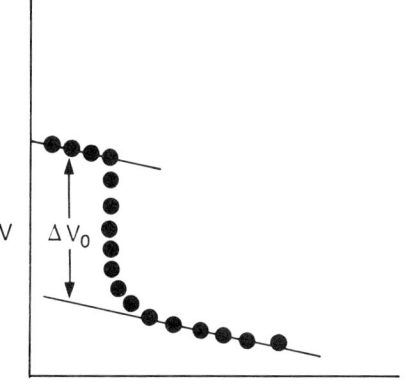

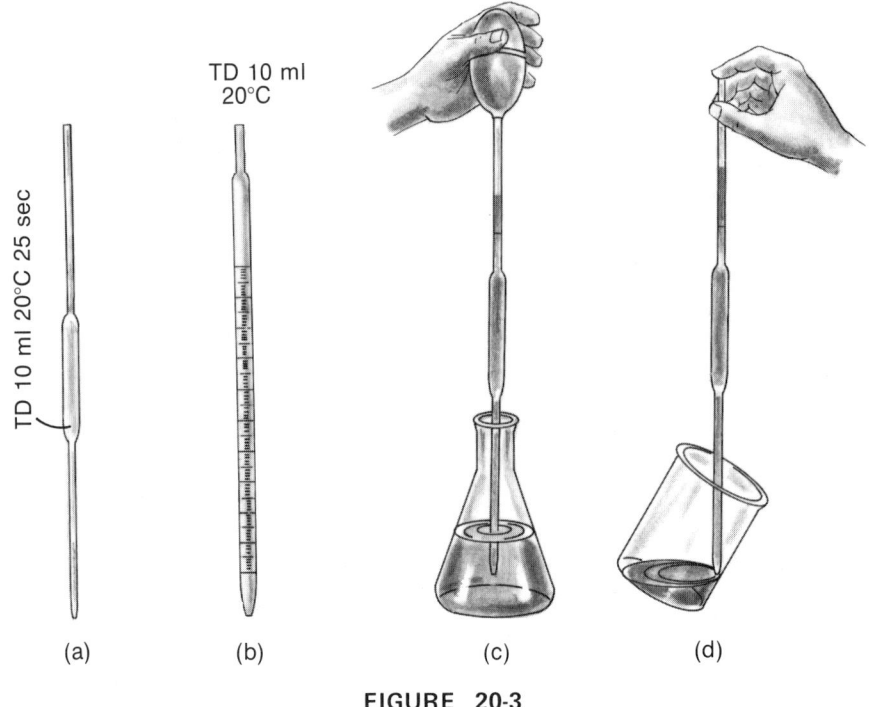

TD 10 ml
20°C

TD 10 ml 20°C 25 sec

(a)   (b)   (c)   (d)

FIGURE 20-3

Pipets: (a) transfer pipet, (b) measuring pipet, (c) filling pipet — liquid drawn above graduation mark, and (d) use of forefinger to adjust liquid level in pipet.

Having adjusted to the mark, the liquid is delivered by allowing it to run out the tip by gravity; do not blow the liquid out. If you are using a transfer pipet (Fig. 20-3) allow the pipet to empty; drain for a further 20-30 seconds, then touch the tip to a wet surface within the reaction vessel to allow capillary action to adjust the height of the liquid remaining in the tip. Do not attempt to remove this remaining liquid; the pipet is calibrated to deliver the stated volume when that bit of liquid remains. If you are using a measuring pipet (Fig. 20-3), keep your finger near the top, allow the liquid to run down until the meniscus reaches the proper mark specifying the volume to be introduced, then stop the flow with the finger. Volumes cannot usually be measured as precisely with a measuring pipet as they can with a transfer pipet.

Pipets must of course be clean, so that a uniform film of liquid, without visible droplets, forms on the inner surfaces. They may be cleaned with hot detergent solution, followed by thorough rinsing. They may be dried before each use, so that no contamination of the solution being pipetted occurs, or they may be rinsed with a small quantity of the solution by filling part way and then turning the pipet horizontally to contact all surfaces with the solution. The solution used for rinsing should be discarded.

PROCEDURE

A. Enthalpy of Formation of Mg^{2+}

Prepare the calorimeter by filling the beaker to the top with a mixture of crushed ice and water. Put the stopper (previously cooled to 0°C) securely in place with the glass rod removed from the rubber tubing. Shake the beaker gently about a dozen times to get rid of air bubbles that may have been trapped on the outside of the test tube. Place the beaker and contents into a 1500-ml. beaker or plastic container and fill this larger container with a mixture of ice and water. Pile ice on top of the stopper as high as the lip of the test tube. Add 2 ml. of 6 M HCl to 4 ml. of H_2O in a graduate, mix and pour the mixture into the test tube and allow about 10 minutes for the calorimeter to cool to the temperature of ice. Use a micro pipet to pour water into the rubber tubing until the water level in the calorimeter rises to the top of the rubber tubing; then place the glass rod plunger into the tubing and drive the water into the 1-ml. pipet as high as possible.

Prepare calorimeter and add acid.

Measure the change in water level in the pipet at 30 second intervals and when this becomes very small (about 5×10^{-3} ml. per minute) take time-volume readings every minute for five minutes. (If the water level changes more rapidly than indicated, examine the apparatus for leaks.)

Read water level at measured time intervals.

Then add to the test tube a weighed quantity of not more than 2.5×10^{-3} moles of magnesium. Take volume readings every 30 seconds until the change in volume is 5×10^{-3} ml. per minute; then take readings every minute for five more minutes. Repeat this experiment to obtain duplicate values for the enthalpy of formation of the metal ion.

Add known weight of Mg; continue readings.

B. Enthalpy of Neutralization

Once again prepare the calorimeter as previously described, but add from a pipet exactly 5 ml. of 3.0M HCl to the test tube. Take time-volume readings every minute for five minutes. Meanwhile, cool 20 ml. of a 3.0M solution of sodium hydroxide to $0°C$. in an ice bath and pipet exactly 5 ml of this solution into the test tube. Take volume readings at 30-second intervals and then at one-minute intervals as previously described. Repeat this experiment to obtain duplicate values for the enthalpy of neutralization of the base.

Prepare calorimeter; take water level readings; add base to acid; continue readings.

NOTEBOOK AND CALCULATIONS

Plot the decrease in volume of water in the pipet against time for the period immediately before and following mixing of the reactants. From the graph determine the change in volume of the ice-water mixture due to evolution of heat. The heat evolved is calculated from the data using the following considerations:

$$\text{Volume of ice melted} = \frac{\text{mass of ice melted}}{\text{density of ice}} \qquad (5)$$

$$\text{Volume of water formed} = \frac{\text{mass of ice melted}}{\text{density of water}} \qquad (6)$$

$$\text{Change in volume} = \text{Equation (5)} - \text{Equation (6)} \qquad (7)$$

$$= \text{mass of ice melted} \left(\frac{1}{\text{density of ice}} - \frac{1}{\text{density of water}} \right)$$

Heat liberated = grams of ice melted × heat of fusion per gram of ice **(8)**

= grams of ice melted × 79.6 cal./gram

Solving equation (7) for mass of ice melted and substituting in equation (8)

$$\text{Heat liberated = change in volume} \left(\frac{\text{density water} \times \text{density ice}}{\text{density water} - \text{density ice}}\right) \times 79.6 \text{ cal./g.}\textbf{(9)}$$

At 0°C the density of water is 0.999 g. per milliliter; the density of ice is 0.917 g. per milliliter.

Calculate the enthalpy change per mole for the two reactions studied as follows:

$$\Delta H = - \frac{\text{heat evolved}}{\text{moles of reactant used}}$$

Combine the results and the data given on p. 238 to calculate the enthalpy changes in reactions (3) and (4)

Part II. Alternative Procedure.

Measurement of the enthalpy change in a reaction can also be made in a calorimeter which does not operate at constant temperature as the ice calorimeter does. In this case we measure the temperature rise (or fall) produced by the reaction in an insulated vessel containing a mass of known heat capacity. For the insulated vessel we shall use a styrofoam coffee cup. To take advantage of the greater precision of the −15°C to +15°C thermometers used in Experiment 14, we shall operate in a temperature range below fifteen degrees.

The ice calorimeter required no calibration; all the heat evolved in the reaction went to melting ice. In a calorimeter which undergoes a change in temperature during use some of the heat is used in heating the calorimeter vessel; hence the calorimeter must be calibrated. In precise work this is carried out by introducing heat from an electric current, which can be measured precisely. In a simpler procedure, we shall calibrate by measuring the heat absorbed by the cooling of hot copper, taking the specific heat of copper as a known constant. The calibration procedure will be to heat a measured weight of copper to a measured temperature, add the hot copper to a measured quantity of water, and measure the temperature rise of the water (and the calorimeter). Knowing the specific heats and weights of copper and water, and the temperature change, the only unknown quantity is the amount of heat used to warm the calorimeter vessel. The determination of this unknown quantity is the purpose of the calibration procedure.

Preliminary for Part B. Before starting part A, prepare for part B by placing a 15 ml sample of 3.0M NaOH in a clean test tube in some ice in a plastic cup. Put a second 5-10 ml sample of 3.0M NaOH in another test tube (referred to later as test tube no. 2).

Cool NaOH solution.

A. Calibration Procedure. Weigh, on the analytical balance, two of the copper cylinders. The weight needs to be known only to the nearest 0.01g, so do not take the time needed to get the weight more precisely. (Observe, however, proper handling of the balance, particularly with respect to support of the balance pan and beam when adding or removing something from the pan.) Place the weighed cylinders in a *dry* test tube and set the test tube into a beaker of boiling water, full enough with water so that the level is well above the height reached by the cylinders in the test tube.

[margin: Weigh Cu cylinders; heat in boiling H_2O.]

Using the analytical balance, weigh to the nearest 0.01g a clean, dry styrofoam coffee cup of about 200 ml capacity. Add to the weighed cup 60 ml of distilled water and a few pieces of clean ice to cool the water to 3°C-5°C as read on the thermometer. Remove any excess ice with a clean spatula and reweigh the cup and water to the nearest 0.01g.

[margin: Determine weight of cooled H_2O in cup.]

Using the thermometer with 0.1°C graduations, read and record the temperature of the cooled water every 30 seconds for 5 minutes. At the next 30-second mark, remove the test tube from the boiling water (use a test-tube holder), immediately wipe the outside with tissue or towel and dump the copper cylinders into the water in the plastic cup. Record the time of addition and continue to read the time and temperature at 30 second intervals until the temperature change within the interval is about what it was before adding the copper, and for 5 minutes longer. Stir the water occasionally with the thermometer (gently, don't bang it against the copper cylinders) during this period.

[margin: Take temperature readings at measured time intervals; add hot Cu, and continue readings.]

Use an ordinary −10° to 110° thermometer (*not* the −15° to +15° thermometer) to measure the temperature of the boiling water, estimating to the nearest 0.1°, and record the value.

[margin: Measure boiling temperature.]

B. Enthalpy of Neutralization. Remove the copper cylinders from the water and add to the water 10 ml of 3.0 M HCl, using a pipet. Add ice and cool below 5°C. Remove excess ice and weigh cup and solution to the nearest 0.01 gram.

[margin: Determine weight of cooled HCl solution in cup.]

Rinse the pipet with water and with 3.0M NaOH solution from the number 2 test tube prepared before you started part A. Rinse and wipe the thermometer, take the temperature of the cold sodium hydroxide solution, and record the value. Rinse and wipe the thermometer again and place it in the hydrochloric acid solution in the plastic cup. Observe and record the temperature of the solution at 30-second intervals for five minutes. Using the rinsed pipet, add 10 ml of the cold 3.0M NaOH solution to the acid in the plastic cup, noting the time of addition. Stir. Observe and record the temperature at 30-second intervals for five minutes.

[margin: Take temperature readings at intervals; add NaOH, and continue readings.]

C. Enthalpy of Formation of Mg^{2+}(aq). Weigh out on clean paper or a watch glass 2.5×10^{-3} moles of magnesium turnings.

Discard the salt solution in the plastic cup, rinse it with distilled water, and add from a graduate 20 ml. of 6 M HCl and 40 ml. of H_2O, and sufficient ice to cool below 5°C. Remove excess ice and weigh the hydrochloric acid solution to 0.01 g. Observe and record the temperature of the solution at 30 second intervals for five minutes. Add the magnesium, noting the time of addition. Stir and record the temperature at 30-second intervals until all the magnesium has disappeared and for five minutes longer.

[margin: Measure temperature rise for known weight of acid; add weighed Mg; continue to measure temperature rise.]

D. Since both parts B and C depend upon the calibration results of part A, if you have laboratory time available, repeat part A and average the calculated results with those of the first set.

CALCULATIONS

Plot the data of Parts A, B and C of Part II as in Figure 20-2, but use temperature as the ordinate. On each graph, draw a vertical line at the time at which the copper, sodium hydroxide or magnesium was added and extrapolate the nearly horizontal portions of the curves to intersect it. Determine the temperature difference, ΔT, resulting from the additions and the corrected final temperature of the calorimeter vessel and contents, T_F. The latter is the temperature corresponding to the intersection of the upper extrapolated curve and the vertical line drawn as above.

For Part A we have, denoting by T_B the temperature of the boiling water in which the copper was heated,

(a) $\left(\begin{smallmatrix}\text{Calories lost}\\\text{by copper}\end{smallmatrix}\right)$ = (grams of copper) $\times$ (0.092 cal. g.$^{-1}$ deg^{-1}) $\times$ ($T_B - T_F$)

(b) $\left(\begin{smallmatrix}\text{Calories gained}\\\text{by water}\end{smallmatrix}\right)$ = (grams of water) $\times$ (1.002 cal. g.$^{-1}$ deg^{-1}) $\times \Delta T$

(c) Calories gained by calorimeter vessel = (X) $\times \Delta T$

Here 0.092 cal. g.$^{-1}$ deg^{-1} is the heat capacity of 1 g of copper near room temperature and 1.002 cal. g.$^{-1}$ deg^{-1} is the heat capacity of 1 g of water in the region near 10°C; the unknown, X, is the amount of heat in calories needed to change the temperature of the calorimeter vessel by 1°C when it holds about 65 g. of H_2O.

Since

$$(a) = (b) + (c)$$

$$X = \frac{(a) - (b)}{\Delta T}$$

For Part B we have, letting T_F represent the final temperature of the calorimeter, ΔT the temperature change, and T_N the temperature of the sodium hydroxide solution.

(a) Calories produced in the neutralization reaction = Y

(b) $\left(\begin{smallmatrix}\text{Calories to warm the}\\\text{solution in the calorimeter}\end{smallmatrix}\right) = \left(\begin{smallmatrix}\text{mass of}\\\text{HCl solution}\end{smallmatrix}\right) \times (\Delta T) \times$ (1.002 cal. g.$^{-1}$ °C^{-1})

(c) Calories to warm the calorimeter = X $\times$ (ΔT)

(d) $\left(\begin{smallmatrix}\text{Calories to warm the}\\\text{sodium hydroxide solution}\end{smallmatrix}\right) = \left(\begin{smallmatrix}\text{mass of}\\\text{NaOH solution}\end{smallmatrix}\right) \times (T_F - T_N) \times$ (1.005 cal.g.$^{-1}$ °C^{-1})

In (b) and (d) we have assumed that the specific heats of the dilute solutions do not differ significantly from that of pure water and in (d) we have used 1.005 cal g^{-1} deg^{-1} as the heat capacity of 1 g of water near 3°C. Also,

$$Y = (b) + (c) + (d)$$

Use 1.083 g cm^{-3} as the density of the sodium hydroxide solution for calculating its mass.

The calculated value of Y represents the number of calories evolved on reaction of $\frac{10}{1000}$ × 3.0 moles of HCl and an equal number of moles of sodium hydroxide. From these data calculate the enthalpy of neutralization for one mole of each, at the dilution represented in your experiment.

All the heat liberated by the reaction of Part C goes to heating the calorimeter and contents.

(a) Calories evolved in forming magnesium ion in solution = Z

(b) Calories to warm the calorimeter = X × ΔT

(c) $\left(\begin{array}{l}\text{Calories to warm the}\\\text{hydrochloric acid solution}\end{array}\right) = \left(\begin{array}{l}\text{mass of}\\\text{HCl solution}\end{array}\right) \times \Delta T \times (0.891 \text{ cal. g.}^{-1}\ ^\circ C^{-1})$

Here 0.891 cal g^{-1} $^\circ C^{-1}$ is the heat capacity for 2 M HCl in the region near 8°C.

Also Z = (b) + (c) for the quantity of magnesium used. Calculate the enthalpy of formation of one mole of magnesium ion (p. 240) from this result.

Use the results of B and C and the data given on p. 238 to calculate the enthalpy changes in reactions (3) and (4).

DISCUSSION

Strictly speaking, the change in enthalpy measured in the reaction of the metal with acid is not the enthalpy of formation of the metal ion M(aq)$^{+2}$. In addition to the measured quantity, the changes in enthalpy for at least two other processes must be taken into account if the enthalpy of formation of the metal ion is to be calculated accurately. Fortunately, the contribution of these additional reactions to the enthalpy of formation is very small – at the most, a few tenths of 1 percent – and well within the experimental error of our measurement. However, because it is important to understand the principles involved in the concept of the enthalpy of formation of ions and to appreciate the approximations made in determining it as we do in this experiment, the following explanation is provided:

For the reaction of a metal with acid, the quantity measured in the ice calorimeter in Part I is the change in enthalpy for the reaction

$$Mg(s) + 2H_3O^+ \text{ (2.0M)} \longrightarrow Mg^{2+} \text{ (0.42M)} + H_2 \text{ (1 atm.)} \qquad \textbf{(10)}$$

and in Part IIC it is

$$Mg(s) + 2H_3O^+ \text{ (2.0M)} \longrightarrow Mg^{2+} \text{ (0.042M)} + H_2 \text{ (1 atm.)} \qquad \textbf{(11)}$$

Since only 2.5 × 10^{-3} moles of metal are used in the experiment, the heat liberated in the calorimeter must be multiplied by 4.0 × 10^2 to get the ΔH for reactions (10) and (11) in which 1 mole of metal is stipulated.

The enthalpy of formation of a divalent metal ion is given by the equation:

$$M(s) \longrightarrow M^{2+} \text{ (aq)} + 2e^- \text{(aq)} \qquad \textbf{(12)}$$

By convention this is measured in the reaction

$$M(s) + 2H_3O^+ (aq) \longrightarrow M^{2+}(aq) + H_2 \text{ (1 atm.)} \tag{13}$$

in which the enthalpy change for the process

$$2e^-(aq) + 2H_3O^+(aq) \longrightarrow H_2 \text{ (1 atm.)} + 2H_2O$$

is taken as zero.

The difference between equation (13) — the equation defining the enthalpy of formation of the ion — and equations (10) and (11) for which the enthalpy changes are measured in this experiment — appears in the different concentrations of the ions $H_3O^+(aq)$ and $M^{2+}(aq)$. The symbolism in equation (13) implies that these ions are present in the solution at infinite dilution; in equations (10) and (11) they are present at the concentrations specified. In taking these ions from the concentrations specified to infinite dilution, the following processes can be imagined to occur:

$$2H_3O^+ \text{ (2.0M)} \longrightarrow 2H_3O^+ \text{ (aq)} \tag{14}$$

$$Mg^{2+} \text{ (0.42M or 0.042M)} \longrightarrow Mg^{2+}(aq) \tag{15}$$

Because the heats of dilution of ions are small — often much less than 4×10^{-2} kcal. per mole — reliable enthalpies of formation of metal ions can be obtained by measuring the change in enthalpy for the reaction of the metal with acid as described in this experiment.

By similar reasoning, it can be shown that the change in enthalpy measured in Parts IB and IIB are not actually the enthalpy of neutralization. Here again the difference between the measured and defined values is small and is associated with the heats of dilution of the ions. This is illustrated by the following equations:

$$H_3O^+ \text{ (3.0M)} + OH^- \text{ (3.0M)} \longrightarrow 2H_2O \tag{16}$$

$$H_3O^+ \text{ (aq)} + OH^-(aq) \longrightarrow 2H_2O \tag{17}$$

Equation (16) represents the reaction in which the change in enthalpy is measured; equation (17) is the equation defining the enthalpy of neutralization.

In both parts of the experiment, the enthalpies of dilution of the ions not involved in the reactions, sodium and chloride ions, should be included in the calculations, but these enthalpies also are very small and can be neglected in the calculations.

QUESTIONS–PROBLEMS, PART I

1. What error would be produced in the ΔH values determined in this experiment by each of the following: (a) failure to place the calorimeter in a larger beaker of ice; (b) addition of more metal than the acid present can oxidize; (c) addition of less metal than the acid present can oxidize; (d) addition of less base than the acid present can neutralize?

2. Write equations for the three reactions for which ΔH could be determined in this experiment if cadmium were used as the metal in the first part and potassium hydrogen carbonate were used as the base in the second part.

3. Could the ice calorimeter be used to measure changes in enthalpy in endothermic reactions? Why or why not?

4. Why is the ice calorimeter suited to measuring changes in enthalpy in fast reactions?

QUESTIONS—PROBLEMS, PART II

5. In each of parts A, B, and C the quantity of liquid in the styrofoam cup has been kept nearly the same. What scientific reason is there for this, other than the practical one that this gives a convenient depth for covering the thermometer bulb with liquid?

6. Was there an increase or decrease in temperature in Part B when the hydrochloric acid solution was added to the water? To what do you attribute this?

7. In the experiment of part C, the magnesium was at room temperature when added to the solution, so that it is as if you had dropped hot pieces of magnesium into the cold water, thus adding additional heat to the calorimeter liquid. Yet no account was taken of this in the calculations. Remembering the Law of Dulong and Petit, what error does neglect of this factor have on the calculated value of the enthalpy of formation of Mg^{2+}?

REFERENCES

1. Shoemaker, D. P., and Garland, C. W.: *Experiments in Physical Chemistry*. 2nd edition, McGraw-Hill, New York, 1967, Chapter 5.
2. Garrett, A. B., Lippincott, W. T., and Verhoek, F. H.: *Chemistry, A Study of Matter*. 2nd edition, Xerox College Publishing, Lexington, Mass., 1972, Chapter 18.
3. Mahan, B.: *J. Chem. Educ., 37:* 634 (1960).

**Experiment 20
REPORT SHEET**

Name _____ Section No. _____

Objective

PART I

A. Weight of magnesium used _____

   ΔV measured   _____

   Heat liberated   _____

   ΔH_f for Mg^{2+}(aq)   _____

B. Volumes and concentrations of acid and base solution used: acid _____ _____

                                                              base _____ _____

   ΔV measured   _____

   Heat liberated   _____

   Molar enthalpy of neutralization   _____

   Enthalpy change for reaction (3) from your data _____

   Enthalpy change for reaction (4) from your data _____

   Attach pages with answers to question 1, 2, 3, 4. Attach graphs of volume against time.

PART II

A. Weight of water in cup   _____

  Weight of copper added   _____

  ΔT _____

  $T_B - T_F$ _____

  Calculated value of X   _____

B. Weight of HCl solution   _____

  Weight of NaOH solution   _____

  ΔT _____

  $T_F - T_N$ _____

  Molar enthalpy of neutralization _____

C. Weight of HCl solution   _____

  Weight of magnesium added   _____

  ΔT _____

  ΔH_f for Mg^{2+} _____

  Enthalpy changes for reactions (3) and (4) from your data: (3) _____

                                                              (4) _____

  Attach pages with answers to questions 5, 6, 7.

  Attach graphs of temperature against time.

CHANGES IN FREE ENERGY, ENTHALPY AND ENTROPY IN THE REACTION: $I_2(aq) + I^-(aq) \rightleftharpoons I_3^-(aq)$

OBJECTIVE

To determine the concentrations at equilibrium in the reaction given and so to determine the equilibrium constant and standard free energy change. To show how the standard enthalpy and entropy changes in the reaction can be obtained from measurements of the equilibrium constant at more than one temperature.

EQUIPMENT NEEDED

Four 25-ml volumetric flasks; 250-ml Erlenmeyer flask; 5-ml pipet; three 13 X 120 mm culture tubes; one 600-ml beaker; rubber bulb for pipetting; aspirator pump and tubing connections; spectrophotometer for visible wavelengths; pair of matched cuvettes.

REAGENTS NEEDED

Decahydronaphthalene; 0.01M I_2 in decahydronaphthalene; 0.100M KI in water; Alconox; acetone for rinsing; aluminum foil or polyethylene film; paper tissues; ice.

INTRODUCTION

In an aqueous solution containing iodine and iodide ion, an equilibrium is established, represented by the equation

$$I^- + I_2 \rightleftharpoons I_3^- \tag{1}$$

The purpose of this experiment is to determine the value of the equilibrium constant for this reaction.

$$K = \frac{[I_3^-]_{eq}}{[I^-]_{eq}\,[I_2]_{eq}} \tag{2}$$

In Part A this constant is determined at room temperature. In Part B the measurements are extended to a higher and a lower temperature, and the results at several temperatures are used to calculate the changes in enthalpy and in entropy for the reaction represented by equation (1).

A. The Triiodide Equilibrium at Room Temperature

In order to determine the equilibrium constant, equation (2), it is necessary to know the equilibrium concentrations of the three species present. A direct determination of these concentrations is not feasible, because analytical methods do not distinguish between I_2 and I_3^-, nor between I^- and I_3^-. The material balance tells us, however, that

$$\text{Total } I_2 = I_2(eq) + I_3^-(eq) \tag{3}$$

$$\text{Total } I^- = I^-(eq) + I_3^-(eq) \tag{4}$$

Hence, if we measure how much iodine and how much iodide we add to the solution initially, and measure the amount of free iodine present at equilibrium, $I_2(eq)$, we can calculate the equilibrium amounts of the other two species. The amount of iodine present at equilibrium may be determined by taking advantage of the fact that iodine, but neither of the ionic species, is soluble in organic solvents such as decahydronaphthalene and it participates in a *distribution equilibrium* between water and decahydronaphthalene if the two solvents are shaken together.

Distribution Equilibria. The establishment of a distribution equilibrium of a solute between two immiscible solvents is similar to the establishment of a vapor pressure equilibrium.

A pure liquid in a confined space at a constant temperature is in equilibrium with its vapor when the rate at which molecules escape from the liquid into the vapor is equal to the rate at which molecules from the vapor phase condense into the liquid. Similarly, a volatile solution is in equilibrium with its vapor when the rate at which its molecules escape from the liquid into the vapor is equal to the rate at which its molecules return to the liquid from the vapor phase.

An analogous situation holds true for the escape of solute molecules from one liquid into a second liquid that is not soluble in the first. Equilibrium is established when the rate at which solute molecules cross from liquid B into liquid A is equal to the rate at which molecules cross from A into B. If, initially, no solute is present in A, the number of solute molecules crossing from liquid B to A is greater than the number moving in the opposite direction in unit time. The concentration of solute in liquid A thus increases until the number of molecules moving from liquid A to liquid B in unit time equals the number moving from liquid B to liquid A in the same time unit, just as the concentration of vapor increases to a certain level in the case of vapor pressure. After this equilibrium point is reached, there is no further change in the concentrations in either liquid A or liquid B.

In general, the concentrations in the two solvents are not the same at equilibrium, because the solute interacts to different extents with the two solvents. (Note that the solvents must themselves be very different from each other; if they were not, they would dissolve in each other and not form two layers.) Nevertheless, the equilibrium concentrations should appear always

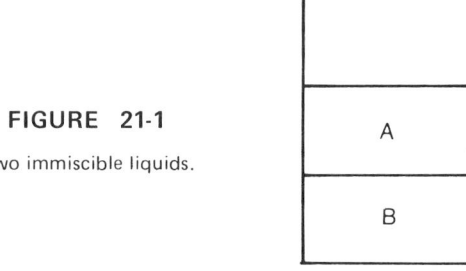

FIGURE 21-1

Two immiscible liquids.

(at a single constant temperature) in a constant ratio to each other, and this ratio of concentrations would be expected to be independent of the total amount of solute or the total volume of the two liquids.

The value of the distribution coefficient for iodine in the iodine, water, decahydronaphthalene system is recorded in Table 21-1. Decahydronaphthalene is less dense than water, and forms the A layer of Fig. 21-1. The distribution coefficient is given as the ratio $\dfrac{\text{conc. in organic layer}}{\text{conc. in water layer}}$.

TABLE 21-1 DISTRIBUTION OF IODINE BETWEEN WATER AND DECAHYDRONAPHTHALENE AT ROOM TEMPERATURE*

$[I_2]_O$ moles per liter	$[I_2]_W$ moles per liter	$K_D = \dfrac{[I_2]_O}{[I_2]_W}$
0.0337	0.000518	65.1
.0500	.000833	60.0
.0386	.000624	61.9
		62.3 Ave.

*Herz, W. and Schuftan, P., Z. physik. Chem., 101, 269 (1922).

Knowledge of K_D makes it possible to determine $[I_2]_{eq}$ in the iodide, iodine, triiodide system. Because the distribution equilibrium of iodine between water and decahydronaphthalene is maintained no matter what other equilibria are involved, we need only establish the iodine distribution equilibrium between the aqueous solution of triiodide and a decahydronaphthalene layer and measure the concentration of iodine in the organic layer. Iodine is the only solute present in that solvent. The equilibrium iodine concentration in the water layer is then simply

$$[I_2]_{eq,W} = \frac{[I_2]_{eq,O}}{K_D} \qquad (5)$$

Since some of the iodine is now in the decahydronaphthalene layer, some in the water layer as iodine, and some in the water layer as triiodide ion, the material balance equation (3) becomes

Total moles of iodine = Moles of iodine in decahydronaphthalene layer at equilibrium + Moles of iodine in water layer at equilibrium + Moles of triiodide ion in water layer at equilibrium   (6)

Equation (4) is unchanged

$$\text{Total moles} = \text{Moles of iodide ion} + \text{Moles of triiodide ion} \tag{7}$$
$$\text{of iodide} \quad\quad \text{at equilibrium} \quad\quad\quad \text{at equilibrium}$$

In this experiment iodine will be introduced into the equilibrum system by adding a measured volume, V_{I_2} ml., of a solution of iodine in decahydronaphthalene of known molar concentration, M_{I_2}. Iodide will be introduced as a measured volume, V_{KI} ml., of an aqueous solution of potassium iodide of known molar concentration, M_{KI}. The volume of the decahydronaphthalene layer is then V_{I_2} and the water layer V_{KI}, and equations (6) and (7) become

$$\frac{V_{I_2}M_{I_2}}{1000} = \frac{V_{I_2}[I_2]_{eq,O}}{1000} + \frac{V_{KI}[I_2]_{eq,W}}{1000} + \frac{V_{KI}[I_3^-]_{eq}}{1000} \tag{8}$$

$$\frac{V_{KI}M_{KI}}{1000} = \frac{V_{KI}[I^-]_{eq}}{1000} + \frac{V_{KI}[I_3^-]_{eq}}{1000} \tag{9}$$

If we measure $[I_2]_{eq,O}$, equations (5), (8) and (9) will represent three simultaneous equations in the three unknowns $[I_3^-]_{eq}$, $[I_2]_{eq,W}$, and $[I^-]_{eq}$ which are needed for the determination of K in equation (2).

The successful calculation of K thus depends upon a measurement of the concentration of iodine in the decahydronaphthalene layer at equilibrium, $[I_2]_{eq,O}$. This concentration will be determined by a spectroscopic procedure.

TECHNIQUE

The techniques of spectrophotometric measurement are described in Experiment 23, and the introductory portions of that experiment should be studied.

PROCEDURE

Preparation of Iodine Solutions.

Thoroughly clean four 25-ml volumetric flasks with Alconox and water, rinse them thoroughly with water and then successively with two 5-ml portions of acetone followed by two 5-ml portions of decahydronaphthalene. Do not discard the decahydronaphthalene into the sink; put it into a 250-ml Erlenmeyer flask (the "recovery flask") for disposal as instructed later. The volumetric flasks need not be dried after rinsing with decahydronaphthalene. Label the flasks 1, 2, 3, 5.

Clean four volumetric flasks and rinse with decahydronaphthalene.

Clean a 5-ml transfer pipet with hot detergent solution, rinse with water and then with acetone, and dry by drawing a slow current of air through it with the aspirator pump.

Clean and dry pipet.

Take the flasks to the buret at the reagent shelf and add 1.00 (±0.02) ml of 0.01M I_2 in decahydronaphthalene to flask 1; 2.00 ml of the solution to flask 2; 3.00 ml to flask 3; and 5.00 ml to flask 5. Use the buret carefully, measuring the volumes used to the nearest 0.02 ml, since the success of your experiment depends upon the precision with which you carry out this initial procedure. Record the exact volume added to each flask in your notebook.

Add decahydronaphthalene to the marks in each of flasks 1, 2, 3, 5 and shake each until the mixture is homogeneous.

Determination of the Appropriate Wavelength for Spectrophotometric Measurements. Choose matched cuvettes as instructed in Experiment 23, and measure the absorption of the solution from volumetric flask 2 as a function of wavelength. Make readings 50 millimicrons apart between 340 and 650 millimicrons. Be sure to reset the instrument for zero transmission and for 100 percent transmission with solvent only for each change in wavelength setting. Plot percentage transmission against wavelength; find the region with minimum transmission and make additional readings at 5-millimicron intervals over a small range in this region. Add these data to the graph of the transmission spectrum, and choose the wavelength at which absorption is greatest as the working wavelength. By choosing the value at minimum transmission, you will have chosen a place at which the curve for transmission vs. wavelength is horizontal, and a slight deviation of wavelength on either side of the minimum will have much less effect than if you chose a wavelength where the curve is steeper.

Determination of the Absorbance-Concentration Relation. You are now ready to establish Beer's law for the iodine solution in decahydronaphthalene.

Read carefully the transmission of the solution used in the preceding paragraph at the working wavelength, taking care to balance the instrument for zero and for 100 percent transmission of pure solvent. Empty the cuvette containing the iodine solution into the recovery flask for later recovery of iodine and solvent; clean the cuvette thoroughly, rinsing it finally with acetone and drying it by inserting a glass tube connected to the aspirator pump and drawing air through it.

Refill the cuvette with one of the other solutions of iodine prepared by the dilution procedure outlined above, and determine its absorption at the working wavelength. Discard the solution into the recovery beaker, clean and dry the cuvette, and determine the absorption of a third solution at the working wavelength.

Repeat this procedure until the transmissions of all four diluted solutions have been measured.

Determination of the Concentration of Iodine in Decahydronaphthalene When Distribution Equilibrium Has Been Reached With Aqueous Triiodide Solution. Pipet 5.0 ml of 0.1M KI solution in water into each of three culture tubes. Using a clean and dry pipet each time, pipet 5.0 ml of solution 1 into one of the tubes, 5.0 ml of solution 3 into another, and 5.0 ml of solution 5 into the third. Cover one of the tubes with a bit of aluminum foil or plastic film, press down firmly with a thumb to close the opening and shake the contents vigorously for not less than one minute. Shake each of the other tubes in a similar way, and set them aside for 10 minutes to allow separation of the layers.

Read a thermometer and record the temperature of the room.

Determine
absorbance of the
organic layers at the
working wavelength.
When the layers have separated, take a sample of the decahydro-naphthalene layer from one of the tubes and add it to the solution cuvette for determination of the absorption. Both the pipet or dropping tube used to take the sample, and the cuvette, should be clean and dry so that no untoward dilution or contamination occurs. Read and record the light transmission of the equilibrated solution at the working wavelength.

In a similar way draw off samples from the other two tubes and determine their absorption.

CALCULATIONS

Make a Beer's Law
plot.
Calculate the concentrations of iodine in solutions 1, 2, 3, 5, using the precise values of the volumes of 0.01M solution used for each.

Calculate the absorbance [equations (3) and (4), Exp. 23] for each of the four solutions and plot this against the concentration of the solutions. Draw the best straight line through the points and determine ϵb of equations (8) and (10), Experiment 23, for solutions of iodine in decahydro-naphthalene.

Calculate the absorbance of the solutions, and the concentration of iodine in them, using the value of ϵb determined earlier or reading from the Beer's law graph. The values obtained represent $[I_2]_{eq, o}$ in equations (5) and (8). Calculate $[I_2]_{eq,w}$, $[I^-]_{eq}$, and $[I_3^-]_{eq}$, using equations (8) and (9), and calculate K in equation (2).

Examine the three values to see whether they show a trend; that is, are the deviations from the average random, or do the sign and values of the deviations seem to depend upon the concentration of the decahydro-naphthalene solution? If such a trend appears, try to account for it.

Report your values of K, and the temperature of the experiment, to the instructor, and calculate, for each trial, the percentage deviation from the class average.

If you are not to continue with Part B, add the remaining contents of flasks 1, 2, 3 and 5 to the recovery flask and take the entire mixture to the recovery bottle as instructed.

B. Standard Enthalpy and Entropy Change in the Triiodide Equilibrium

The two thermodynamic relations

$$\Delta G^\circ = -2.303RT \log K \qquad (10)$$

and

$$\Delta G^\circ = \Delta H^\circ - T \Delta S^\circ \qquad (11)$$

make it possible, from measurements of K at more than one temperature, to calculate ΔH° and ΔS° for reaction (1).

That this is the case is readily seen by substituting equation (10) into equation (11) and dividing by 2.303RT.

$$-2.303RT \log K = \Delta H^\circ - T \Delta S^\circ$$

$$\qquad (12)$$

$$\log K = -\frac{\Delta H^\circ}{2.303RT} + \frac{\Delta S^\circ}{2.303R}$$

Equation (12) shows that if K is measured at several temperatures, a plot of log K against $\frac{1}{T}$ is a straight line of slope $-\frac{\Delta H^\circ}{2.303R}$ and intercept $\frac{\Delta S^\circ}{2.303R}$. Such a plot thus permits us to determine the standard enthalpy change in the reaction, just as a similar plot of log (vapor pressure) against $1/T$ enabled us to calculate the enthalpy of vaporization in Experiment 12.

To conserve time, we shall measure K at one additional temperature only, so that our graph will have only two points on it.

PROCEDURE

Repeat the portion of Part A outlined in the paragraph headed "Determination of the Concentration of Iodine in Decahydronaphthalene when Distribution Equilibrium has been Reached With Aqueous Triiodide Solution" using a well-stirred bath of ice and water to maintain a constant temperature. Wrap a large beaker with a towel and place it on a towel, fill it with small pieces of ice, and add water. Shake the tubes containing the iodine solution of decahydronaphthalene and the potassium iodide solution as before, and place them in the ice water for 10 minutes. Stir the ice and water bath containing the tubes occasionally during this period. Remove each tube from the ice water, shake it for one minute, and replace it for 10 additional minutes. Shake each tube again for 30 seconds and return the tube to the ice water for separation of the layers. With the tubes in the cold bath, extract samples of the organic layers with a pipet as before, and put them into clean cuvettes. Allow the cuvettes to warm to room temperature before reading the transmission (or warm them to room temperature by dipping them into lukewarm water) and wipe the outsides carefully with tissue to prevent scattering of the light beam by condensate on the outsides of the cuvettes.

CALCULATIONS

Calculate the value for K [equation (2)] for each trial. The value for K_D is nearly independent of temperature over the range from 0° to $50^\circ C$, and the room temperature value may be used for both calculations. Plot the logarithm of each K value against the reciprocal of the corresponding absolute temperature. Draw the best straight line through the points and calculate ΔH° from the slope of the line, using equation (12).

Take the average of your values for K at room temperature and calculate ΔG° for that temperature using equation (10). Using this value and the value for ΔH° determined from the graph, calculate ΔS° from equation (11).

Using the calculated values for ΔH° and ΔS°, evaluate ΔG° and K for zero degrees and compare the results with the average value for that temperature. How well do the calculated K values agree with the directly measured values?

QUESTIONS – PROBLEMS

1. What would be the effect on the error in your results at room temperature if the initial concentration of your stock decahydronaphthalene solution was 10 percent less than the value indicated on the label?

2. What would be the effect on the value of K of using 0.2M KI in place of 0.1M KI?

3. What would be the effect on the values of K of using 10 ml portions of solutions 1, 3, or 5 to 5 ml of aqueous KI solution?

4. What would have been the error in your K values at room temperature if you had used two pipets in making up the mixtures, one for the organic solution and one for the water solution, both of which were marked 5 ml but the one for the organic liquid actually contained 5.5 ml?

5. Iodine reacts with water to establish the equilibrium

$$I_2 + 2H_2O \rightleftharpoons HOI + I^- + H_3O^+$$

This reaction was ignored in your determination of K. What effect might this have on the precision of your result?

6. In measurements of the distribution of benzoic acid between benzene and water, it was found that a distribution coefficient calculated as

$$K_D = \frac{[C_6H_5CO_2H]_{H_2O}}{[C_6H_5CO_2H]_{C_6H_6}}$$

was not constant with changes in the total amount of benzoic acid present, but that a coefficient calculated as

$$K_D = \frac{[C_6H_5CO_2H]^2_{H_2O}}{[C_6H_5CO_2H]_{C_6H_6}}$$

was constant. Suggest an explanation for this result.

7. The standard enthalpy of formation of I_2(aq) at $298°K$ is 5.4 kcal. per mole. The standard enthalpy of formation of I^-(aq) at $298°K$ is -13.19 kcal. per mole. From your data, calculate the enthalpy of formation of I_3^-(aq).

8. From your data and the following values for standard free energies of formation, calculate the equilibrium constant for the reaction

$$I_3^- + 2Fe^{2+} \longrightarrow 2Fe^{3+} + 3I^-$$

For I^-(aq), $\Delta G_f° = -12.33$ kcal. per mole. For Fe^{2+}(aq), $\Delta G_f° = 20.30$ kcal. per mole. For Fe^{3+}(aq), $\Delta G_f° = -2.52$ kcal. per mole. For I_2(aq), $\Delta G_f°$ is $+3.92$ kcal per mole.

REFERENCES

1. Davies, M., and Gwynne, E.: *J. Am. Chem. Soc., 74:*2748 (1952).
2. *Selected Values of Chemical Thermodynamic Properties.* National Bureau of Standards Circular No. 500. U.S. Department of Commerce, Washington, D.C., 1952.

**Experiment 21
REPORT SHEET**

Name _____ Section No. _____

A. Wave length of maximum absorption _____

Value of ϵb from your data _____

	Flask 1	Flask 3	Flask 5
Concentration of iodine in organic solvent before mixing with KI solution	_____	_____	_____
Concentration of KI solution	_____	_____	_____
Absorbance of I_2 in organic layer after equilibrium has been established	_____	_____	_____
Concentration of I_2 in organic solvent $[I_2]_{eq, O}$	_____	_____	_____
Concentration of I_2 in water $[I_2]_{eq, W}$	_____	_____	_____
Concentration of I_3^-, $[I_3^-]_{eq}$	_____	_____	_____
Concentration of I^-, $[I^-]_{eq}$	_____	_____	_____
Value of K	_____	_____	_____
Temperature of the equilibration experiment	_____	_____	_____

Repetition of the above at $0°$

Calculated values of $\Delta H°$, $\Delta S°$, $\Delta G°$

Value of K at zero degrees, calculated from ΔH° and ΔS°.   _____

Attach Beer's Law graph and graph for determining ΔH°.

Attach Pages with Answers to Assigned Questions.

INORGANIC SYNTHESES

PREPARATIONS OF PHOSPHORUS, SULFUR, AND CHLORINE OXYANIONS

OBJECTIVE

To prepare representative oxygen anions of the elements in groups V, VI, and VII of the Periodic Table.

EQUIPMENT NEEDED

A 6-inch porcelain casserole; suction filter and trap bottle; filter paper to fit 3-inch Büchner funnel; 4-inch crystallizing dish; 3-inch Büchner funnel; 400-ml. beaker; iron ring and ring stand; Meker burner; two 125-ml. Erlenmeyer flasks; medicine dropper; simple desiccator; pH paper; crucible with lid; spatula; 100-ml. beaker; hydrogen sulfide source and generator in hoods; rubber policeman, watch glass, buret; a 250-ml. Erlenmeyer flask.

REAGENTS NEEDED

30 g. bone ash, 80 percent $Ca_3(PO_4)_2$, or tribasic calcium phosphate, $Ca_{10}(OH)_2(PO_4)_6$; 18M H_2SO_4; solid anhydrous sodium carbonate; phosphoric acid wash solution; saturated ammonium molybdate solution; 0.1 M KH_2PO_4; 0.1 M Na_2HPO_4; 0.1 M Na_3PO_4; Kodak D-72 developer; calcium chloride 6-hydrate, $CaCl_2 \cdot 6H_2O$; ethanol; methyl orange and phenolphthalein indicators; potassium sulfite; powdered sulfur; ice; iodine, 95 percent C_2H_5OH; 50 percent C_2H_5OH; crystallized barium hydroxide, $Ba(OH)_2 \cdot 8H_2O$; Aich-Two-Es, solid source of hydrogen sulfide; dilute hydrochloric acid; carbon disulfide; potassium chlorate; 8 M nitric acid.

INTRODUCTION

All the elements of Groups VA, VIA, and VIIA form strong bonds with oxygen. In all these oxides except that containing fluorine, the element other than oxygen is assigned a positive oxidation state. This experiment illustrates some of the chemical reactions and preparative routes to these oxygen compounds of the non-metallic elements. The experiment is divided into three subsections, one concerning each of the elements phosphorus, sulfur, and chlorine. You may be asked to prepare one or more compounds for a given element, or you may prepare one compound of each element.

Read the entire experiment to obtain experience with the general preparative chemistry, but follow your instructor's directions regarding the assignment and preparation of specific compounds.

A. Group VA – Phosphorus

Each of the Group VA elements has an outer shell electronic configuration of $ns^2 np^3$. Since each p orbital is half-filled and may form a covalent bond, a common molecular formula in this family is EX_3 (where E equals either N, P, As, Sb, or Bi.). In addition to utilizing the three p electrons, the elements of Group V also form compounds with a formal oxidation state of +3 and +5 (for example, +5 is illustrated by NO_3^-, N_2O_5, As_4O_{10}, Sb_2O_5, PO_4^{3-}, and AsO_4^{3-}).

Orthophosphoric acid, H_3PO_4, a hydrolysis product of P_4O_{10}, can be represented by the Lewis representation, A, or by the structural formula, B.

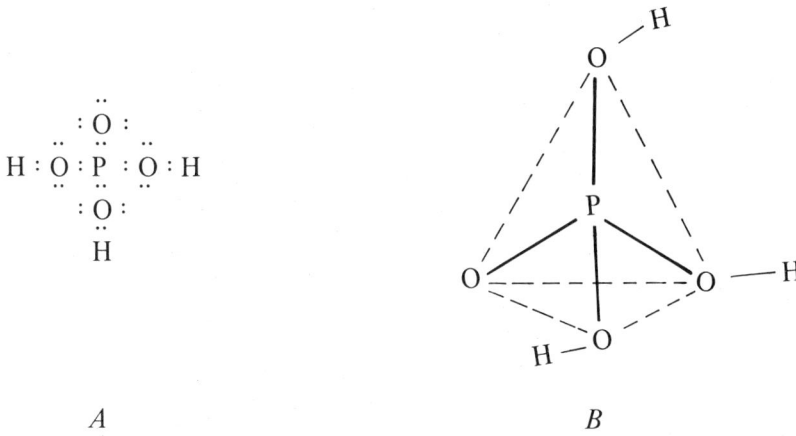

A B

All structures of orthophosphoric acid, the condensed phosphoric acids, and their salts are based upon the PO_4^{3-} tetrahedral unit.

In dilute aqueous solution, orthophosphoric acid is a triprotic acid in which the three protons do not show equal acid strengths but exhibit successive ionization constants, which decrease by a factor of $\sim 10^{-5}$. The values for the three ionization steps at 25°C are:

$$H_3PO_4 + H_2O \rightleftharpoons H_3O^+ + H_2PO_4^- \qquad K_{a_1} = 7.5 \times 10^{-3}$$

$$H_2PO_4^- + H_2O \rightleftharpoons H_3O^+ + HPO_4^{2-} \qquad K_{a_2} = 6 \times 10^{-8}$$

$$HPO_4^{2-} + H_2O \rightleftharpoons H_3O^+ + PO_4^{3-} \qquad K_{a_3} = 3.6 \times 10^{-12}$$

Note that the value for the third ionization constant is so low that appreciable concentrations of PO_4^{3-} may exist only in very basic solutions. As a result, normal orthophosphate salts (salts containing PO_4^{3-} ions) are difficult to obtain. The normal orthophosphate compounds which may be obtained pure are those that exhibit only a slight solubility in water (for example, $NH_4MgPO_4 \cdot 6H_2O$, Ag_3PO_4, and $FePO_4 \cdot 4H_2O$). The hydrogen orthophosphate, HPO_4^{2-}, and dihydrogen orthophosphate, $H_2PO_4^-$, anions appear in well characterized series of salts, especially those of the alkali metals.

The species present in solution is strongly dependent upon the pH as illustrated by the following equilibria:

$$H_3PO_4 \rightleftharpoons H_2PO_4^- \rightleftharpoons HPO_4{}^{2-} \rightleftharpoons PO_4{}^{3-}$$

increasing pH, decreasing acidity

decreasing pH, increasing acidity

These equilibria are examined in more detail in Experiment 28, which is concerned with determining the concentrations of orthophosphoric acid, H_3PO_4, and dihydrogen orthophosphate, $H_2PO_4^-$, present in an unknown sample. In the present experiment, we shall (1) extract orthophosphate from a natural source, bone ash, or from a chemical source, tribasic calcium phosphate, $Ca_{10}(OH)_2(PO_4)_6$; (2) isolate the orthophosphate compound $Na_2HPO_4 \cdot 12H_2O$; and (3) use the phosphate to prepare a complex anion of phosphorus. The extraction of the orthophosphate provides a pure reagent to use in the subsequent synthesis. Thus this experiment should illustrate some of the steps often followed by a synthetic chemist: (1) preparation of an intermediate compound, (2) analysis or characterization of a product, and (3) utilization of the intermediate compound for tests or for preparation of another compound.

B. Group VIA – Sulfur

Oxygen and sulfur belong to the same family of the periodic table. Consequently, each element has an electron configuration of $ns^2 np^4$, and the chemistry of the two elements is often quite similar. However, in some cases the chemistry of oxygen is quite different from that of sulfur. In this experiment some similarities in oxygen and sulfur are illustrated by the chemical bonding and structures that are formed. However, part of the experiment is chosen to illustrate an important difference (the tendency for catenation), because sulfur forms stable $-S-S-S-S-$ chains, whereas a chain containing more than two oxygen-oxygen linkages is *unknown.*

One of the objectives of this portion of the experiment is to prepare sulfur analogs of the familiar sulfate and carbonate anions. As a possible extension of this experiment, you may determine the purity of your potassium thiosulfate sample by an iodine titration of the thiosulfate ion.

The term *thio* is used in the name of a compound when an oxygen atom in a chemical species is replaced by a sulfur atom. The thio analogs to be prepared in this experiment are the thiosulfate and the trithiocarbonate ions. These species and their oxygen analogs may be represented as follows:

Sulfate ion	Thiosulfate ion	Carbonate ion	Trithiocarbonate ion

For each of the ions, only one of the resonance structures is pictured. The two carbonate ions are planar (using sp^2 hybridization at carbon), whereas the two sulfate ions are tetrahedral.

The thiosulfate ion can be synthesized by a reaction between the sulfite ion and sulfur that can be represented as follows:

$$8 \; |\overline{O} - \underset{|\underline{O}|}{\overset{|\overline{O}|}{S}}| \;^{2-} + S_8 \rightarrow 8 \; |\overline{O} - \underset{|\underline{O}|}{\overset{|\overline{O}|}{S}} - \overline{\underline{S}}| \;^{2-} \tag{1}$$

Sulfite ion + Sulfur → Thiosulfate ion

The trithiocarbonate ion can be synthesized by a combination of sulfide ion with carbon disulfide as:

$$|\overline{\underline{S}}| \;^{2-} + |\overline{S} = C = \overline{S}| \rightarrow |\overline{S} = C \underset{|\underline{S}|}{\overset{|\overline{S}|\;^{2-}}{<}} \tag{2}$$

Sulfide ion + Carbon disulfide → Trithiocarbonate ion

The thiosulfate ion can be converted into the tetrathionate ion $S_4O_6{}^{2-}$ (part 2) by oxidation with iodine (Eq. 3).

$$2 \; O - \underset{O}{\overset{O}{S}} - S \;^{2-} + I_2 \rightarrow O - \underset{O}{\overset{O}{S}} - S - S - \underset{O}{\overset{O}{S}} - O \;^{2-} + 2I^- \tag{3}$$

C. Group VIIA – Chlorine-Iodine

The oxyanions of the halogen family, containing the positive oxidation states of the halogen, are mild to very strong oxidizing agents. This oxidation chemistry is illustrated by the preparation of the iodate ion, $IO_3{}^-$, by oxidizing iodine (which is a mild oxidizing agent itself, as illustrated by Equation 4) with the chlorate ion, $ClO_3{}^-$.

$$I_2 + 2KClO_3 \rightarrow 2KIO_3 + Cl_2 \tag{4}$$

SYNTHESES OF PHOSPHORUS, SULFUR, AND CHLORINE COMPOUNDS

A. Syntheses of Phosphorus Compounds

1. Disodium Hydrogen Orthophosphate·12-Hydrate, $Na_2HPO_4 \cdot 12H_2O$

The raw material from which phosphorus and its compounds are prepared industrially is calcium phosphate, either as natural phosphate rock or as bone ash. For laboratory preparations, bone ash is preferable because it is nearly free of fluorides and iron compounds. The bone ash may be assumed to be ~80 percent $Ca_3(PO_4)_2$, 10 percent $CaCO_3$, and 10 percent inert material. Upon dissolution of bone ash in sulfuric acid, the following reactions may occur:

$$CaCO_3 + H_2SO_4 \longrightarrow CaSO_4 \downarrow + CO_2 + H_2O \tag{5}$$

$$Ca_3(PO_4)_2 + 3H_2SO_4 \longrightarrow 3CaSO_4 \downarrow + 2H_3PO_4 \text{ (soluble)} \tag{6}$$

$$Ca_3(PO_4)_2 + 2H_2SO_4 \longrightarrow 2CaSO_4 \downarrow + Ca(H_2PO_4)_2 \text{ (soluble)} \tag{7}$$

$$Ca_3(PO_4)_2 + H_2SO_4 \longrightarrow CaSO_4 \downarrow + 2CaHPO_4 \downarrow \tag{8}$$

The inert material does not react with the acid and can be filtered off with the insoluble calcium salts.

It can be seen from equations (5) to (8) that if all the phosphate is to be dissolved, a sufficient amount of sulfuric acid must be used to react with all the calcium carbonate and to provide more than 2 moles of sulfuric acid per mole of calcium phosphate. However, there must not be more than 3 moles of sulfuric acid per mole of calcium phosphate, as shown in equation (6), because the excess acid would remain in solution and cause the product to be contaminated with sodium sulfate after neutralization with sodium carbonate. A quantity of sulfuric acid is chosen, therefore, that is more than sufficient to complete equation (7) but insufficient to exceed equation (6).

After removal of the insoluble calcium sulfate, the filtrate is neutralized with sodium carbonate, which reacts according to equations (9) and (10).

$$H_3PO_4 + Na_2CO_3 \longrightarrow Na_2HPO_4 + CO_2 + H_2O \tag{9}$$

$$Ca(H_2PO_4)_2 + 2Na_2CO_3 \longrightarrow 2Na_2HPO_4 + CaCO_3 \downarrow + CO_2 + H_2O \tag{10}$$

An excess of sodium carbonate does not convert disodium hydrogen phosphate into trisodium orthophosphate, because the orthophosphate ion, $PO_4{}^{3-}$, is a stronger base than the carbonate ion, $CO_3{}^{2-}$. The ionization constants of the conjugate acids, $HCO_3{}^-$ and $HPO_4{}^{2-}$ are 4.8×10^{-11} and 3.6×10^{-12}, respectively. Because the position of equilibrium is always toward the weaker acid, the reaction represented by equation (11) is favored instead of the reverse reaction.

$$NaHCO_3 + Na_3PO_4 \longrightarrow Na_2HPO_4 + Na_2CO_3 \tag{11}$$

Procedure. Place 30 g. of bone ash (~80 percent $Ca_3(PO_4)_2$, labeled "for Exp. 22") in a 6-inch porcelain evaporating dish, add 30 ml. of distilled water, and stir the mixture until a thick paste is obtained. Stir it with a porcelain spatula and add 14.1 ml. 18 M H_2SO_4 *as rapidly as is possible without causing excessive heating.* Continue to stir the mixture vigorously, until it begins to stiffen. This usually requires 10 to 15 minutes. Add 150 ml. of cold distilled water and stir the mixture until a thin paste, free from lumps, is obtained. Filter the cold mixture on a Büchner funnel, using suction (Note 1). Wash the solid in the funnel with 15 ml. of cold water and then discard the precipitate. Combine the filtrates and add to them solid anhydrous sodium carbonate in small portions, with stirring, until effervescence no longer takes place and a drop of the solution turns phenolphthalein indicator solution pink (Note 2). This should require 16.5 to 18 g. of Na_2CO_3. Filter and evaporate the solution to about 80 ml. Transfer the clear liquid (filter while hot, if necessary) to a 4-inch crystallizing dish, cover it so that dust cannot fall into the dish, and allow it to stand until the next laboratory period so that a satisfactory crop of crystals is obtained. Decant the liquid from the crystals and rinse them with 5 to 10 ml. of distilled water.

Disodium hydrogen phosphate is efflorescent (Note 3). Such crystals should be dried as quickly as possible and protected from the carbon dioxide of the air. To do this, quickly press the disodium hydrogen phosphate crystals between pieces of filter paper until as much of the liquid as possible is absorbed. Then enclose the crystals in a tightly folded package of three or four layers of fresh filter paper and leave it in the desk for not more than 48 hours (Note 4). The layers of filter paper permit the water to be soaked up and also minimize diffusion of air through the paper with consequent contamination of the crystals by carbon dioxide absorption. When the product is dry, place it in a closed bottle until needed for the second part of this experiment.

Test 0.1 M KH_2PO_4, 0.1 M Na_2HPO_4, and 0.1 M Na_3PO_4 solutions with litmus paper. Explain and write equations for your observations. How can a compound be an acid salt and react alkaline to litmus? Check the pH of each solution with pH paper. Which is more basic? Why?

2. Synthesis of a Heteropoly Anion of Phosphorus

As the pH is lowered the oxyanions of Subgroups VB and VIB undergo self-condensation to form condensed (or *isopoly*) anions. For example, ammonium molybdate, the chief source of molybdenum compounds in the laboratory, has the composition $(NH_4)_6Mo_7O_{24}$. The highly symmetric anion has a central molybdenum atom in an octahedral hole of a "cage" formed by condensation of six molybdate ions so that each molybdenum atom has an octahedral MoO_6 unit sharing edges (Fig. 22-1). The central molybdenum atom, but apparently no others, may be replaced by other electronegative elements, such as iodine, iron, chromium, or cobalt, which can sustain the six-coordination with oxygen. The latter condensed species are referred to as *heteropoly* anions.

Just as it is possible to copolymerize organic compounds, e.g., 1,3-butadiene and styrene in synthetic rubber, one can also copolymerize some inorganic oxyanions. For example, condensed anions (and acids) of characteristic yellow colors are formed by the condensation of simple anions, such as AsO_4^{3-}, PO_4^{3-}, and SiO_4^{4-}, with a large but definite amount of molybdenum trioxide. Analogous heteropoly anions of vanadium and tungsten are also known.

The heteropoly anions (and acids) are usually described in terms of the number of molybdenum (or tungsten or vanadium) atoms associated with the other anionic atom, such as phosphorus, in the empirical formula. Three main classes of heteropoly species are now recognized: the 6-, the 9-, and the 12-polyacids. Apart from their uses as (1) good precipitating agents for high molecular weight cations, such as proteins, (2) agents for producing intense and excellent quality color toners for the printing industry, (3) catalysts for

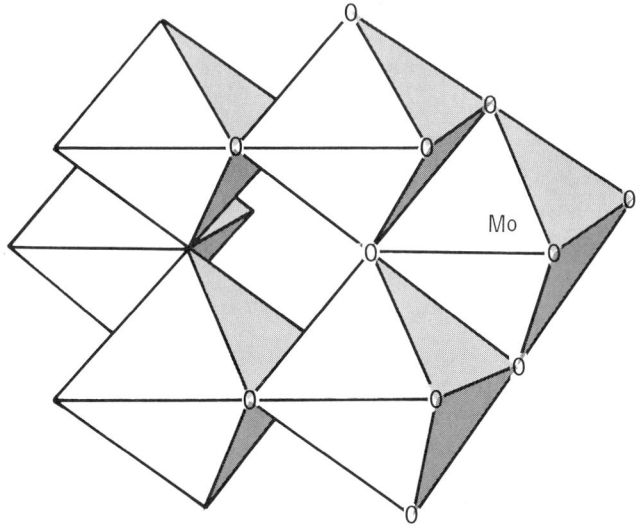

FIGURE 22-1

Arrangement of MoO_6 octahedra in a 6-acid.

hydrocarbon reactions, and (4) analytical reagents, the structures of these heteropoly species pose many problems. X-ray crystallography has recently provided a basis for systematically classifying the different series of isopoly and heteropoly compounds. Figure 22-2 shows the structure of 12-phospho-tungstic acid. The hetero-atom phosphorus occupies a tetrahedral hole in a "cage" formed by twelve WO_6 bridged octahedra. Figure 22-3 depicts how each oxygen atom of the PO_4^{3-} ion is bridged equally to three tungsten atoms. The $[PMo_{12}O_{40}]^{3-}$ salts have completely analogous structures.

In this experiment we shall prepare a yellow crystalline 12-phospho-molybdate salt, using the sodium dihydrogen orthophosphate. We shall also qualitatively test the effect of reducing agents on the color of the $[PMo_{12}O_{40}]^{3-}$ ion.

Procedure. Dissolve 6.0 g. of $Na_2HPO_4 \cdot 12H_2O$ in 30 ml. of 8 M nitric acid contained in a 125-ml. Erlenmeyer flask, and cool the solution to room temperature if necessary. Add 22.7 ml. of saturated ammonium molybdate solution (Note 5) an eyedropperful at a time, while stirring

Dissolve
$Na_2HPO_4 \cdot 12H_2O$.

Add molybdate
solution.

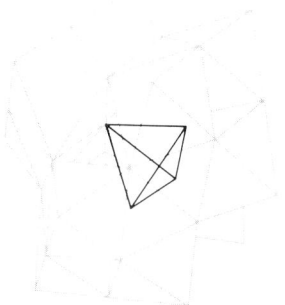

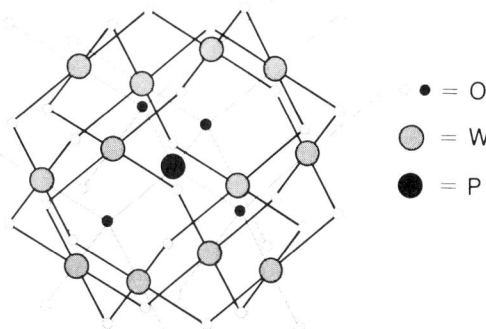

• = O

◉ = W

● = P

FIGURE 22-2

The structure of the $[PW_{12}O_{40}]^{3-}$ anion in the 12-polyacids. The PO_4 tetrahedron is shown with black lines, whereas the WO_6 or MoO_6 octahedra are drawn with gray lines.

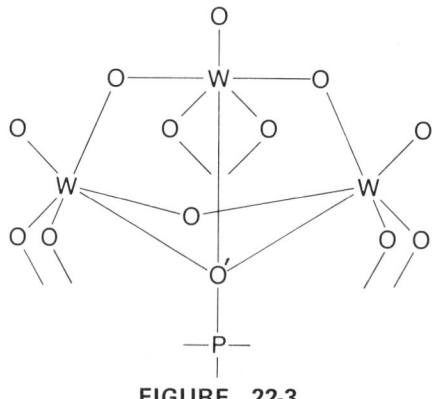

FIGURE 22-3

Illustration of how each oxygen of PO_4^{3-} is bridged to three WO_6 octahedra. (MoO_6 is analogous.)

Collect and dry the yellow compound.

with a glass rod. A deep yellow, microcrystalline salt should precipitate immediately. Collect the compound on a Büchner funnel (Note 1) and then wash it twice with 10-ml. portions of hot water to which a few drops of concentrated nitric acid have been added. Dry the crystals in air for 5 minutes and then dry them in the desiccator until the next laboratory period. Transfer the dry materials to a weighed sample vial and determine the yield.

Label the bottle and give it to your instructor, along with the remainder of your sample from Part 1 of this experiment.

Test for phosphate.

Qualitative Test for Phosphate. Dissolve a small crystal of disodium hydrogen phosphate in ~0.5 ml. of H_2O, make it acidic with a drop of nitric acid, and place two or three drops of the solution on a piece of filter paper. Add a drop each of ammonium molybdate and D-72 developer. A blue stain, with the intensity depending on the phosphate content, indicates the presence of phosphate.

The reduction of the simple and heteropoly anions of molybdenum and tungsten leads to bright blue colors, which have intense electronic absorption bands in the region of 7000 to 8000 Å. It is thought that the color results from reduction of approximately four of the 12 molybdenum atoms in the heteropoly anion, permitting some charge transfer to occur in which an electron on a Mo(V) atom jumps to a Mo(VI) atom within the "cage" structure of the heteropoly anion. Reducing agents include salts of Cu(I), Sn(II), Sb(III), Fe(II), Hg(I), and Tl(I).

Possible Research Extensions. If you are interested and have laboratory time available, several interesting additional studies could be undertaken with these materials.

1. You could investigate other reducing agents to obtain the "molybdenum blue" color.

2. You could construct an absorbance *vs.* concentration curve by using different volumes of a carefully prepared standard solution of $Na_2HPO_4 \cdot 2H_2O$ as a standard with a spectrophotometer after developing the blue color.

3. You may wish to obtain the free acid $H_3[PMo_{12}O_{40}]$ by ion-exchange of the ammonium salt (see ref. 2, p. 1701).

4. You may wish to prepare the nearly colorless phosphotungstate ion $[PW_{12}O_{40}]^{3-}$ (see ref. 2, p. 1720).

Check with your instructor before undertaking any of these suggested extensions.

B. Syntheses of Sulfur Compounds

1. Preparation of Potassium Thiosulfate, $3K_2S_2O_3 \cdot 5H_2O$

The sulfite ion, which contains sulfur in oxidation state IV, can be oxidized easily by either oxygen or sulfur to form the sulfate ion, SO_4^{2-}, or the thiosulfate ion, $S_2O_3^{2-}$, respectively, in which the formal oxidation state of the central sulfur atom is VI. The two sulfur atoms in $S_2O_3^{2-}$ are quite different chemically.

Procedure. Dissolve 15 g. of K_2SO_3 crystals in 60 ml. of distilled water contained in a 125-ml. Erlenmeyer flask. Add 3 g. of powdered sulfur and gently boil the suspension for about 45 minutes or until the solution is no longer alkaline to litmus paper. Add additional water if necessary.

Combine the reagents and heat to boiling.

Remove any undissolved or suspended material by filtration* and then concentrate the filtrate in a 100-ml. beaker to a volume of 15 ml. Cool the beaker in ice to induce crystallization. Potassium thiosulfate often forms supersaturated solutions; hence, scratching the bottom of the beaker with a glass rod or adding a small seed crystal of potassium thiosulfate may be necessary to obtain crystals. Collect the cold crystals by suction filtration as rapidly as possible to prevent redissolving of the crystals. Concentrate the filtrate to 5 ml. and cool it in ice to obtain an additional crop of crystals. The total sample should be 10 to 12 g. (Note 6).

Filter the mixture and evaporate the filtrate.

Collect the resultant crystals.

Remove the absorbed water from the crystals by pressing them between two pieces of filter paper; then dry the crystals on a watch glass in your desiccator until the next laboratory period.

Dry the crystals.

Determine the total weight of crystals and place each crop in a separate labeled bottle. Use the first crop for determination of purity. Calculate the percentage yield in the preparation on the basis of the amount of potassium sulfite used.

Determine % yield.

Dissolve about 0.5 g. of the product in 5 ml. of water and add 2 ml. of dilute hydrochloric acid. Carefully observe the odor of the gas and the color of the precipitate. What is the free acid corresponding to the salt, potassium thiosulfate? What can you conclude about the stability of this acid? Write equations in your laboratory notebook to represent the reaction of the thiosulfate ion with hydrochloric acid.

Test the compound in HCl.

2. Preparation of Potassium Tetrathionate, $K_2S_4O_6$

Procedure. Dissolve 7.9 g. $3K_2S_2O_3 \cdot 5H_2O$ in so little water that the solution is saturated; cool in ice. Dissolve 5.2 g. I_2 in 30 ml. of 95% C_2H_5OH and add the iodine solution drop by drop, with stirring, to the cold thiosulfate solution in a 100 ml. beaker. *Very vigorous stirring is needed during the addition.* The reaction is instantaneous. The tetrathionate, which is insoluble in ethyl alcohol, separates as small crystals or a lumpy solid. When the addition is complete, collect the solid on a Büchner funnel and wash it with 5-ml. portions of alcohol until the wash solution is free of iodine and iodide. (How can you detect the absence of iodine and iodide?) To purify the salt, redissolve it in a minimum of water at room temperature,

Dissolve the reagents and combine dropwise.

---

*The Büchner funnels must be washed *very* thoroughly if any work was done above with PO_4^{3-} solutions.

and reprecipitate it by stirring in three volumes of ethyl alcohol. The precipitate (small, shiny crystals) should be pure anhydrous potassium tetrathionate. Dry the crystals by pressing them between filter paper and then place them on a watch glass in the desiccator until the next laboratory period (Note 7).

Collect and dry the crystals.

Weigh the potassium tetrathionate crystals and calculate the percentage yield on the basis of the amount of potassium thiosulfate used. Place the sample in a labeled bottle, and give the sample to your instructor.

3. Preparation of Barium Trithiocarbonate, $BaCS_3$

Procedure.   Dissolve 15 g. of barium hydroxide in approximately 75 ml. (or less) of warm water in the following manner: add the barium hydroxide to 10 ml. of H_2O and heat slightly with stirring; add another 10 ml. of H_2O, heat and stir; repeat this process until all (or nearly all) the barium hydroxide dissolves. Divide this solution into two equal parts.

Dissolve barium hydroxide.

Hydrogen sulfide generators, like that shown in Figure 22-4, are in the hood, and they are to be used in the hood (Note 8). Warm the tube gently to liberate the hydrogen sulfide. Saturate one portion of the barium hydroxide solution with hydrogen sulfide (be sure the generator tube extends nearly to the bottom of the flask), and then add to it the other portion of the barium hydroxide solution. This gives a solution of barium sulfide.

Add H_2S to one-half of the barium hydroxide.

Filter the solution, using suction filtration, to remove undissolved barium hydroxide, barium carbonate (Note 9), and any elemental sulfur that may have formed. Wash the residue with a little distilled water until it is nearly colorless; let the washings run into the filtrate. Discard the residue.

Filter the solution.

At the hood, add 3 ml. of carbon disulfide, CS_2, (Note 10) to the barium sulfide solution, stopper the flask and shake vigorously for about ten minutes. *CAUTION*—the flask *must* be unstoppered *frequently* to prevent the build-up of pressure from vaporized carbon disulfide. During this process

Carefully add CS_2.

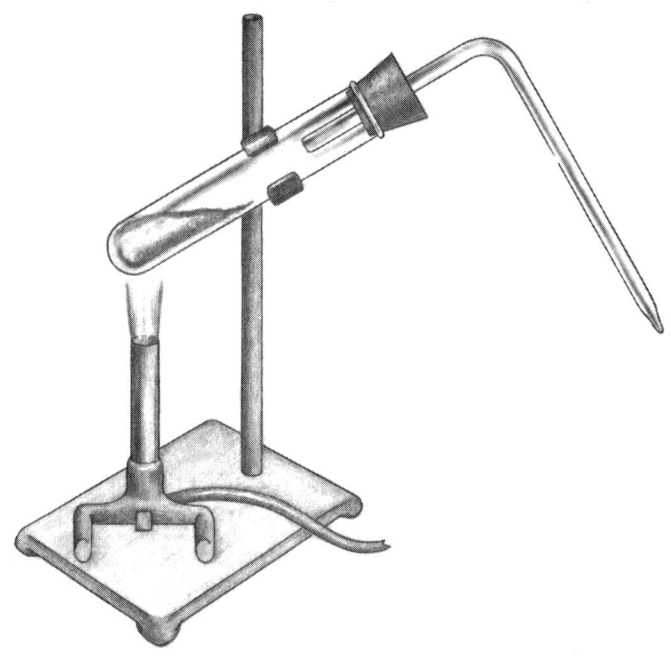

FIGURE   22-4

Apparatus for generating hydrogen sulfide.

barium trithiocarbonate will precipitate as a yellow crystalline powder. All the carbon disulfide should react.

Collect the product by filtration and return the filtrate to the flask. A second crop of crystals may be obtained by adding absolute alcohol to the filtrate (Note 11). Add approximately 10 ml. at a time until no more precipitate appears to form. Add these crystals to the first batch on the filter and wash them first with a few ml. of cold distilled water, then with 50% alcohol, and finally with absolute alcohol. The first washing removes unreacted barium sulfide; the alcohol wash removes water so the product can be dried more easily. Dry the yellow crystals between filter paper until the next lab period. Turn in the product in a labeled bottle.

Collect, wash, and dry the compound.

C. Synthesis of an Iodide Compound

Preparation of Potassium Iodate, KIO_3

Procedure. In a 250 ml. Erlenmeyer flask dissolve 6.75 g. of potassium chlorate in 25 ml. of water containing a few drops of concentrated nitric acid and heat to 45°C. Add five grams of iodine to the flask and warm the mixture gently *in the hood* until a reaction just begins. Cover the mouth of the flask loosely with an inverted beaker to prevent loss of iodine; provide a bath of ice water in case the reaction becomes too vigorous. The reaction is complete in about 15 minutes, as shown by the disappearance of the iodine color.

Weigh the reagents. Combine them and heat in the hood gently.

Filter the warm solution to remove any solids, and reduce the volume of solution to one-half by heating it in a beaker to *ca.* 50°C and blowing a stream of air across the surface of the liquid (Note 12). Set the solution aside to crystallize and cool in ice if necessary to obtain a good crop of crystals.

Filter the solution.

Collect the crystals on a filter and return the filtrate to the beaker. A second crop of crystals may be obtained by again evaporating the filtrate to one-half of its volume. Wash the crystals with 5 ml. of *cold*, distilled water. Dry the crystals in your desiccator until the next laboratory period. Turn in the crystals in a properly labeled bottle.

Collect and dry the crystals.

NOTEBOOK AND REPORT

Record in your notebook the quantities of reagents used for each synthesis and any observations that differed from the procedure described in this manual. Prepare a report on the experiment that includes a brief statement of the purpose and procedure, complete and balanced equations for all reactions, a statement about any difficulties encountered, and any modifications that you made in the procedure.

For *one* of the three syntheses, explain in terms of sound chemical principles and procedures why it was necessary to do each of the things specified in the procedure. For example, in the preparation of potassium thiosulfate: Why were potassium sulfite crystals dissolved in water? Why was the mixture boiled? Was the length of heating chosen the proper length of time? Why could sulfur separate from the solution? Why is the solution evaporated to 15 ml.? There are several other points to consider.

If you are not able to answer these questions, try to find the answers by thinking carefully about each step of the reaction, by going to the library (books on inorganic preparations may be useful), and by finally asking your instructor. Your instructor will not supply answers to the questions when you have enough information to arrive at a satisfactory answer without help.

CALCULATIONS

Indicate in your notebook and report the data used and the arithmetic steps needed to determine the moles of reagents and products and the percentage yield of the product.

NOTES

Note 1. See Experiment 2 for the set-up of the suction filtration apparatus.

Note 2. A 0.5 to 1 percent phenolphthalein indicator solution is used.

Note 3. An efflorescent substance is subject to changing throughout or on the surface to a powdery substance upon exposure to air, as a crystalline substance changing through loss of water of crystallization.

Note 4. If you do not intend to return to the laboratory within two days, store the solid in your desiccator.

Note 5. Commercial ammonium molybdate is generally $(NH_4)_6Mo_7O_{24} \cdot 4H_2O$; the saturated solution contains 430 g. per liter.

Note 6. Potassium thiosulfate is very soluble in water at room temperature; thus, the filtration must be done while the mixture is cold and as rapidly as possible.

Note 7. The potassium tetrathionate crystals are colorless and platelike or prismatic in shape; the pure dry material is stable for a very long time without change, but it decomposes if potassium thiosulfate or occluded mother liquor is present, producing a characteristic odor. When strongly heated, potassium tetrathionate decomposes to potassium sulfate, sulfur dioxide, and sulfur.

The compound potassium tetrathionate is readily soluble in water: 12.60 g. per 100 g. H_2O at 0°C; it is insoluble in ethyl alcohol. An aqueous solution of potassium tetrathionate undergoes disproportionation slowly to potassium trithionate, $K_2S_3O_6$, and potassium pentathionate, $K_2S_5O_6$.

Note 8. Hydrogen sulfide is toxic. The reaction should be performed in the hood.

Note 9. Barium carbonate may be formed (due to carbon dioxide, CO_2, from the air). This must be removed by filtration before continuing.

Note 10. CARBON DISULFIDE IS EXTREMELY FLAMMABLE; be sure there are no flames near where carbon disulfide is poured or near your work bench.

Note 11. Beware of careless technique when C_2H_5OH and flames are not far apart. The alcohol has been denatured — it is poisonous (and it *cannot* be made fit for consumption).

Note 12. The set-up for the apparatus for directing a stream of air across the surface is given on p. 17.

QUESTIONS–PROBLEMS

1. Using the preparative conditions, what controls the pH of the solution so that you can isolate the $Na_2HPO_4 \cdot 12H_2O$ compound rather than the $H_2PO_4^-$ or PO_4^{3-} ions in part A-1?

2. If the reduced heteropoly anion of molybdenum absorbs in the region of 6000 to 8000 Å, why is it blue?

3. What can you deduce about the absorption region in the yellow heteropoly anions? Is the energy of the absorption higher or lower than for the blue form? Why?

4. What is the average oxidation state of sulfur in the sulfite ion, thiosulfate ion, tetrathionate ion, and trithiocarbonate ion?

5. From a practical consideration, why is it preferable to use a slight excess of polysulfide rather than an excess of potassium sulfite in the synthesis of potassium thiosulfate?

6. What type of experiment could one do to prove that the central sulfur atom in $S_2O_3^{2-}$ did not come from the elemental sulfur used in the synthesis?

7. How do you explain the yellow color of barium trithiocarbonate, whereas barium carbonate, potassium thiosulfate, and potassium tetra-thionate are white crystals?

8. *Balance* the following equation that represents the reaction for the formation of the molybdenum heteropoly anion:

$$\_HNO_3 + \_Na_2HPO_4 + \_(NH_4)_6Mo_7O_{24} \cdot 4H_2O \rightarrow \_(NH_4)_3PMo_{12}O_{40} +$$

$$\_NaNO_3 + \_NH_4NO_3 + \_H_2O.$$

9. Acid decomposes barium trithiocarbonate to give an unstable oily substance, trithiocarbonic acid (H_2CS_3), which dissociates into carbon disulfide and hydrogen sulfide. Write the equations.

REFERENCES

1. Schlessinger, G.G.: *Inorganic Laboratory Preparations.* Chemical Publishing Co., New York, 1962, pp. 40, 50, and 66-68.
2. Brauer, G.: *Handbook of Preparative Inorganic Chemistry.* Academic Press, New York, 1963, pp. 399-400, 543-552, and 1698-1735.

Experiment 22
REPORT SHEET

Name _____   Section _____

Synthesis of   _____

A. State concisely what each step in the procedure accomplished in terms of physical transformations or chemical reactions.

B. Balanced Equation(s) for the Synthesis

C. Properties of the Product (color, crystal shape, etc.)

D. Calculation of yield(s)

   1. Theoretical yield (show calculations) _____

   2. Weight obtained (g) _____

   3. Percentage yield obtained _____

QUANTITATIVE ASPECTS
OF COORDINATION
CHEMISTRY

DETERMINATION OF THE FORMULA OF A COMPLEX ION BY A SPECTROPHOTOMETRIC METHOD

An Iron(III)-Thiocyanate Complex by Job's Method of Continuous Variation

OBJECTIVE

To use spectrophotometry to determine the presence and formulas of complex ions in solution.

EQUIPMENT NEEDED

Three burets; a pair of matched cuvettes or cells for the spectrophotometer; spectrophotometer; 10 small bottles or 25-ml. test tubes, fitted with stoppers.

REAGENTS NEEDED

Solution A, 3×10^{-3} M in $NH_4Fe(SO_4)_2$; solution B, 3×10^{-3} M in NH_4SCN; buffer solution, 0.25 M in $(NH_4)_2SO_4$ and 0.125 M in H_2SO_4.

INTRODUCTION

Suppose that two substances react to form a compound of unknown formula

$$A + nB \longrightarrow AB_n \qquad (1)$$

If the compound AB_n is stable, it is a simple matter to prepare it, analyze it, and determine its formula by fundamental stoichiometric methods, as in Experiments 4 and 5. Often, however, the compound is unstable and exists only in solution, and always in equilibrium with its components A and B. This is the case with many complex ions. Such ions cannot be isolated; they must be studied in solution, in the equilibrium mixture. To determine the formula of the complex ion, advantage may be taken of the fact that equilibrium is established.

The determination of the formula of the complex ion as it exists in solution is particularly simple if the compound is colored and its two components are colorless and if *one* of the possible complexes is considerably more stable than the others. (In such a case, the stable species is the only compound present in appreciable concentrations.) Thus, for the reaction

$$A \quad + \quad nB \;\rightleftharpoons\; AB_n \qquad\qquad (2)$$
$$\text{(colorless)} \quad \text{(colorless)} \qquad \text{(colored)}$$

the concentration of AB_n can be measured by the amount of light it absorbs. The change in concentration of AB_n, as the relative amounts of A and B are changed, can be used to determine the coefficient n in equation (2).

In this experiment the formula of the deep red iron(III)-thiocyanate complex will be determined by the method of continuous variations. This method involves measuring the absorbances of a series of solutions of varying composition at the wavelength at which the complex has maximum absorptivity. The *total number of moles* of Fe^{3+} and SCN^- in a given volume of the solution is kept constant while the *molar ratios* of the two reactants are varied. The absorbances are plotted against the fraction of the total moles represented by one reagent and the values of n are determined from the fraction at which the maximum absorbance occurs. For example, in Figure 23-1 the maximum absorbance is at 0.75; thus, the formula of the complex is MX_3 since this solution contains 0.25 moles of M^{m+} ion for each 0.75 moles of X^{x-} ion. If the complex is somewhat unstable, the plots are curved near the maximum as in Figure 23-1. In this case, one should extend the straight-line portions of the graph until they cross to determine the mole fraction at maximum absorbance.

Principles of Spectrophotometry[1]

We shall discuss briefly some important principles of spectrophotometry and equipment for the spectrophotometric technique.

Laws of Absorption. A property possessed by a beam of radiation is its *intensity* or *radiant power,* which is defined as the number of photons that strike a unit area per unit time. If the radiation is partly absorbed, the radiant power (P) measured by detectors, such as photocells or phototubes, is less than the incident power (P_o). The ratio of the radiant power transmitted through a sample to the radiant power incident on the sample is the transmittance, T.

$$T = P/P_o \qquad\qquad (3)$$

The transmittance is often expressed as a percentage by multiplying by 100,

$$\%T = T \times 100$$

---

[1]The terminology and abbreviations used in this section have been endorsed by the advisory board members of *Analytical Chemistry* in an attempt to obtain consistency in the field.

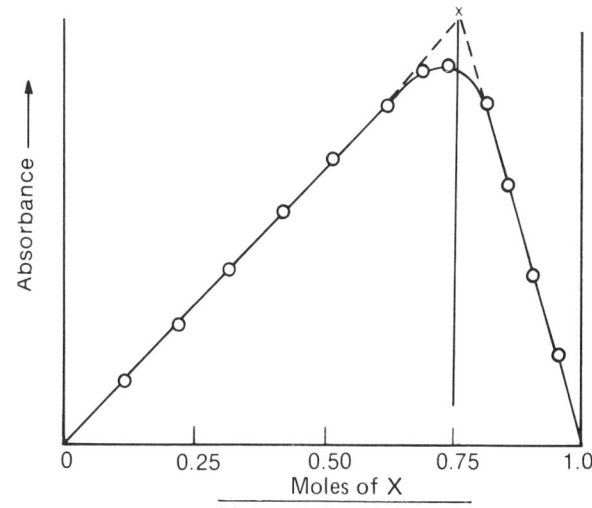

FIGURE 23-1

Absorbance as a function of mole fraction.

The logarithm to the base 10 of the reciprocal of the transmittance is the absorbance.

$$A = \log_{10}(1/T) = \log_{10}(P_0/P) \tag{4}$$

The absorbance is more useful than the transmittance for quantitative work because it is directly proportional to concentration. If a monochromatic beam of radiation enters an absorbing medium, the rate of decrease in radiant power is proportional to the power of the incident radiation; that is, the light intensity is diminished in a geometric (not arithmetic) or exponential progression. Thus,

$$\frac{-dP}{P} = k\,db \tag{5}$$

Upon integrating and changing to logarithms of base 10, and putting $P=P_0$ when $b=0$, one obtains

$$2.303 \log(P_0/P) = kb \tag{6}$$

This is equivalent to stating that the radiant power of the unabsorbed light decreases exponentially as the thickness (b) of the absorbing medium increases arithmetically.

Beer's Law. This relationship indicates that the radiant power of a beam of monochromatic radiation decreases in a similar manner as the concentration of the light-absorbing constituent increases. Thus,

$$2.303 \log(P_0/P) = k'c \tag{7}$$

Equations (6) and (7) may be combined and written as a single equation:

$$\log(P_0/P) = \epsilon bc \tag{8}$$

$$P = P_0 10^{-\epsilon bc} \tag{9}$$

where P = power or intensity of radiation after passage through a sample.

P_o = power or intensity of radiation before passage through a sample. (It is assumed here that the solvent does not absorb radiation in the region discussed.)

ϵ = a constant for a given absorbing species, solvent, and temperature, dependent principally upon the intrinsic nature of the absorbing species, and having a value dependent upon the wavelength, called the molar absorptivity.

c = the (stoichiometric) concentration of the absorbing species (moles/liter).

b = the radiation path length (centimeters) through the sample.

Absorbance is the product of the molar absorptivity, the optical path length, and the concentration of the absorbing species.

$$A = \epsilon bc \tag{10}$$

From equation (10) it is evident that for a given absorbing species at a specific wavelength and for a given sample holder *(cuvette),* of fixed b, a plot of A against c is a straight line if Beer's law is obeyed. Hence, the applicability of Beer's law can be determined by measuring the absorbance as a function of concentration, using the same cuvette for all measurements, and examining the linear character of the plot. Furthermore, after the slope of the line, ϵb, has been determined, the concentration in any unknown solution can be calculated by measuring the absorbance and dividing this value by the slope. This is essentially the procedure of spectrophotometric analysis.

The Spectrophotometer. A spectrophotometer is a device for producing electromagnetic radiation (light) of rather precisely defined frequency (or wavelength), this frequency being continuously variable over a range, and for measuring the intensity of the resulting radiation, usually after its passage through a sample, the absorption characteristics of which are under investigation. Thus, the basic components of a spectrophotometer are a radiation source, a dispersion element, a selection element, a sample holder, and a detection element. These are briefly described here for the visible and infrared regions of energy.

Radiation Source. The source of radiation must have *continuous* output in the range of frequencies (or wavelengths) of interest, and it must be relatively constant in intensity over the range of interest. The output must be relatively stable, i.e., not vary measurably with time, and the output must be appropriately intense. A tungsten filament or a mercury vapor lamp is commonly used as the source in the visible region and a solid rod (globar) heated to glowing is useful for the infrared region.

Dispersion Element. This element resolves the continuous source output into narrow ranges of essentially monochromatic radiation, i.e., radiation of a single wavelength or frequency. There are two prominent types of dispersion elements: (a) prisms made from materials selected because of their absorption properties in the range of wavelengths to be dispersed, e.g., rock salt or other alkali halide prisms for the infrared region and glass prisms for the visible region, and (b) diffraction gratings (with spacings of rulings depending upon the range of wavelengths to be dispersed). The dispersion by prisms is due to *refraction,* which occurs because the velocity of radiation in a material medium depends upon its frequency (or wavelength), whereas dispersion by gratings depends upon *diffraction* followed by constructive and destructive interference of the diffracted beams.

Selection Element.   This element selects successively the narrow ranges of essentially monochromatic radiation produced by the dispersion element and transmits the selected radiation to the sample, which is fixed in space by the sample holder. A scan by the selection element should be approximately uniform in rate of passage through the spectral region covered. The rate of scan should be variable and conveniently repeatable and reproducible; i.e., the dial reading giving the selected wavelength must always be the same for any given wavelength. (It may be noted that a unit combining a dispersion element and a selection element is called a monochromator.)

Sample Holder.   This device positions the sample in the selected beam of radiation and consists of a cell holder and a cell. Clearly, the cell must be quite transparent to the radiation of interest, it should not have strong absorption peaks of its own, and it should provide a definite path length for the radiation to follow (particularly for quantitative analytical determinations) through the sample, because the extent of absorption depends, among other things, upon how many molecules the beam encounters. The material for cell "windows" thus varies with the spectral region under study. For example, for the visible region the cells are made of glass, whereas for the ultraviolet region they are made of quartz. In order that the radiation path be well defined, and for other reasons as well, these cells are built with plane, parallel faces. For less precise work, selected cylindrical cells may be used.

Detection Element.   This is a composite element, consisting of the transducer, e.g., a thermocouple or a photocell, and an amplification element, which converts the weak detected signal into a much stronger derived signal.

In the Bausch & Lomb Spectronic 20, 70, and 100 and the Coleman Models 6A and 6C spectrophotometers, the *radiation source* is an incandescent tungsten filament. In the Spectronic 20 the light is focused on the objective lens by the field lens. It is then reflected and dispersed by the diffraction grating *(dispersion element).* In order to obtain the desired wavelength, the wavelength cam *(selection element)* is turned. Light of the selected wavelength passes through the exit slit, and this monochromatic light passes through the *sample holder* (cell) and strikes the *detection element,* a phototube detector in which the light energy is converted to an electric signal for an amplifier, which increases the strength of the signal for reading on the meter. (See Fig. A-II-2 for a schematic diagram of the Spectronic 20.)

Analysis by Spectrophotometry. Instruction will be given in the laboratory concerning the proper operation of the spectrophotometer. *Read Appendix II before class.* The instrument is a carefully constructed (and expensive) optical and electronic device. MAKE NO ADJUSTMENTS THAT YOU DO NOT THOROUGHLY UNDERSTAND.

Techniques of Spectrophotometry

First select a pair of cuvettes that have approximately the same absorption and scattering characteristics, so that a proper comparison of the absorption of the iron(III)-thiocyanate solutions and the absorption of the solvent can be made. Examine several of them and select two that, when filled with solvent and placed in the properly set instrument, each show 100 percent transmittance when the aperture in front of the light source is adjusted so that the meter reading is 100 percent for one of them. In using these cuvettes during the experiment, always carefully place them in the sample holder in the same orientation (etch mark toward the front). Precautions of cleanliness, no fingerprints, and so forth, are, of course, necessary. It is advisable to wipe each cuvette with lens paper just before inserting it in the sample holder. It is probably wise also to check the pair of cuvettes for likeness with the same solution in each, using a solution to give a transmittance of about 30 percent to see that both read the same. Thereafter, always use one as the reference cell and the other as the solution cell. *Mark them* high on the neck above the area where the light will pass through the cell.

The instrument gives the least error in the concentration of solute for a given error in the measurement of %T (the percentage transmittance on the meter) when %T is in the middle range, e.g., 20 to 60 or 70 percent. Hence, to take advantage of the precision of the instrument, solutions should be diluted so that their absorbance values are in this range.

Having chosen matched cuvettes, it is then normally necessary to choose a wavelength for the absorbance measurements. The wavelength chosen should be one for which the molar absorptivity is high, i.e., ϵ large in equation (10), so that a slight change in concentration makes a great change in the absorbance, and it should be one for which the molar absorptivity changes as little as possible with a slight change in wavelength, so that the unavoidable mis-setting of the wavelength dial in repeated experiments does not result in a significant change in ϵ. Both these criteria can be satisfied by choosing the wavelength that results in maximum absorbance by the chemical species being studied. At the maximum, the curve of absorbance versus wavelength is broad. For this reason, a slight error in setting the wavelength dial results in only a slight change in ϵ. But in Part A of this experiment, you do not know the identity of the light absorbing species, and in fact the solutions may contain more than one light-absorbing species. Thus, in order to simplify your experimental problems, the wavelength resulting in maximum absorbance by the predominant iron(III)-thiocyanate complex has been found for you and is 450 millimicrons.

Quantitative Basis of the Experiment

The equilibrium constant for the reaction that is represented by equation (2) is

$$K = \frac{[AB_n]_{eq}}{[A]_{eq}[B]_{eq}^n} \tag{11}$$

Suppose that a series of experiments has been arranged in which the sum of the initial concentrations (before reaction) of A and B is always the same, and equal to some value c. Let the initial concentration of A be x; then the initial concentration of B is c−x. When equilibrium is established, let the concentration of AB_n be y. Using information from the chemical equation (2), we then have

$$[AB_n]_{eq} = y \tag{12}$$

$$[A]_{eq} = x-y \tag{13}$$

$$[B]_{eq} = c-x-ny \tag{14}$$

and the equilibrium constant becomes

$$K = \frac{y}{(x-y)(c-x-ny)^n} \tag{15}$$

The concentration of AB_n, that is, y, is proportional to the absorbance of the solution, by Beer's law,

$$A = \log \frac{P_o}{P} = \epsilon b y \tag{10}$$

if AB_n is the only absorbing substance. Hence, we can measure changes in the concentration of AB_n, even though we may not know the value for ϵ, needed to determine its actual concentration, by equation (16).

$$y = \frac{A}{\epsilon b} \tag{16}$$

Equation (16) shows that when the absorbance is zero, y is zero, and when the absorbance is a maximum, y is a maximum. A plot of the absorbance of various solutions against the values of x (the initial concentration of reactant A) for those solutions will, thus, have the same shape as a plot of y against x.

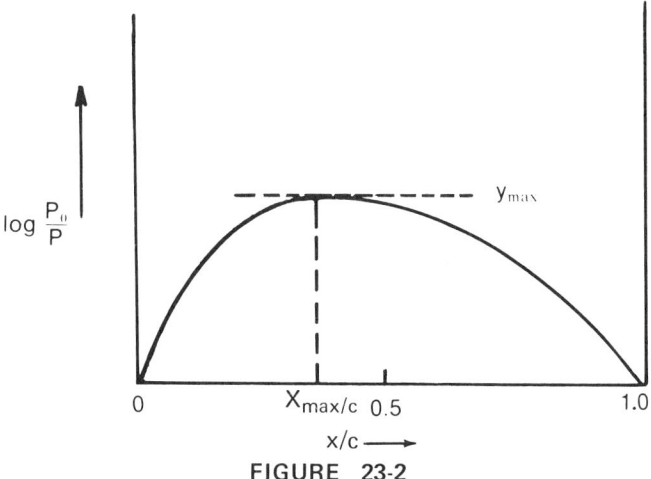

FIGURE 23-2

Absorbance as a function of concentration of A for a mixture of A and B which forms only one complex AB_n.

It can be shown that the maximum on such a plot appears at the point where the value of x is given by

$$x_{max} = \frac{c}{n+1} \tag{17}$$

regardless of the value for K (Fig. 23-2).

PROCEDURE

Prepare the ten solutions.   Make 10 solutions by mixing reagents in the following quantities, (to 0.02 ml.) using the three burets,

Solution Number	1	2	3	4	5
Solution A	0 ml.	1 ml.	3 ml.	5 ml.	7 ml.
Solution B	16 ml.	15 ml.	13 ml.	11 ml.	9 ml.
Buffer	4 ml.	4 ml.	4 ml.	4 ml.	4 ml.

Solution Number	6	7	8	9	10
Solution A	9 ml.	11 ml.	13 ml.	15 ml.	16 ml.
Solution B	7 ml.	5 ml.	3 ml.	1 ml.	0 ml.
Buffer	4 ml.	4 ml.	4 ml.	4 ml.	4 ml.

Determine absorbance of each solution.   and determine their absorbance in the spectrophotometer at 450 millimicrons,[1] using a solution of 4 ml. of buffer diluted to 20 ml. with distilled water in the reference cell. Rinse the sample cell two or three times with small amounts of each new solution before taking the absorbance reading. Discard the wash solutions. This procedure should assure that the concentration of the solution in the cuvette is the same as that in the bottle.

Note that solution 10 has slight absorbance at this wavelength, so that any unreacted iron(III) ion in the other solutions will also contribute to the absorption (i.e., the assumption in equation (2), that reagent A is colorless, is not correct). To correct for this, we may make use of the fact that the total absorbance of the solution is the sum of the absorbances of the individual absorbing species, i.e.,

$$\text{Absorbance} = \epsilon_i c_i b + \epsilon_j c_j b + \epsilon_k c_k b + \ldots \tag{18}$$

If the absorbance of solution 10 (where [B] = 0 and the complex ion is, of course, not present) is A_{10}, we have

$$A_{10} = \epsilon_{Fe} c_{10} b = \epsilon_{Fe} \left(\frac{MV_{10}}{20} \right) b \tag{19}$$

where c_{10} is the concentration of iron(III) in solution 10, V_{10} is the volume (milliliters) of solution A in solution 10, and M the molarity of solution A.

---

[1] Recent nomenclature recommendations include the designation that wavelength be measured in meters, i.e., 450 millimicrons would be 450 nanometers. Since chemists are more accustomed to the term millimicrons, we shall continue using it in this experiment.

For any other solution the absorbance due to iron(III) may then be approximated as $A_{10} \dfrac{V_x}{V_{10}}$ where V_x is the volume (milliliters) of solution A, used in solution x.

Note that this is an overcorrection, because in all solutions other than solution 10, some iron(III) has been removed from the solution by reaction (2). How much is thus removed depends upon the (unknown) magnitude of K, but the absorbance in the other solutions due to the complex should be approximately equal, from equation (18), to

$$A_{complex} = A_{measured} - \frac{A_{10} V_x}{V_{10}} \qquad \textbf{(20)}$$

Correct the measured absorbances according to equation (20), and plot $A_{complex}$ against x, or against some variable that is proportional to x. If more data are needed in order to define the maximum clearly, interpolate other solutions near the maximum. Find x_{max} and calculate the formula of the complex

$$Fe(NCS)_n^{(3-n)+}$$

NOTEBOOK AND REPORT

Write a brief report of the experiment, showing the tabulated data and representative calculations. Include the graph showing the absorbance plotted against mole fraction. Include sufficient discussion to answer any of the following questions assigned by the instructor.

QUESTIONS—PROBLEMS

1. What are some of the problems that would be encountered if you attempted to determine the formulas of the other iron(III)-thiocyanate complexes in equilibrium with $Fe(NCS)_n^{(3-n)+}$ by this method of continuous variations?

2. Although we have written the formula of the iron(III)-thiocyanate complex as $Fe(NCS)_n^{(3-n)+}$, in considering the normal coordination number of Fe^{3+} and the fact that it is in water, what is probably a more correct formula for the complex?

3. Another spectrophotometric technique for determining the formula of a complex is the *mole-ratio* method, involving a series of solutions in which the concentration of one reagent, such as Fe^{3+}, is constant and that of the other, such as SCN^-, is varied. The absorbances are plotted against the *ratio* of the moles of SCN^- to Fe^{3+}. What type of plot would you expect in this iron(III)-thiocyanate case? What difficulties could also be incurred with this method owing to the other iron(III)-thiocyanate complexes in equilibrium?

294

292 Experiment 23

4. Derive equation (17), starting with equation (15). Procedure: Consider y as a function of x. Remember that at the maximum, the derivative of y with respect to x is zero. Write equation (15) in the form

$$y = K(x-y)(c-x-ny)^n \tag{21}$$

and differentiate implicitly. Set the derivative equal to zero, and solve the resulting equation for n in terms of c and x_{max}.

REFERENCES

1. Day, R.A., and Underwood, A.L.: *Quantitative Analysis.* 2nd edition. Prentice-Hall, Englewood Cliffs, N.J., 1967, Chapter 12.
2. Carmody, W.R.: *J. Chem. Ed., 41:* 615 (1964).

Experiment 23
REPORT SHEET

Name _____ Section _____

Initial Concentrations and Absorbances of Solutions (450 millimicrons)

Solution No.	1	2	3	4	5
$x = [A]_{initial}$					
Total Absorbance					
Absorbance due to Fe^{3+} $(A_{10} \cdot V_x / V_{10})$					
Absorbance due to complex ion					

Solution No.	6	7	8	9	10
$x = [A]_{initial}$					
Total Absorbance					
Absorbance due to Fe^{3+} $(A_{10} \cdot V_x / V_{10})$					
Absorbance due to complex ion					

Sample Calculations

Formula of the Complex Ion

DETERMINATION OF THE FORMULA OF A COMPLEX ION BY A TITRATION PROCEDURE

A Silver-Ammonia Complex

OBJECTIVE

To determine the formula of a complex ion using the principles of chemical equilibrium and a titration procedure.

EQUIPMENT NEEDED

Five 250-ml. Erlenmeyer flasks; three burets; 10-ml. pipet; black paper or other dark background for observing the formation of a white precipitate (cloudiness) in the solution; a source of bright light (incandescent lamp) to aid in observing the cloudiness.

REAGENTS NEEDED

Solution A, 0.020M $AgNO_3$; solution B, 2.00M aqueous NH_3; solution C, 0.010M KBr.

INTRODUCTION

The background information needed for this experiment is similar to that needed in Experiment 23, and the first paragraph of the introduction to that experiment is repeated here.

Suppose that two substances react to form a compound of unknown formula

$$A + nB \longrightarrow AB_n \tag{1}$$

If the compound AB_n is stable, it is a simple matter to prepare it, analyze it, and determine its formula by fundamental stoichiometric methods, as in Experiments 4 and 5. Often, however, the compound is unstable and exists only in solution, and always in equilibrium with its components A and B. This is the case with many complex ions. Such ions cannot be isolated; they must be studied in solution, in the equilibrium mixture. To determine the formula of the complex ion, advantage may be taken of the fact that equilibrium is established.

We have, as the equilibrium constant for reaction (1),

$$K = \frac{[AB_n]_{eq}}{[A]_{eq}[B]_{eq}^n} \qquad (2)$$

If we make a solution by adding known amounts of A and B to water, and measure the equilibrium concentration of A, we can calculate n as the integer that makes K a constant when the equilibrium concentrations of B and AB_n satisfy the stoichiometry of the chemical equation, (1).

The equilibrium concentration of A may be determined in several ways. One method, which we shall use in this experiment, is to add a third substance, C, which reacts with A to produce an insoluble substance, as shown in equation (3).

$$A + C \rightleftharpoons AC(s) \qquad (3)$$

We must determine the least concentration of C that will precipitate A from its solution in equilibrium with B and the complex AB_n. In particular, if both A and C are ions, the equilibrium, equation (3), is represented by a solubility product constant. If the value for this is known and if the concentration of C just sufficient to establish equilibrium (3) is measured, $[A]_{eq}$, and hence K, equation (2), can be calculated. The determination of n, however, is a simpler task than the determination of K, and this can be done even when K_{sp} is not known.

In this experiment we shall determine the formula of the complex formed between silver ion (A), and ammonia (B), using bromide ion as the precipitant (C). The basic reactions are:

$$Ag^+ + nNH_3 \rightleftharpoons [Ag(NH_3)_n]^+$$

and

$$Ag^+ + Br^- \rightleftharpoons AgBr(s)$$

Quantitative Relationships. Consider the equilibria, equations (1) and (3), and the equilibrium constant K, equation (2), and define the solubility product constant K_{sp} by

$$K_{sp} = [A]_{eq}[C]_{eq} \qquad (4)$$

Prepare a solution initially of concentration a in A and b in B, with the concentration of B very large compared to that of A. In such a case the position of equilibrium in equation (1) is far to the right. It is mathematically convenient to presume that reaction (1) goes to completion, and that the concentration a of AB_n thus formed dissociates to form a small but unknown concentration d of free A. From the stoichiometry of reaction (1) we have

$$[AB_n]_{eq} = a-d \qquad (5)$$

$$[A]_{eq} = d \qquad (6)$$

$$[B]_{eq} = b-n(a-d) \qquad (7)$$

If the reaction had gone to completion the a moles of A added would have formed a moles of AB_n, reacting with na moles of B in so doing, and the solution would have had the concentration a of AB_n and (b−na) of B. But AB_n dissociates to produce d moles of A and leave a−d moles of AB_n. Production of d moles of A results in nd moles of B, and the equilibrium concentration of B is [(b−na) + (nd)] as in equation (7).

To the solution at equilibrium we add reagent C until the solubility product constant, equation (4), is just exceeded and precipitation begins. Let the concentration of C at this point be c, so that, substituting in equation (4)

$$d \times c = K_{sp} \tag{8}$$

It is readily shown that, under the condition b ≫ a, plot of the logarithm of [C] at the start of precipitation against the logarithm of b for each of several experiments with constant a must give a straight line of slope n, whatever values K and K_{sp} may have.

This conclusion is justified as follows: One can write equation (2) in terms of a, b, and d, using equations (5), (6), and (7). Thus,

$$K = \frac{[AB_n]_{eq}}{[A]_{eq}\,[B]_{eq}^n} \quad \text{becomes} \quad K = \frac{a-d}{d[b-n(a-d)]^n} \tag{9}$$

However, because the position of equilibrium in equation (1) is far to the right (large excess of B), d ≪ a. Furthermore, from the initial composition of the solutions, a ≪ b. Using these inequalities, we approximate:

$$a - d \cong a$$

$$b - n(a-d) \cong b - na \cong b$$

and equation (9) becomes

$$K = \frac{a}{d \times b^n} \tag{10}$$

but from equation (8)

$$d = \frac{K_{sp}}{c}$$

and

$$K = \frac{c}{b^n} \times \frac{a}{K_{sp}} \tag{11}$$

If the concentrations are chosen so that concentration a has the same value in all experiments, a/K_{sp} is a constant. Taking the logarithm of both sides of equation (11), one obtains

$$\log K = \log c - \log b^n + \log \frac{a}{K_{sp}} \tag{12}$$

Log K and log $\dfrac{a}{K_{sp}}$ are both constants and may be combined to form another constant, log K'. Thus,

$$\log K' = \log c - \log b^n$$

$$\log K' = \log c - n \log b \qquad \textbf{(13)}$$

$$\log c = n \log b + \log K' \qquad \textbf{(14)}$$

If we let log c equal some number y, log b equal some number x, and log K' equal some constant q, we obtain equation (14) in the following form:

$$y = n x + q \qquad \textbf{(15)}$$

This is the form of an equation for a straight line where n is the slope of the line. Thus, a plot of log c against log b for a series of experiments at constant a should give a straight line with slope that is the subscript in the general chemical formula AB_n. In this experiment it is the number of ammonia molecules coordinated (chemically bound) to the silver ion.

PROCEDURE

The experimental procedure involves performing each titration twice. In the first set, you have a total volume of 100 ml., i.e., 10 ml. of solution A, 10 ml. of B, and 80 ml. of H_2O, before adding any potassium bromide solution. In the second set, you are to adjust the amount of water added with reagent B so that the *total volume of the solution after titration* is as close to 100 ml. as possible. How will you deduce the amount of water to add?

Carefully measure reagent solutions.

1. Into each of five marked Erlenmeyer flasks, pipet 10 ml. of solution A. Add to these, from separate burets, solution B and distilled water in the amounts specified (measured to 0.05 ml.).

Solution Number	1	2	3	4	5
Solution B	10 ml.	15 ml.	20 ml.	30 ml.	40 ml.
Water	80 ml.	75 ml.	70 ml.	60 ml.	50 ml.

Titrate the solutions.

Mix thoroughly by swirling the flask, and titrate the contents of each flask with solution C until the first *persistent cloudiness* (due to the silver bromide precipitate) appears; record the volume of solution C needed. You should judge when the solution becomes cloudy by comparing it to 100 ml. of H_2O in a 250-ml. Erlenmeyer flask under similar lighting and background conditions.

Repeat titrations to arrive at 100 ml. total volume.

2. Repeat the experiment by adding 10 ml. of solution A and the indicated volumes of solution B as before, but adding to each flask only enough water so that the final volume *after* titration will be as close as possible to 100 ml. You must estimate from the previous titrations the amount of water needed in each flask so that the total volume will be

approximately 100 ml. after adding solution C. Titrate with solution C again to the first persistent cloudiness and record the volume of solution C.

CALCULATIONS

For the first experiment, plot the logarithm of the volume of solution C against the logarithm of the volume of solution B used and draw a smooth curve through the points (Graph 1).

For the second experiment, calculate the initial concentrations of silver ion and ammonia [a and b, equation (7)] and the concentration of bromide ion [c, equation (8)] for each flask. Plot log c against log b and draw a smooth curve through the points (Graph 2). Determine n from the slope.

If the solubility product constant of silver bromide is 5.2×10^{-13}, estimate, from the information given on Graph 2 and other data from this experiment, the value for K for the formation of the silver ammonia complex ion.

NOTEBOOK AND REPORT

Write a report of the experiment, showing the graphs, tabulated data, and representative calculations, including answers to any of the following questions assigned by the instructor. Report the value for n to three significant figures and the equilibrium constant K to two significant figures. Comment on similarities or differences in the appearance of Graphs 1 and 2.

QUESTIONS—PROBLEMS

1. How do you interpret the observation that more potassium bromide is required to cause precipitation as more ammonia is added to each successive flask?

2. Is Le Chatelier's principle illustrated in this experiment? How?

3. If you used a 6 M instead of a 2 M solution of ammonia, would the titration require more or less bromide solution to reach the point of cloudiness? Why?

4. Would you expect silver bromide to be more (or less) soluble in aqueous ammonia solutions than in pure water?

5. Ammonia reacts with water to form ammonium ion and hydroxide ion; thus, the ammonia concentration is less than that calculated from the concentration indicated on the bottle. Use the ionization constant K_b of ammonia to calculate the effect of this on your results.

6. Silver also forms a 1:1 complex of formula $[AgNH_3]^+$. Would you expect the concentration of that complex to be greater in dilute (0.1 M) or in concentrated (2 M) ammonia solutions? Explain.

REFERENCES

1. Derr, P.F., Stockdale, R.M., and Vosburgh, W.C.: *J. Am. Chem. Soc., 63:*2670 (1941).
2. Vosburgh, W.C., and McClure, R.S.: *J. Am. Chem. Soc., 65:*1060 (1943).
3. Wolfenden, J.H.: *J. Chem. Ed., 36:*490 (1959).

**Experiment 24
REPORT SHEET**

Name _____ Section _____

Set I: Volume of Titrant Added

Flask No.	1	2	3	4	5	
Volume of Titrant Added						

Set II: Amount of Water, Concentration of NH_3 and Volume of Titrant Added

Flask No.	1	2	3	4	5	
ml of H_2O added						
b = $[NH_3]$ total						
Volume of titrant added						

Sample Calculations

Formula of the Complex Ion

Attach Answers to Problems

ELECTRONIC SPECTRA OF SOME
NICKEL(II) COMPLEXES
Determination of the Ligand Field
Parameters Dq and B

OBJECTIVE

To examine the spectral effects of changing the ligands bonded to a metal ion and to establish a spectrochemical series that reflects an increasing ligand field.

EQUIPMENT NEEDED

One 5-ml. pipet; four 25- or 50-ml. Erlenmeyer flasks; spectrophotometer; cuvettes for absorbance measurements.

REAGENTS NEEDED

Solution of 0.05M aqueous $Ni(NO_3)_2$; 0.05M $Ni(NO_3)_2$ dissolved in dimethyl sulfoxide; concentrated ammonium hydroxide; ethylenediamine.

INTRODUCTION

The concepts of crystal field and ligand field theory have been used successfully during the past 20 years in discussing the magnetic and spectral properties of transition metal complexes. In crystal field theory the complex is treated as though the only interaction between the central metal atom and its set of nearest molecules or ions — both of which we shall call *ligands* — is purely electrostatic. To a first approximation the orbitals of the central ion are considered as separated from the ligand orbitals.

To appreciate the effect of a crystal field, imagine that a symmetrical group of ligands is brought up to a charged ion from a far distance. First, the electrostatic repulsions between the ligand electrons and those in the *d*-orbitals of the metal increases the energy of all five *d*-orbitals equally. Then, as the ligands approach to within bonding distances, the repulsion interactions take on a directional character and vary with the particular *d*-orbitals under consideration, because of the different shapes and orientations of the five *d*-orbitals in space along a Cartesian coordinate system (ref. 1).

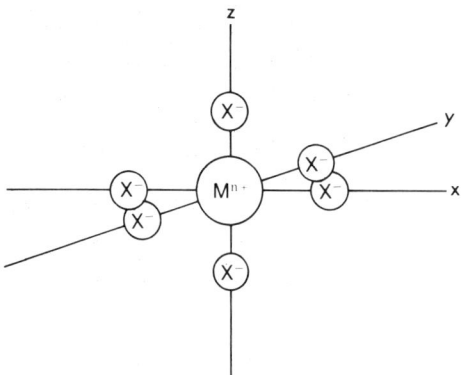

FIGURE 25-1

Six X⁻ ions arranged octahedrally around a central M^{n+} ion.

Consider the specific case of a metal ion, M^{n+}, at the center of an octahedral set of anions, X^- (Fig. 25-1). The orbitals that point directly along the Cartesian coordinate axes toward the ligands ($d_{x^2-y^2}$ and d_{z^2}) experience a greater repulsion interaction than do the orbitals that point between the ligands (d_{xy}, d_{xz}, d_{yz}). In terms of orbital energy levels, the original set of five degenerate orbitals is split into two sets; the higher energy orbitals ($d_{x^2-y^2}$, d_{z^2}) are conventionally labeled e_g (or sometimes $d\gamma$) and the lower energy orbitals (d_{xy}, d_{xz}, d_{yz}) are labeled t_{2g} (or sometimes $d\epsilon$). *Within each set*, the orbitals are all of equal energy; i.e., the t_{2g} level is triply degenerate and the e_g level is doubly degenerate. These results are shown in Figure 25-2.

The crystal-field splitting is the energy difference between the t_{2g} and e_g orbital levels, and it is frequently measured in terms of the parameter Dq (or sometimes Δ; 10 Dq = Δ). The magnitude of the splitting depends upon the nature of the ligands and has been designated 10 Dq. The value for Dq in a given complex can be determined experimentally from its electronic absorption spectrum. Light absorption by the complex corresponds to the excitation of electrons from the lower to the higher levels, and the frequency

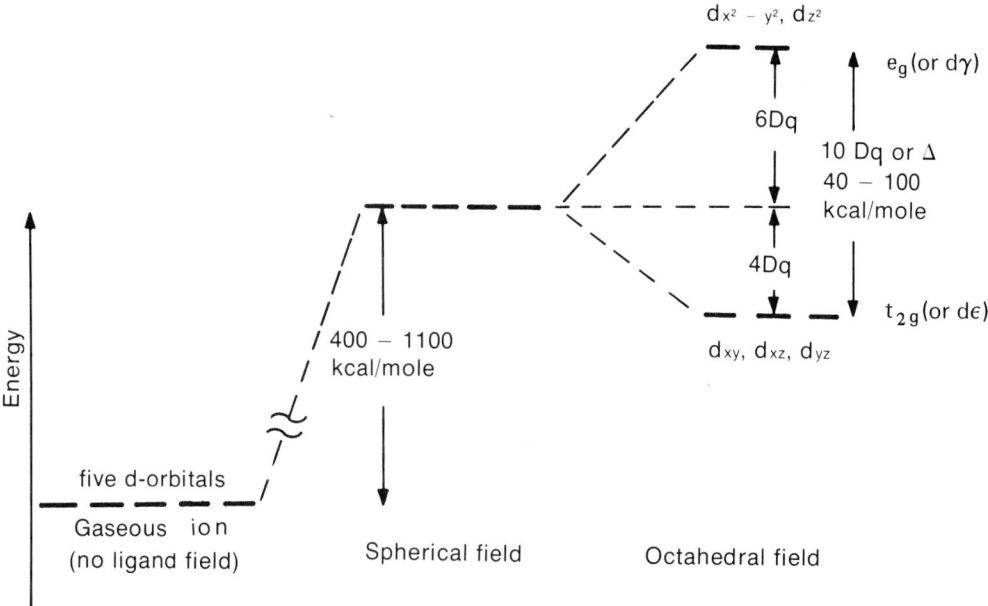

FIGURE 25-2

An energy level diagram illustrating the splitting of the five *d*-orbitals in an octahedral ligand field.

of the light absorbed is related to the energy-level difference by Planck's equation

$$h\nu = E_2 - E_1$$

The spectrum of a d^1 ion, e.g., Ti^{3+}, in a regular octahedral field should exhibit only one absorption band corresponding to excitation of the electron from the t_{2g} to e_g orbitals; therefore, the energy change corresponding to this absorption represents 10 Dq. The spectrum of $[Ti(H_2O)_6]^{3+}$ (d^1) has a maximum at 20,000 cm.$^{-1}$; thus, Dq = 2000 cm.$^{-1}$

The purely electrostatic concept, in which there is no mixing of metal and ligand orbitals or sharing of electrons, is never strictly true. In fact, recent studies of electron spin resonance and nuclear magnetic resonance have shown appreciable delocalization of the central metal's d-electrons into the ligand orbitals. The crystal field theory modified to take account of the existence of moderate amounts of overlap between the metal and ligand orbitals is called *ligand field theory*. When overlap is excessive, e.g., in metal complexes of carbon monoxide or the isocyanides, the *molecular orbital theory* gives a more complete explanation of the metal-ligand bonding.

Correlations of the electronic spectra of a large number of complexes containing various metal ions and various ligands have demonstrated that ligands generally can be arranged in a series according to their capacity to affect the d-orbital splitting. Such a *spectrochemical series*, for some common ligands, is

$$CN^- \gg NO_2^- > dipyridyl > pyridine > OH^- > F^- > Cl^- > Br^- > I^-$$

When an electron undergoes a transition from the ground state, *e.g.*, the t_{2g} level, to an excited state, *e.g.*, the e_g level, the complex must absorb energy. In the case of transitions involving the d-orbitals, e.g., $t_{2g} \rightarrow e_g$ electron transitions, the absorption bands occur in the near infrared, visible, or near ultraviolet region. One might infer from the scale in Figure 25-3 that as the colors of a series of complexes of a given metal shift from red to violet, the $t_{2g} \rightarrow e_g$ splitting increases. However, we must recall that the color observed by the eye is the complementary color of that which is absorbed. Thus, a complex that absorbs strongly at $\sim$14,500 cm.$^{-1}$ appears blue and one that absorbs at $\sim$20,000 cm.$^{-1}$ appears red to the observer. The shape of the bands and colors of specific complexes are given in Figure 25-4.

Octahedral Nickel(II) Complexes. The splitting diagram in Figure 25-2 implies that all transition metal complexes would exhibit one d-d absorption band. Most complexes, in fact, exhibit two or three d-d electronic transitions because ions with more than one but less than nine d-electrons experience electron-electron interactions that remove the equal energy representation (i.e., degeneracy) of the e_g and t_{2g} levels and give rise to a series of energy

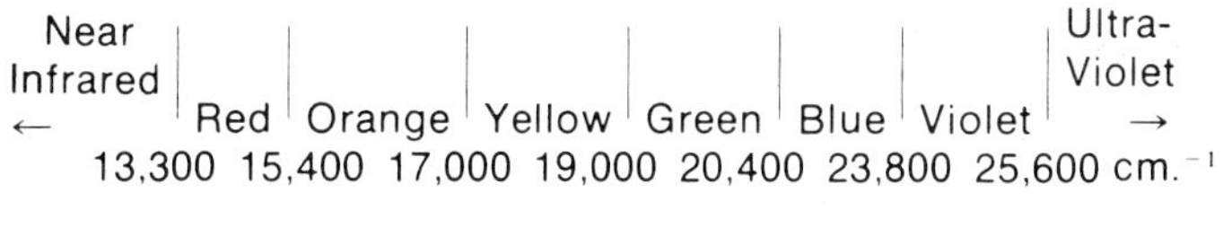

Near Infrared | Red Orange Yellow Green Blue Violet | Ultra-Violet
13,300 15,400 17,000 19,000 20,400 23,800 25,600 cm.$^{-1}$

FIGURE 25-3

The visible region of the spectrum.

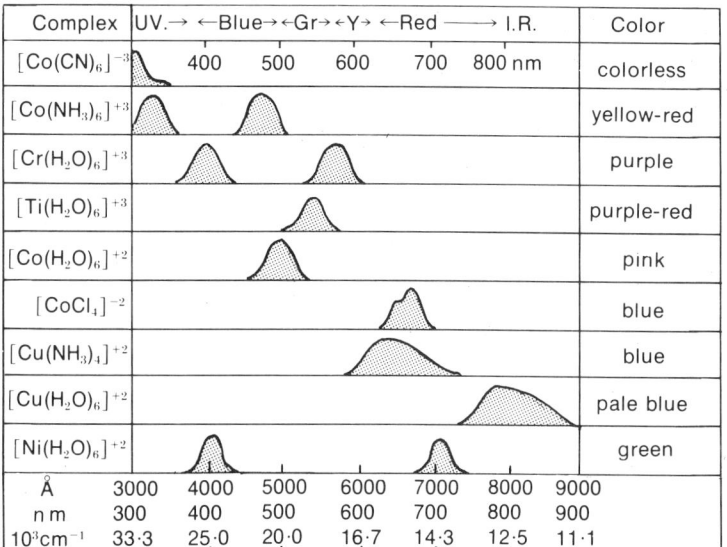

FIGURE 25-4

An illustration of the relationship between the colors and the positions of the electronic absorption spectra of some inorganic complexes. The peaks in the spectra represent absorbance as a function of wavelength. (Courtesy of Oxford University Press.)

states. For example, the free ion Ni^{2+}, a d^8 configuration, has ^3F, ^1D, ^3P, ^1G, and ^1S states (ref. 2); here the left superscript numbers on the capital letters represent the multiplicity of the states: triplet (^{3}) for two unpaired electrons per Ni^{2+} or singlet (^{1}) for zero unpaired electrons per Ni^{2+}.

A spectral selection rule makes electronic transitions between states of different multiplicity much less probable than those between states of equal multiplicity; that is, singlet ↔ triplet transitions usually have low intensities of absorption. Because the lowest energy state of an octahedral nickel(II) complex is derived from the ^3F state, we need to consider only the triplet states in order to assign the observed spectral bands. The lines in Figure 25-5 represent different energy states for octahedral nickel(II) complexes. The lowest energy state in a nickel(II) complex may be repre-

sented by the electronic distribution $-\!\!\!\begin{smallmatrix} \uparrow \;\; \uparrow & e_g \\ \uparrow\downarrow \; \uparrow\downarrow \; \uparrow\downarrow & t_{2g} \end{smallmatrix}$, and is given

the label $^3A_{2g}$. The excited states $^3T_{2g}$, $^3T_{1g}(F)$, and $^3T_{1g}(P)$ result from excitation of an electron from the t_{2g} level to the half-filled e_g orbitals. Figure 25-5 shows how the energies of the ^3F and ^3P states of a gaseous Ni^{2+} ion vary with the nature of the ligands; as the degree of interaction increases, the energy levels undergo greater separation. Electronic transitions from the $^3A_{2g}$ ground state to the three excited states are expected for octahedral d^8 complexes. The assignments and energies (in terms of the parameters Dq and B) of these absorption bands are

Transitions			**Energy Difference Between $^3A_{2g}$ and the Excited State**	
$^3A_{2g} \longrightarrow {}^3T_{2g}$	ν_1		$10Dq$	(1)
$^3A_{2g} \longrightarrow {}^3T_{1g}(F)$	ν_2		$7.5B + 15Dq - \tfrac{1}{2}(225B^2 - 180DqB + 100Dq^2)^{1/2}$	(2)
$^3A_{2g} \longrightarrow {}^3T_{1g}(P)$	ν_3		$7.5B + 15Dq + \tfrac{1}{2}(225B^2 - 180DqB + 100Dq^2)^{1/2}$	(3)

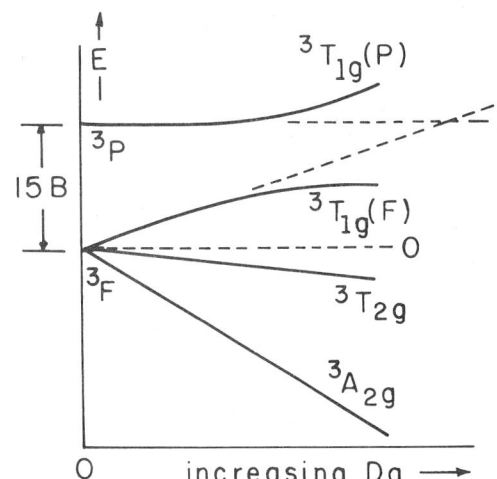

FIGURE 25-5

An energy level diagram that represents the effect of increasing ligand interaction in an octahedral nickel(II) complex. Zero Dq corresponds to a noncomplexed Ni^{2+} ion.

In the absence of any ligand field, e.g., gaseous Ni^{2+}, the energy separation between the ^3F and the ^3P terms corresponds to $15B = 15,836$ cm.$^{-1}$. In the complexes we make the assumption that the center of gravity of the $^3A_{2g}$, $^3T_{2g}$, and $^3T_{1g}(F)$ states maintains this correspondence relative to the $^3T_{1g}(P)$ state. Equations (1)–(3) may be used to solve for the Dq and B values. When all three transitions are observed, note that ν_1 gives 10Dq directly and the calculation of Dq and B is simple, but the wavelength range of simple spectrophotometers often prevents the observation of the near infrared band (ν_1). You will be provided the values of ν_1 on the Report Sheet.

In this experiment we shall examine the electronic absorption spectra of a series of nickel(II) complexes and determine a spectrochemical series involving the ligands ammonia, ethylenediamine, water, and dimethyl sulfoxide.

TECHNIQUE

Review the principles of spectrophotometry and the instructions for operating the instrument that are given in Experiment 23 and Appendix II, respectively.

NOTEBOOK AND REPORT

Your notebook should contain, in legible tabular form, all raw data, e.g., concentration of complex solution, dilutions made, absorbance and wavelength values, and color to the eye, required in the Procedure. Each absorption spectrum should be plotted on graph paper and the results of the team tabulated and analyzed as indicated in the Calculations section. Each student should write a separate report, utilizing all data collected by his team.

Include at the end of the experiment the answers to the questions that are assigned by your instructor.

PROCEDURE

Obtain about 16 ml. of aqueous 0.05M $Ni(NO_3)_2$ solution and divide it into 5 ml. portions in small Erlenmeyer flasks. At the hood add 5 ml. of concentrated NH_4OH to one flask and 5 ml. of ethylenediamine $(H_2NCH_2CH_2NH_2)$ to the second flask. IF ANY ETHYLENEDIAMINE OR AMMONIUM HYDROXIDE IS SPILLED ON YOUR SKIN, WASH IT OFF IMMEDIATELY WITH WATER. Shake the flasks to mix the solutions, and visually compare the colors of the resulting solutions with that of the original solution in the third flask. What can you conclude about the relative energies of the visible absorption bands? (Remember the complementary colors.)

To examine the spectra of a series of complexes, each person of a four-member team will measure the absorption of one of the following complexes: $[Ni(H_2O)_6]^{2+}$, $[Ni(NH_3)_6]^{2+}$, $[Ni(DMSO)_6]^{2+}$, and $[Ni(en)_3]^{2+}$ (where DMSO is dimethyl sulfoxide, $(CH_3)_2SO$, and en is ethylenediamine, $H_2NCH_2CH_2NH_2$). Two members of the team should obtain 0.05M aqueous $Ni(NO_3)_2$ and 0.05M $Ni(NO_3)_2$ dissolved in DMSO (both marked "for Exp. 25") and measure their spectra directly, using the same techniques as in Experiment 23. The third and fourth team members should measure the spectra of the $Ni(NO_3)_2-NH_4OH$ and $Ni(NO_3)_2-en$ solutions that were prepared in the above paragraph.

Each person is to measure the absorbance values of his complex at every 20 millimicrons from 650 to 340 nanometers (Note 1); in the regions of a peak maximum, obtain the absorbance every 10 nanometers. Prepare a graph by plotting the absorbance against wavelength (nanometers) and compare the spectrum with those obtained by the other team members. To make comparisons easier, each person of the team should use the same nanometer and absorbance scales for the graphs.

CALCULATIONS

The range of the simple spectrophotometers, *e.g.*, the Spectronic 20's, permit measurement of only one of the three absorption bands for octahedral nickel(II) complexes. The wider range (325-925 nm.) of the Spectronic 70's will permit measurement of the ν_2 and ν_3 bands. If you have the latter spectrophotometers determine these two bands, calculate B from the position of maximum absorption and the value of ν_1 given on page 313, and check the position of ν_3 or ν_2 against the value that you would calculate from equation (3) or (2) using the values of Dq and B obtained from the other equation (Note 2).

If you are unable to determine two absorption maxima, due to the limited range of your spectrophotometer, consider that the maximum observed near 400 nm. in the $[Ni(H_2O)_6]^{2+}$ and $[Ni(DMSO)_6]^{2+}$ complexes is ν_3 and that the maximum observed at approximately 550 nm. in the $[Ni(NH_3)_6]^{2+}$ and $[Ni(en)_3]^{2+}$ complexes is the ν_2 absorption.

Using the appropriate energy values obtained from Table 25-1 (p. 313) and the experimental absorption maxima, derive the Dq and B values for each of the four ligands, and arrange the ligands in a spectrochemical series with increasing value of Dq.

In the cases where the absorption maxima for two of the three bands are available (*e.g.*, ν_1 from Table 25-1 and your observed band), we can solve for Dq and B by use of simultaneous equations, given that

$$\nu_3 + \nu_2 - 3\nu_1 = 15B \qquad (4)$$

for nickel(II) complexes (Note 2). Thus, Dq may be substituted back into one of the quadratic equations (1)–(3) as a function of B. Use energy values (cm^{-1}) for ν_1, ν_2, and so forth.

The somewhat cumbersome calculations can be avoided by making use of solutions of equation (1) that have been prepared by others (Table 25-2). In making this table, various ratios of Dq/B have been chosen, and ν_3/B, which is equal to

$$7.5 + 15\frac{Dq}{B} + \frac{1}{2}\left(225 - \frac{180Dq}{B} + 100\left(\frac{Dq}{B}\right)^2\right)^{\frac{1}{2}}$$

has been calculated in terms of these chosen ratios. The ratios ν_2/ν_1, ν_3/ν_1, and ν_3/ν_2 can similarly be expressed in terms of Dq/B, and the calculated values are recorded in the table for each chosen ratio Dq/B. It is necessary, then, only to calculate experimental values of ratios such as ν_3/ν_2 and look down the last column of the table until that ratio appears. Suppose, for example, that the experimental ν_3/ν_2 ratio is 1.800. In the table that ratio corresponds to Dq/B = 0.9 and to ν_3/B = 27.0. Since ν_3 is known from the experiment, B = $\frac{\nu_3}{27.0}$ and Dq = $\frac{0.9\nu_3}{27.0}$. In the lower part of the table, the same value for ν_3/ν_2 corresponds to more than one value for Dq/B; the choice of the correct Dq/B is then made by considering the experimental value for ν_3/ν_1 or ν_2/ν_1. Experimental values that do not occur explicitly in the table may be obtained by interpolation.

TABLE 25-2

Dq/B	ν_3/B	ν_2/ν_1	ν_3/ν_1	ν_3/ν_2
		A₂ Ground State		
0.05	15.60	1.794	31.21	17.39
0.1	16.21	1.789	16.21	9.062
0.15	16.83	1.783	11.22	6.291
0.2	17.45	1.777	8.723	4.909
0.25	18.07	1.771	7.230	4.083
0.3	18.71	1.764	6.236	3.535
0.35	19.35	1.757	5.529	3.147
0.4	20.00	1.750	5.000	2.857
0.45	20.66	1.743	4.591	2.634
0.5	21.32	1.735	4.265	2.458
0.55	22.00	1.727	4.000	2.316
0.6	22.69	1.719	3.781	2.199
0.65	23.38	1.711	3.597	2.102
0.7	24.08	1.703	3.440	2.021
0.75	24.80	1.694	3.306	1.952
0.8	25.52	1.685	3.190	1.893
0.85	26.26	1.676	3.089	1.843
0.9	27.00	1.667	3.000	1.800
0.95	27.76	1.657	2.922	1.763
1.0	28.52	1.648	2.852	1.731
1.1	30.08	1.629	2.735	1.679

TABLE 25-2 *(Continued)*

Dq/B	v_3/B	v_2/v_1	v_3/v_1	v_3/v_2
		A$_2$ Ground State		
1.2	31.69	1.610	2.610	1.640
1.3	33.33	1.590	2.563	1.612
1.4	35.00	1.571	2.500	1.591
1.5	36.70	1.553	2.447	1.576
1.6	38.45	1.535	2.403	1.566
1.7	40.21	1.517	2.365	1.559
1.8	42.00	1.500	2.333	1.556
1.9	43.81	1.484	2.306	1.554
2.0	45.64	1.468	2.282	1.554
2.1	47.49	1.453	2.261	1.556
2.2	49.35	1.439	2.243	1.559
2.3	51.22	1.425	2.227	1.563
2.4	53.11	1.412	2.213	1.567
2.5	55.00	1.400	2.200	1.571
2.6	56.90	1.388	2.189	1.577
2.7	58.82	1.377	2.178	1.582
2.8	60.74	1.367	2.169	1.587
2.9	62.66	1.357	2.161	1.593
3.0	64.59	1.347	2.153	1.599
3.1	66.53	1.338	2.146	1.604
3.2	68.47	1.329	2.140	1.610
3.3	70.42	1.321	2.134	1.616
3.4	72.37	1.313	2.128	1.621
3.5	74.32	1.305	2.123	1.627
3.6	76.27	1.298	2.119	1.632
3.7	78.23	1.291	2.114	1.638
3.8	80.19	1.284	2.110	1.643
3.9	82.16	1.278	2.107	1.648
4.0	84.12	1.272	2.103	1.653
4.1	86.09	1.266	2.100	1.658
4.2	88.06	1.261	2.097	1.663
4.3	90.03	1.255	2.094	1.668
4.4	92.00	1.250	2.091	1.673
4.5	93.97	1.245	2.088	1.677
4.6	95.95	1.240	2.086	1.682
4.7	97.93	1.235	2.084	1.686
4.8	99.90	1.231	2.081	1.691
4.9	101.9	1.227	2.079	1.695
5.0	103.9	1.223	2.077	1.699

From A. B. P. Lever: *J. Chem. Ed., 45,* 711 (1968).

Another way of using the data in Table 25-2 is to construct a graph by plotting various ratios of v_3/v_1, v_3/v_2, and v_2/v_1 against the Dq/B values. A second graph may be obtained by plotting the various ratios against the v_3/B values given in Table 25-2. By using Table 25-2 or these graphs, derive the Dq and B values for each of the four complexes, and arrange the ligands in a spectrochemical series of increasing Dq.

For at least one ligand compare the value for Dq obtained by this procedure to that obtained by using simultaneous equations involving equation (4) and one of the quadratics of equations (1)–(3).

NOTES

Note 1. The spectrophotometer cuvettes (or cells) should be covered (or stoppered) during the absorbance measurement to minimize odors and absorption of water from the atmosphere. The latter is especially important in the case of the hygroscopic dimethyl sulfoxide solution.

Note 2. The absorption maxima of the bands are usually reported with units of cm.$^{-1}$, which is the reciprocal of the wavelength expressed in centimeters, i.e., 500 nanometers = 500×10^{-9} m. = 20,000 cm.$^{-1}$. Until recently the wavelength in the visible region was expressed in millimicrons (1 millimicron = 1 nanometer).

QUESTIONS–PROBLEMS

1. By considering where the absorption maxima of the four complexes occur, how do you rationalize their observed colors? Take advantage of Figure 25-3 and the fact that one or both of the ν_2 and ν_3 bands may contribute to the observed color.

2. If the path length of your cell is known or you have a method to determine it, calculate the absorptivity A (A = ϵbc; terms defined in Exp. 23) of the complexes at their band maxima.

3. Even if a quantitative determination of A is not practical, consider that the path lengths of the cuvettes were similar and qualitatively determine which of the four complexes have (has) more intense bands than the others. (Remember that some solutions were diluted before measurement.) What significance can you attach to the observed intensity trend?

4. The ^{3}F to ^{3}P energy separation, i.e., 15B, in the free nickel(II) ion is 15,836 cm.$^{-1}$. From your data calculate 15B for each complex and arrange the complexes according to increasing values. Is the order in this series the same as in the Dq series? In general the values for B decrease as the diffuse character of the ligand orbitals increases; some chemists have tried to relate an increasing degree of covalent bonding to the lower B values. Do your data permit any such judgment about the nitrogen ligands (*i.e.,* NH_3 and en) as compared to the oxygen ligands, (*i.e.,* H_2O and DMSO)?

REFERENCES

1. Harvey, K.B., and. Porter, G.B.: *Introduction to Physical Inorganic Chemistry.* Addison-Wesley, Reading, Mass., 1963, p. 68.
2. Hochstrasser, R.M.: *Behavior of Electrons in Atoms.* W.A. Benjamin, New York, 1964, Chapter 5.

3. Kilner, M., and Smith, J.M.: *J. Chem. Ed., 45:*94 (1968).
4. Lever, A.B.P.: *J. Chem. Ed., 45:*711 (1968).
5. Meek, D.W., Drago, R.S., and Piper, T.S.: *Inorg. Chem., 1:*285 (1962).

Experiment 25
REPORT SHEET

Name _____   Section _____

TABLE 25-1 SUMMARY OF SPECTRAL DATA

Complex	ν_1 nm.	ν_1 cm.$^{-1}$	ν_2 nm.	ν_2 cm.$^{-1}$	ν_3 nm.	ν_3 cm.$^{-1}$
$[Ni(H_2O)_6]^{2+}$	1165	_____	_____	_____	_____	_____
$[Ni(DMSO)_6]^{2+}$	2190	_____	_____	_____	_____	_____
$[Ni(NH_3)_6]^{2+}$	943	_____	_____	_____	_____	_____
$[Ni(en)_3]^{2+}$	875	_____	_____	_____	_____	_____

Calculated Ligand Field Parameters

	Dq	B
H_2O	_____	_____
DMSO	_____	_____
NH_3	_____	_____
en	_____	_____

Spectrochemical Series of Ligands in Terms of Dq

Ligand Series in Terms of B

Attach sheets with answers to assigned questions.

SOLID STATE

STRUCTURE OF IONIC CRYSTALS AS ANALYZED BY X-RAY DIFFRACTION

EQUIPMENT NEEDED

None.

REAGENTS NEEDED

X-ray powder photographs of sodium chloride, taken in a camera of known dimensions with x-rays of known wavelength.

INTRODUCTION

Diffraction of a beam of x-rays by powdered samples of solid materials produces diffraction patterns characteristic of the nature and position of the scattering units (atoms or ions) in the powdered solid.

A common method of recording these diffraction patterns is to photograph the diffracted radiation, using a diffraction camera, which is schematically shown in Figure 26-1. The x-ray beam enters at X and passes through a hole in a strip of photographic film wrapped in a circle about the diffracting powder in a capillary at C, and the diffracted beams impinge on the film in both forward and backward directions, as shown.

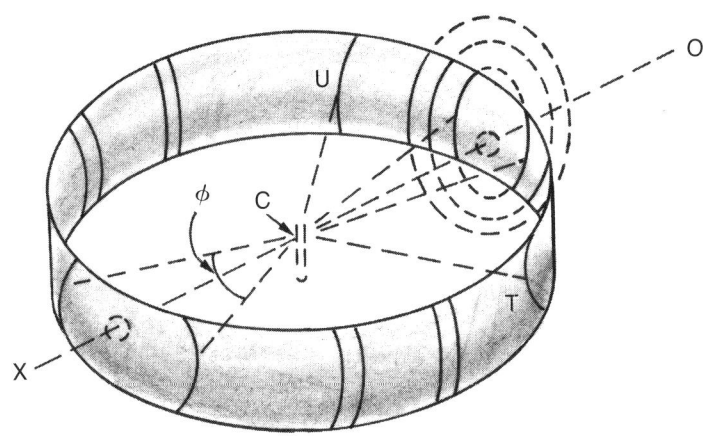

FIGURE 26-1

Geometry of an x-ray powder camera. The powdered sample at C is rotated in the beam.

The experimental measurements confirm the hypothesis that the external symmetry of crystalline forms results from an internal symmetry in which the ultimate particles of the substance — atoms, molecules, or ions — are at symmetrically spaced lattice points. Such symmetrically spaced lattice points permit planes to be drawn through them in such a way that some planes are parallel to other planes at fixed distances, symbolized as d, from each other; examination of Figure 26-9, for example, shows that other planes through lattice points parallel to those shown can readily be drawn. If the ultimate particles are situated at the lattice points, the x-ray photons interact with them and produce diffracted waves. The diffracted x-rays are reinforced when the relationship between the angle that the incident and "reflected" beams make with a plane through the diffracting particles and the distance between that plane and a similar plane parallel to it is that expressed by the Bragg relation

$$n\lambda = 2d \sin \theta \qquad \textbf{(1)}$$

where n is an integer and λ the wavelength of the x-ray. When the angle is such that its sine is not equal to $n\lambda/2d$, interference occurs, and no diffracted beam is seen at such an angle.

The diffracted beams appear as cones about the powdered sample, as shown in Figure 26-2, and thus form arcs of the circumferences of the bases of the cones on the strip of photographic film in the camera. The arcs appear upon developing the film. The relative number of x-ray quanta falling on particular arcs determines the relative blackness of these arcs when the film is developed.

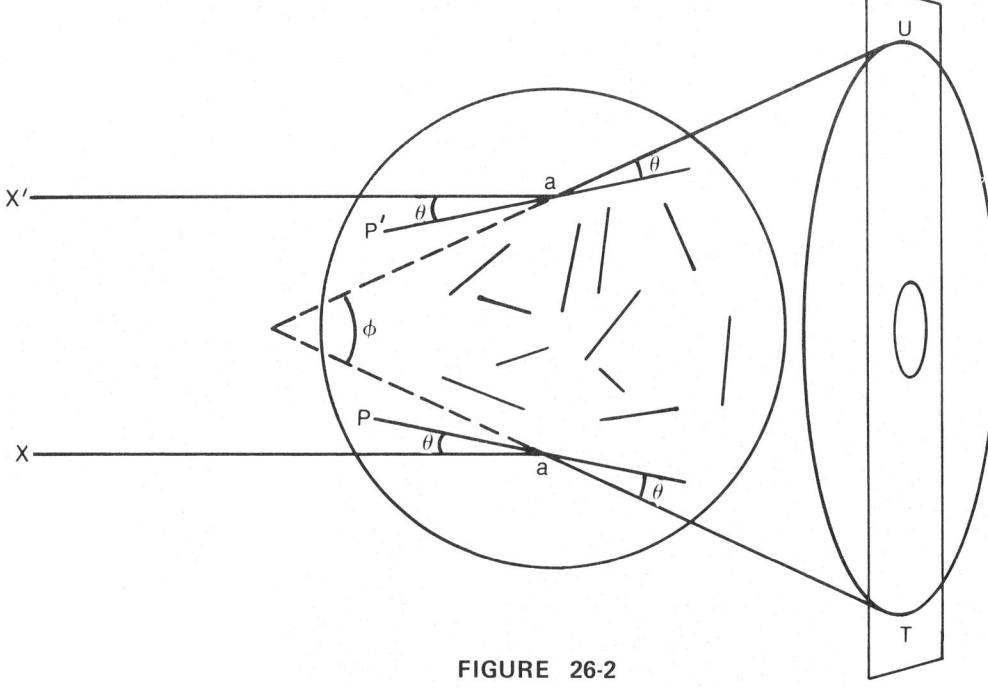

FIGURE 26-2

Formation of a diffraction cone. The x-ray beam of width XX' meets sets of parallel planes in the randomly oriented crystallites of the powdered sample, those parallel to P and P', for example, which make such an angle with the beam that $\sin \theta = n\lambda/2d$. Reinforced diffracted beams thus appear at T and U. Other sets of planes above and below the plane of the paper, similarly oriented with respect to the incident beam, appear at other points above and below to make a cone base as shown. Rotation of the cylindrical sample about the axis of the cylinder helps to insure the randomness of orientation of the planes.

Other cones may appear at an angle θ' if parallel planes at a spacing d' are present so that $\sin \theta' = n\lambda/2d'$.

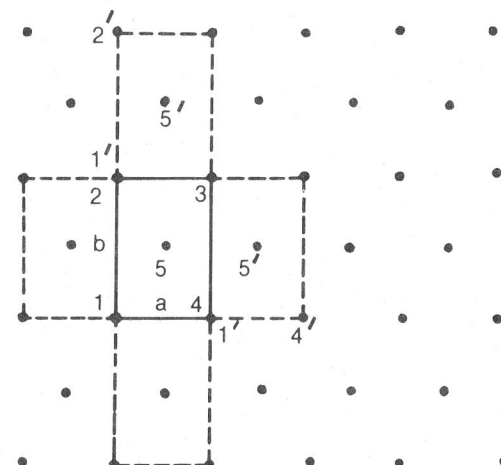

FIGURE 26-3

Reproduction of a two-dimensional lattice by translation of the two-dimensional unit cell (1,2,3,4).

Such photographs have been made for use in this experiment, the purpose of which is the determination of the crystal structure of sodium chloride from such *x-ray diffraction powder photographs.*

A. The Unit Cell of Sodium Chloride—Its Dimensions, and the Calculation of Avogadro's Number

The Unit Cell. In a symmetrical array of lattice points there are repeated examples of certain groups of points that are in identical relation to each other. Consider, for example, the two-dimensional lattice of Figure 26-3 and the group outlined in solid lines. Groupings identical to that of these five points appear throughout the figure. In fact, the whole figure can be regenerated by repeatedly lifting this outlined group off the paper, moving it horizontally or vertically by appropriate distances, putting it down on the paper again, and marking the points. The distances the group is to be moved are indicated as *a* for the horizontal and *b* for the vertical distance so that points 1,4 take up new positions at 1′,4′ and points 1,2 take up new positions at 1′,2′. In each case 5 appears at 5′ after the translation, and continued translation by distances *a* and *b* reproduces the whole set of lattice points.

A grouping of points having the symmetry of the lattice that by translation can reproduce the whole set of lattice points is known as the *unit cell,* and the translation distances by which the unit cell must be moved to reproduce the lattice are the dimensions of the unit cell. Thus, the two-dimensional unit cell of Figure 26-3 has the dimensions *a, b;* for a three-dimensional unit cell these would be denoted as *a, b, c.* The *directions* of the translations need not be at right angles, but they are at right angles for cubic, orthorhombic, and tetragonal systems.

Miller Indices and the Law of Rational Indices. Many sets of parallel planes can be drawn through a three-dimensional set of lattice points. To distinguish one set from another, a system of describing a plane has been agreed upon. This is most easily understood for cubic, orthorhombic, and tetragonal systems in which three mutually perpendicular lines can be drawn

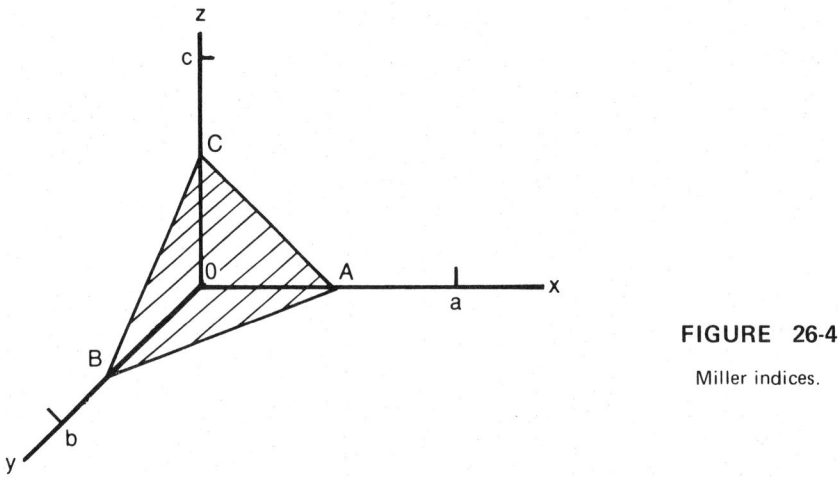

FIGURE 26-4

Miller indices.

from one lattice point O to three other lattice points at the unit cell distances a, b, and c (Fig. 26-4). Any arbitrary plane ABC may then be described in terms of its intercepts on these three lines. Thus, if the plane cuts the line OX at a distance $\frac{a}{h}$, line OY at $\frac{b}{k}$, line OZ at $\frac{c}{l}$, it may be catalogued as the plane $\frac{a}{h}, \frac{b}{k}, \frac{c}{l}$.

It is an experimental fact, recorded in the *law of rational indices*, that if the plane ABC passes through three noncolinear lattice points, the quantities h, k, and l are integers, usually small, or zero. These numbers rather than the fractions $\frac{a}{h}, \frac{b}{k}, \frac{c}{l}$ are, therefore, convenient indexing quantities, and they are called the *Miller indices*. The Miller indices are thus the reciprocals of the intercepts, in terms of the unit distances a, b, and c, of the planes on the three axes. Thus, for example, if the intercepts of the plane ABC were $\frac{a}{2}, \frac{b}{2}, \frac{c}{2}$, it would be called the 222 plane; if the intercepts were $\frac{a}{2}$, b, and c, it would be the 211 plane. A plane through a that is parallel to the bc plane would intercept OY and OZ at infinity (a, ∞, ∞) and would be called the 100 plane $\left(\frac{a}{1}, \frac{b}{0}, \frac{c}{0} \right)$.

Relationship Between the Miller Indices and the Distance Between Planes.   In crystal structure analysis, the distance between planes is important, because it indicates the spacing of the lattice points in relation to each other. The distance, d, between a plane ABC with Miller indices h, k, l and a plane parallel to it passing through the origin O is given by

$$\frac{1}{d^2} = \frac{h^2}{a^2} + \frac{k^2}{b^2} + \frac{l^2}{c^2} \tag{2}$$

For a cubic crystal, the lattice points are the same distance from the origin in each of the three directions, and $a = b = c$. Thus, for such a cubic lattice

$$\frac{a^2}{d^2} = h^2 + k^2 + l^2 \tag{3}$$

Geometry of the Diffraction Camera. The purpose of the diffraction study is to determine d, and from d the unit cell size a, b, c (or, if cubic, a). To determine d we need to know θ in equation (1). What is the relationship between the spacing of arcs in the powder photograph and the value for θ? This relation depends upon the geometric dimensions of the camera.

In many cameras, the circumference of the circle on which the film is placed is 360 mm. or some simple multiple of that number. If it is 360 mm., the radius of the circle is

$$\text{Radius} = \frac{360}{2\pi} \text{ mm.} \qquad (4)$$

The length of a piece of this circumference, as from T to U in Fig. 26-1, is measured by the product of the radius and the angle ϕ (measured in radians). Thus

$$\overline{TU} = (\text{Radius}) \times \phi = \frac{360\,\phi}{2\pi} \qquad (5)$$

Now $\overline{TU}$ is the distance between the arcs traced by the base of the cone in Figure 26-3, and a measurement of $\overline{TU}$ when the film is flattened makes it possible to calculate the central angle ϕ in Figure 26-1.

$$\phi = \frac{2\pi\,(\overline{TU})}{360} \text{ radians} \qquad (6)$$

But as shown in Figure 26-3, ϕ is also the apex angle of the diffraction cone.

It can be shown that ϕ is equal to 4θ, where θ is the angle of incidence of the x-ray beam with the diffracting plane, as indicated in Figure 26-3. Substituting, we have

$$4\theta = \frac{2\pi\,(\overline{TU}) \text{ radians}}{360} \times \frac{180 \text{ degrees}}{\pi \text{ radians}} \qquad (7)$$

where the conversion factor for degrees to radians is included to give θ in degrees. If $\overline{TU}$, the distance between arcs on the film, is measured in millimeters,

$$\theta = \frac{1}{4}\left(\frac{2\pi\,(\overline{TU} \text{ mm.})}{(360 \text{ mm.})}\right) \times \frac{180}{\pi} \qquad (8)$$

it is seen that, for a film circumference of 360 mm., θ in degrees is numerically equal to one-fourth the distance between arcs on the film measured in millimeters. For some camera dimension other than 360 mm., another factor will appear.

PROCEDURE

Proof That the Photographs Represent Cubic Unit Cells. From the external symmetry of alkali halide crystals, we suspect that the internal symmetry should also be cubic. For a cubic structure, a combination of equations (1) and (3) gives

$$\sin^2 \theta = \left(\frac{n\lambda}{2a}\right)^2 (h^2 + k^2 + l^2) \tag{9}$$

and the square of the sine of θ as measured from the films should be directly proportional to $(h^2 + k^2 + l^2)$ where h, k, and l are, by the law of rational indices, integers. To find out whether this proportionality exists (1) make* a table of increasing values for $h^2 + k^2 + l^2$ for integral values of h, k, and l, starting with h = 1, k = 0, l = 0. Cover the range in the sum $h + k + l$ from 1 to 8. For convenience maintain the sequence $h \geq k \geq l$, e.g., 321 not 213.

Now (2) measure* the distance between corresponding arcs on the diffraction photograph, and (3) calculate* θ and the square of its sine. (Neglect the smallest circle, which represents a hole in the film to let the incident beam through the film without fogging and scattering.) If cubic, squares of all the measured values of $\sin \theta$ must be proportional to values for $h^2 + k^2 + l^2$ listed in your table, with the same proportionality constant. However, because some planes may not be populated, or may not reflect, in the crystal lattice, not all values for $h^2 + k^2 + l^2$ may appear as $\sin^2 \theta$. One can test the symmetry and index the planes by a graphical procedure in which one notes that, from equation (9), $\sin^2 \theta$ is a linear function of $h^2 + k^2 + l^2$ and passes through the origin, so that all experiment values for $\sin^2 \theta$, plotted against appropriate values for $h^2 + k^2 + l^2$, must be on the same straight line. (4) Make* such a graph, accepting the hint that the second arc in the sodium chloride structure has the index 200, with $h^2 + k^2 + l^2 = 4$. From the graph, (5) index* each line with its proper hk$l$ value. Note that neither a nor λ needs to be known in order to make this graph.

Calculation of the Unit Cell Dimension. Having confirmed that the unit cell is cubic, (6) calculate* from the slope of the curve or by using equation (9) and several hk$l$ values for each salt, the value of a (the cube edge), using the given value for the wavelength of the x-rays used (e.g., $\lambda = 1.5374$ Å for a copper source).

Apparent Value of Avogadro's Number. Precise measurements of the unit cell size can be combined with precise measurements of density to determine Avogadro's number. The argument is as follows: From the density, we can calculate the volume of solid that would contain 1 mole of substance, because

$$\text{Volume containing 1 mole} = \frac{\text{formula weight}}{\text{density}}$$

Dividing this volume by the volume of a unit cell (which is a^3 for a cubic unit cell) gives the number of unit cells per mole of substance. If each unit cell contains one molecular unit, e.g., one sodium atom and one chlorine atom for sodium chloride, the number of molecular units in one mole, or Avogadro's number, is immediately given.

---

*Asterisks require an entry in the notebook; see "Notebook and Report," page 328.

Using the density of sodium chloride as 2.165 g./cm.^{3}, (7) calculate*
Avogadro's number, assuming that the unit cell contains one molecular unit.
(8) What do you conclude?*

B. Structure of the Unit Cell of Sodium Chloride

In Part A of this experiment we identified the cubic nature of the unit
cell and its dimensions for sodium chloride. We have not yet determined the
location, within the unit cell, of the centers which diffract the x-rays. In Part
A, we used only the *spacings* of the diffracted arcs on the photographs,
making no use of additional data that might be obtained from the variation
in the *intensities* of the arcs. These are not the same; for example, the first
arc in the sodium chloride spectrum is only 13 percent as intense as the
second – only 13 percent as many quanta are deflected to the position of
the first arc as are deflected to the position of the second.

The intensity of the diffracted line is a measure, for otherwise constant
conditions, of the number of electrons in the scattering center, or, more
precisely, a measure of the density of electrons in the scattering plane. If all
scattering centers were identical and had the same number of electrons, all
the diffracted lines would show a uniform decrease in intensity as $h^2 + k^2 + l^2$
increases. In the photographs, the change in intensity in going from arcs of
smaller to arcs of larger diameter is not uniform. (For sodium chloride, for
example, the change in relative intensity is 13, 100, 55, 2, 15, 6, 1, 11, 7.)
This shows that the scattering centers, though fixed in *position*, differ in
composition, particularly in the density of electrons. If each scattering
center were a molecule of sodium chloride, each scattering center would be
the same, and we would observe a steady decrease in intensity at greater
values of θ; because the intensities do not so decrease, the centers cannot be
the same and we cannot be dealing with identical sodium chloride molecules.
Perhaps we are observing scattering from sodium and chlorine atoms or ions,
with different numbers of electrons (11 and 17 or 10 and 18).

If the atoms or ions are not combined as molecules, but are separate,
where can they be placed in the unit cell so that the cubic pattern is
retained, while satisfying the requirement of chemistry that there be equal
numbers of sodium and chlorine atoms? One can suggest particular cubic
models and examine them to see whether they are physically and chemically
reasonable and whether the diffraction data predicted by them would
correspond to those observed.

The Face-Centered Cubic Unit Cell. One such arrangement consists in
putting chlorine atoms in the corners of a cube and at the centers of each
face, and then putting sodium atoms at the center of the cube and at the
center of each edge of the cube. The resulting structure has the appearance
of Figure 26-5. Note that an atom at the center of a face belongs both to its
own unit cell and to the adjacent unit cell, so that we count only half an
atom as belonging to the cell. In the figure there are six such face-centered
atoms; these correspond to $6 \times \frac{1}{2} = 3$ atoms. The atoms at the corners of the
cube are of the same kind as those in the faces; each of these belongs
simultaneously to the eight unit cells that meet at the corners, $\frac{1}{8}$ to each.

---

*Asterisks require an entry in the notebook; see "Notebook and Report," page 328.

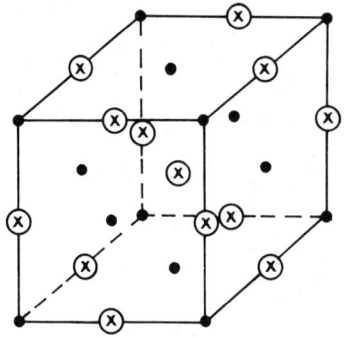

FIGURE 26-5

The sodium chloride structure.

Since there are eight corners, the corner atoms correspond to $8 \times 1/8 = 1$ atom; thus, a total of four atoms of one kind are located at the corners and faces of the face-centered cubic arrangement. (9) Discover* how many cubes an edge atom belongs to and (10) calculate* the total number of atoms of the kind not in the faces and corners that are to be assigned to a unit cell. Chemistry demands that there be a 1:1 ratio between the two kinds.

Using the number of "molecules" per unit cell calculated in the preceding paragraph, (11) recalculate* Avogadro's number for sodium chloride from the unit cell size observed in Part A, and (12) calculate* the percentage error from the accepted value.

X-ray Diffraction from a Face-Centered NaCl Structure. The agreement between the calculated and accepted value of Avogadro's number for the face-centered cubic structure suggests that it is the correct one. Let us see if this suggestion is corroborated by the diffraction data.

(a) Diffraction from Planes Parallel to the 100 Plane. Figure 26-6 shows the 100 plane. (13) Calculate* the number of chlorine atoms and

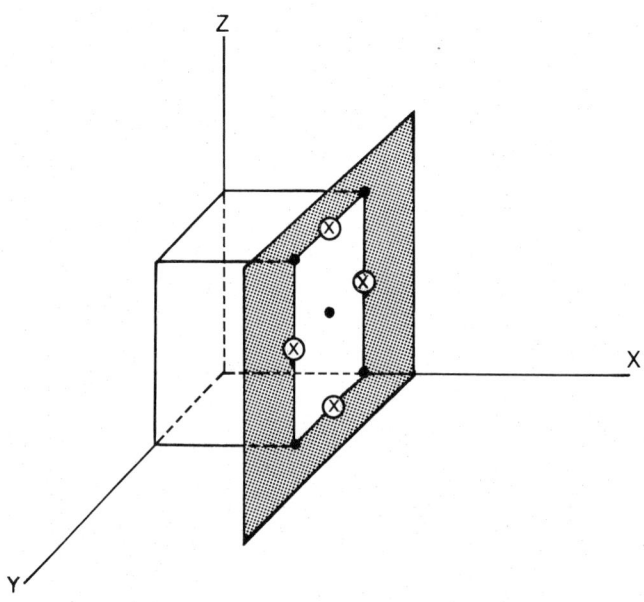

FIGURE 26-6

The 100 plane.

---

*Asterisks require an entry in the notebook; see "Notebook and Report," page 328.

sodium atoms in this plane which belong to the unit cell and the ratio of chlorine atoms to sodium atoms. (14) On a piece of white paper, reproduce* Figure 26-5 (you may place the paper over Figure 26-5 and trace the figure if you wish) and draw in on this reproduction, a 200 plane in the way the 100 plane is drawn in Figure 26-6. (15) Calculate* the ratio of chlorine atoms to sodium atoms in the 200 plane. It is evident from your calculation that the 100 and 200 planes are identical, and hence have the same electron density. Thus, as can be seen from the derivation of the Bragg relation (ref. 1), when reflections from the 100 planes are in phase, the reflections from the halfway between plane will be exactly out of phase and will cancel the 100 reflections. Hence, the greatest distance between identical planes parallel to the YZ plane is the distance $a/2$, the distance to the 200 plane. That a 200 line, but no 100 line, is observed suggests that the face-centered cubic structure is correct.

(b) Diffraction from Planes Parallel to the 110 Plane.   (16) Show* from Figure 26-7 and a reproduction of Figure 26-5 that the situation is the same for the 110 planes and 220 planes, so that there should be a 220 line but no 110 line.

(c) Diffraction from Planes Parallel to the 111 Plane.   Examination of the 111 plane of the sodium chloride structure drawn in Figure 26-8 shows that it contains only chlorine atoms and that halfway between the origin and the 111 plane is a parallel plane containing only sodium atoms, and reflections from it would be expected to interfere with the reflections from the 111 plane. However, because sodium atoms have fewer electrons than do chlorine atoms, although the *atom* densities in these two parallel planes are the same, the *electron* densities are different. Hence, the scattering from the

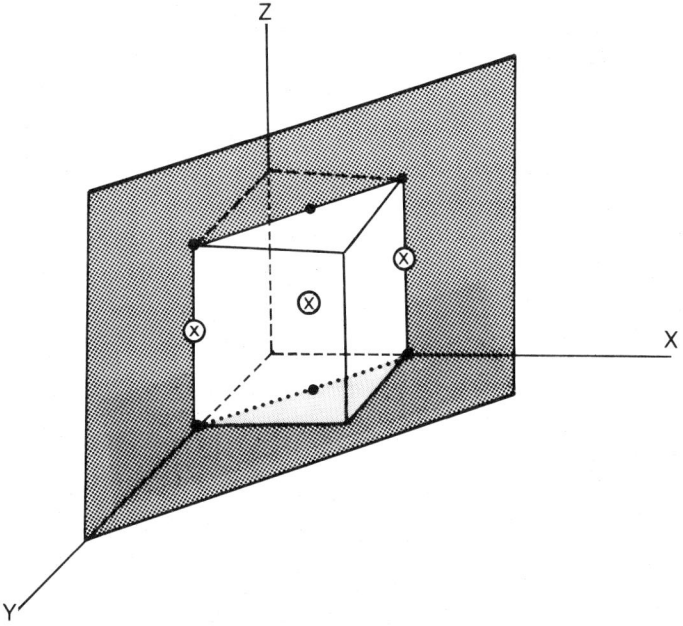

FIGURE 26-7

The 110 plane.

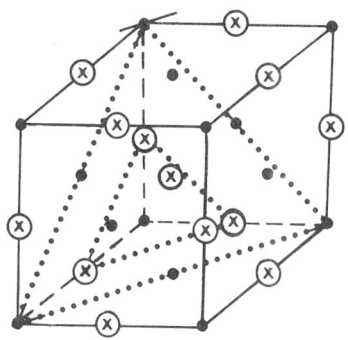

FIGURE 26-8

The 111 plane.

many-electron chlorines overpowers that from the fewer-electron sodiums, and reflection from the 111 planes, though reduced in intensity by interference from the planes containing sodium, is still observable.

(d) Diffraction from Other Planes. The procedures of the last three paragraphs have shown that the structure corresponding to Figure 26-5 fits the chemical data and the measurements on the first three lines in the photograph for sodium chloride. The remaining lines in the photograph, as indexed on your graph, must now be examined to see whether there is a correspondence between the hk*l* indices of the lines and reflecting planes in the crystal. This requires geometric construction that is a little difficult when all indices are different from zero, but simple enough for us to attempt when the last index is zero. The last index zero means that there is no intercept on the Z axis, and that the plane in question is parallel to the Z axis as in Figures 26-6 and 26-7. We can then represent the XY plane as the plane of a sheet of paper, and represent planes h00 or hk0 as planes perpendicular to the plane of the paper. The position of the h00 or hk0 plane then shows on the paper as traces of the intersection of these planes with the plane of the paper. The symbolism is shown in Figure 26-9. The XY plane is shown as it would appear if one were looking down, from a height on the Z axis, at the top of a set of regular unit cells in a crystal structured as in Figure 26-5. The top of one unit cell is outlined in solid lines; the origin is taken at 0. The tops of these unit cells is taken as the XY plane, and traces of the 200 and 220 planes appear on this plane. (17) Construct*, on graph paper so that the distances between atoms is to scale, two diagrams of tops of unit cells as in Figure 26-9. Use one of them (18) to show* traces of h00 planes for h = 1, 2, 3, 4 and the other (19) to show* traces of the 110, 220, and 420 planes. Note the occupancy of these planes by atoms of sodium and chlorine, and imagine a plane drawn parallel to each of them but passing through the origin. Confirm the fact that a Bragg reflection should appear corresponding

---

*Asterisks require an entry in the notebook; see "Notebook and Report," page 328.

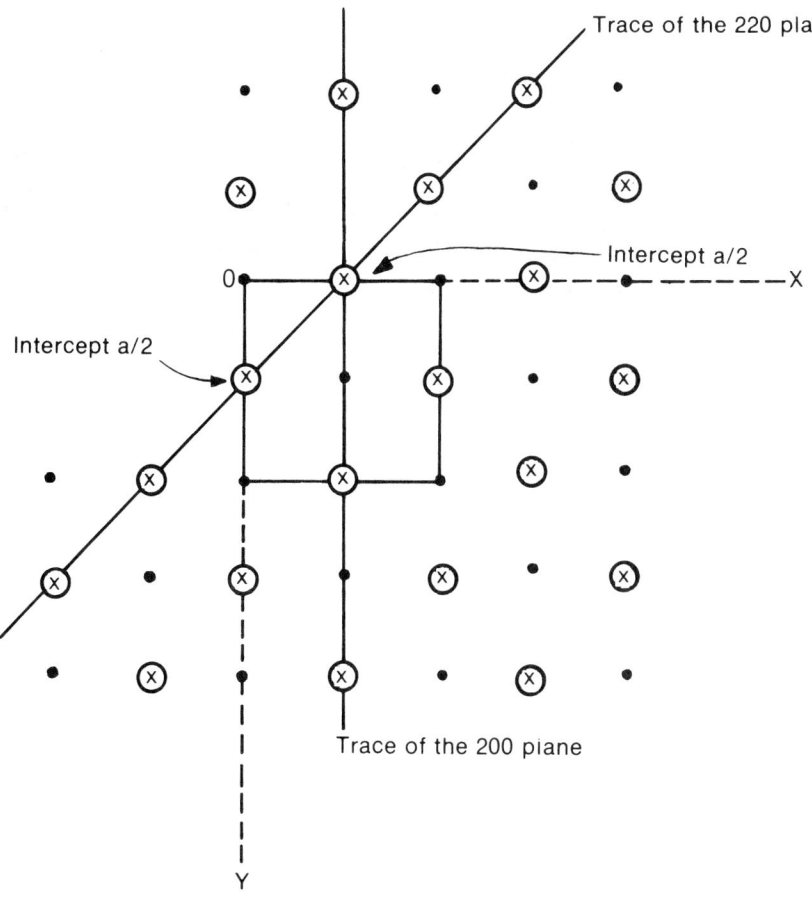

FIGURE 26-9

The XY plane, with traces of the planes h = 2, k = 0, l = 0, and h = 2, k = 2, l = 0.

to the perpendicular distance between the 220 plane and the origin plane, between the 420 plane and the origin plane, and between the 200 plane and the origin plane. (20) Would you predict a Bragg reflection corresponding to the 300 plane? (21) The 400 plane?* (22) Are such reflections shown on your graph?* (23) Explain* (Hint: In equation (1), n is any integer.)

Is It Sodium Atoms and Chlorine Atoms, or Sodium Ions and Chloride Ions? The answer to this question comes from a detailed consideration of the intensity of the lines corresponding to diffraction from the 111 planes discussed above. The reduction in intensity corresponds to interference between planes with electron densities in the ratio 10:18, not 11:17. This is confirmed by the diffraction data from potassium chloride crystals, for, although the spacing, density and Avogadro number calculations correspond to a structure similar to that of sodium chloride (differing only in the size of the unit cell), potassium chloride shows no line corresponding to diffraction from the 111 planes. (24) Show,* by discussion of Figure 26-9, that the absence of a 111 line is definite proof of the ionic nature of potassium chloride.

---

*Asterisks require an entry in the notebook; see "Notebook and Report," page 328.

NOTEBOOK AND REPORT

As you have read through the preceding pages, you will have noted many asterisks* appearing after words such as make*, consider*, calculate* preceded by a number in parentheses. Your notebook should contain the results of carrying out the indicated instructions, in the form of tables, figures, graphs, discussions, calculations, and so forth. Each result should be carried over, by number, to your final report.

QUESTIONS—PROBLEMS

1. Choose the origin at a sodium atom in Figure 26-5 and draw a diagram of a unit cell for sodium chloride with sodium at the origin. Discuss differences or similarities of the resulting diagram as compared to Figure 26-5.

2. Show that a simple cubic unit cell of ions A with an ion B at the center of the cube contains a 1:1 ratio of A to B. Cesium chloride adopts this structure. Using Avogadro's number and the experimental value of a for cesium chloride, 4.110Å, calculate the density of crystalline cesium chloride.

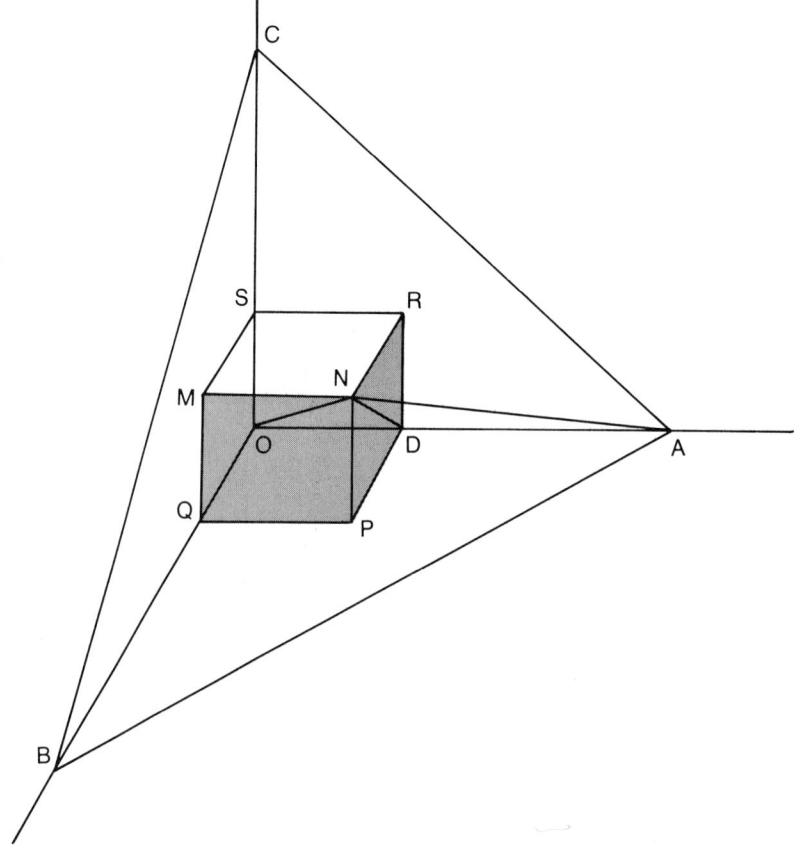

---

*Asterisks require an entry in the notebook; see "Notebook and Report," page 328.

3. Derive equations (2) and (3). To do so proceed as follows: The distance required is the perpendicular distance O to N in the figure on page 328 where N is a point on the plane ABC. Using the three coordinate planes as three sides, construct the rectangular parallelpiped MNPQODRS in such a way that N and O are corners. Note that ON is a diagonal of this parallelepiped.

Put a plane through the three points O, N, and A. The intersection of this plane on the plane ABC is the line AN; its intersection with the parallele-piped face PDRN is DN. Furthermore, ON is perpendicular to AN because ON is perpendicular to the plane ABC and, hence, perpendicular to any line in that plane passing through N. The triangle ONA is thus a right triangle and OA is its hypothenuse. Also, ND is perpendicular to OD because the parallel-epiped was constructed as a rectangular one; hence, the triangle OND is a right triangle and ON is its hypothenuse. Since ONA and OND are right triangles with the common angle DON, they must be similar triangles, and corresponding sides must be proportional to each other. Using this informa-tion, ON may be expressed in terms of OD and OA, and OA is known (it is $\frac{a}{h}$).

By a corresponding plane through B, N, and O, express ON in terms of OQ and OB, where OB is $\frac{b}{k}$.

A plane through C, O, and N, relates ON to OS and OC, where OC is $\frac{c}{l}$.

Combine these results with a relation between ON (which is the inter-planar spacing, d) as the diagonal of the parallelepiped that has edges OD, OQ, and OS to express ON in terms of $\frac{a}{h}, \frac{b}{k}, \frac{c}{l}$.

4. Prove that $\phi = 4\theta$, equations (6) and (7). Start with Figure 26-2, and draw a line aa perpendicular to Xa. Remember that vertical angles are equal and that the sum of the angles in a triangle is 180 degrees.

REFERENCES

1. Shoemaker, D.P., and Garland, C.W.: *Experiments in Physical Chemistry.* McGraw-Hill Book Co., New York, 1967, pp. 348-363.
2. Henry, N.F.M., Lipson, H., and Wooster, W.A.: *Interpretation of X-ray Diffraction Photographs.* Macmillan, London, 1960, Chapters 11-13.
3. Azaroff, L.V., and Buerger, M.J.: *The Powder Method.* McGraw-Hill, New York, 1958, Chapters 1-7.

INORGANIC
QUALITATIVE ANALYSIS

QUALITATIVE ANALYSIS OF
IONIC MIXTURES

OBJECTIVE

To determine the presence or absence of certain cations and anions in aqueous solutions. To study some of the methods used by chemists to make such analyses.

EQUIPMENT NEEDED

Centrifuge; micro burner; heating bath; 10-ml. graduate; dropping tubes or micropipets; semimicro spatula; 4-ml. test tubes; small casserole; 7.5-cm. watch glass; test tube brush; test tube holder; glass stirring rods; pH paper.

REAGENTS NEEDED

6 M and 12 M hydrochloric acid, HCl; 6 M and 16 M nitric acid, HNO_3; 6M acetic acid, CH_3CO_2H; 3 M and 6 M sodium hydroxide, NaOH; 6 M and 15 M aqueous ammonia, NH_3 or NH_4OH; 5% aqueous thioacetamide solution, CH_3CSNH_2; 3% aqueous hydrogen peroxide solution, H_2O_2; 3 M solutions of ammonium acetate, $CH_3CO_2NH_4$, ammonium chloride, NH_4Cl, ammonium sulfide, $(NH_4)_2S$, ammonium thiocyanate, NH_4SCN; 0.2 M dimethylglyoxime solution in 95% ethanol; 0.10 M mercuric chloride, $HgCl_2$; 0.5 M potassium dichromate, $K_2Cr_2O_7$; 0.05 M potassium ferrocyanide, $K_4Fe(CN)_6$; 0.5 M tin(II) chloride, $SnCl_2$; acetone; ether; clean iron tacks; solid ammonium sulfate, $(NH_4)_2SO_4$; solid oxalic acid, $H_2C_2O_4$; solid ammonium chloride, NH_4Cl; aluminon test solution (0.1% aqueous solution of $C_{19}H_{11}(CO_2NH_4)_3$).

Reagents for anion analysis are given in Experiment 3.

INTRODUCTION

As indicated in earlier experiments, qualitative analysis is the process of determining the substances or chemical species present in a sample of matter. The procedures used in such an analysis depend, of course, on the nature of the sample. For example, the analysis of blood for certain drugs or drug residues requires procedures and techniques different from those used for analysis of automotive exhausts for air pollutants. In this experiment you will be asked to determine the presence or absence of certain cations (and anions, in some cases) in various unknown mixtures. The analytical scheme and procedures will be made available to you, but you may not have to use all portions of the scheme in analyzing a given unknown. Once you have learned the analytical scheme, only a little thought and some common sense will enable you to select those portions and procedures that are appropriate for the particular unknown you are about to analyze.

The analytical scheme provided below will enable you to analyze for the following ions in aqueous solution:

Cations: Ag^+, Pb^{2+}, Hg_2^{2+}, Hg^{2+}, Bi^{3+}, Cu^{2+}, Sb^{3+} or Sb^{5+}, Sn^{2+} or Sn^{4+}, Fe^{2+} or Fe^{3+}, Al^{3+}, Cr^{3+}, Co^{2+}, Ni^{2+}, Zn^{2+}

Anions: CO_3^{2-}, SO_4^{2-}, F^-, Cl^-, Br^-, I^-

Schemes are available for the analysis of a much larger group of ions. However, the present scheme contains many of the important features of traditional methods of ion analysis in aqueous solutions.

Before describing the scheme for the ions we have chosen, let us first consider how we can approach the overall problem of analyzing an aqueous mixture containing any or all of a rather large group of ions. For simplicity, let us assume that all of the ions present in the sample are dissolved in the aqueous solution. How then do we proceed to ascertain the identity of these ions? Ideally, of course, it would be desirable to have a chemical probe that when placed in the solution would simply indicate which ions were present, or failing this, to have a unique chemical test for each ion that could be performed on the solution. Unfortunately no such chemical probe has been developed, and the presence of the other ions often interferes with the established tests for a given ion. Most tests for ions are valid only when that ion is separated from most other ions. Hence an important part of qualitative analysis of ions in aqueous solution is the separation of the ions from one another. Once an ion has been separated — and isolated — specific chemical tests can then be performed.

Perhaps the simplest way to separate and isolate an ion from a complex mixture is to add a reagent that will cause this ion — and none of the others in the original mixture — to precipitate from the solution. The precipitate can then be physically separated from the solution and later caused to dissolve forming a new solution — one in which the ion is said to be isolated. This new solution can then be tested to confirm the presence of the ion.

In practice it often is convenient to precipitate and separate not single ions but groups of ions. The groups of precipitated ions can then be treated chemically so as to effect further separations and ultimate isolation and testing.

In designing procedures for separation, isolation and testing of the ions, the chemist makes use of the known properties of the ions, such as the solubility or insolubility of certain of their compounds, their ability to form complex ions, to oxidize or not to oxidize with certain reagents, etc. In this experiment, each of the separations of one ion from another, or of groups of ions, is made on the basis of a difference in solubility, but this difference in solubility is arrived at by making use of a variety of chemical facts. As you proceed with the experiments, you should look for the following chemical principles:

a) Although most chlorides are quite soluble in water, a very few are insoluble, and advantage can be taken of this gross difference in solubility.

b) Some compounds, although insoluble in the cold, are sufficiently soluble in hot water to make it possible to dissolve them away from other compounds which remain insoluble even at higher temperatures.

c) Compounds of some ions may be dissolved by causing the cation present to form a complex ion, thus making it possible to separate the cation from other cations which do not form complex ions.

d) In a way similar to (c), some ions are amphoteric while others are not, and advantage of this may be taken to dissolve compounds of the amphoteric ones while leaving insoluble the compounds of the ions which do not show the property of amphoterism.

e) Salts, M^+, A^- of a weak acid HA tend to be soluble in acidic solutions. If the equilibrium $M^+A^- + H_3O^+ \rightleftharpoons HA + H_2O + M^+$ can be pushed far enough to the right to bring $[A^-]$ to a sufficiently low value, the salt M^+A^- will dissolve, making it possible to separate it from a less soluble salt of a different metal ion.

f) Some cations undergo oxidation to form a soluble salt in the oxidized state, thus making it possible to separate the oxidizable cation from another which is not oxidized by the reagents used.

g) Some compounds dissolve more rapidly than others, so that a timed procedure can separate the rapidly dissolving ones from slowly dissolving compounds.

While the sequence of reactions which lead to the *separation* of one kind of ion from other kinds often produce definite indications that that particular kind is present, it is usually desirable to perform a special test for *identification* of the ion. Such an identifying test *must* be carried out if the separation leads to a colorless solution of the ion, since it is then necessary to show that the ion is indeed present, and that the "solution" of the ion is not simply water containing only previously added reagents. As in the separation procedures discussed above, the identification tests involve several chemical procedures:

a) A precipitate of characteristic color or form may be produced without a change in oxidation state of the ion.

b) The metallic ion present may be (i) reduced to the metal, or (ii) reduced or oxidized to some insoluble compound, often of some characteristic color or form.

c) The ion may be (i) oxidized or (ii) reduced to a soluble form which imparts a characteristic color to its solution.

d) The characteristic precipitate or colored solution may result from oxidation or reduction of an added reagent, such oxidation or reduction being possible only if the ion to be identified is present in the solution. In this case the identifying substance does not contain the element being identified at all.

e) A complex ion of the ion to be identified is formed. This complex ion may give a characteristic color to the solution, or it may form an insoluble precipitate.

f) The precipitate formed for identification as in (a) above may be of such a nature as to be almost invisible under ordinary conditions. To make it more visible, a dye may be added to the solution. Adsorption of the dye on the precipitate gives a color to it which makes it more readily seen.

IMPORTANT LABORATORY OPERATIONS

The procedures you will use in this experiment employ a laboratory technique known as semimicro — or very small scale — analysis. Semimicro

FIGURE 27-1

Transfer of a precipitate using a small spatula.

methods were used in Experiment 5; they are faster and more conservative of materials than are the macro methods you have been using for most of the experiments. Small (4-ml.) test tubes are used for most of these reactions. Reagents are added dropwise using a dropping tube (similar to a medicine dropper) or a 15-μl pipet. Filtering is replaced by first centrifuging the reaction mixture thereby causing the precipitate to settle to the bottom of the test tube, and then removing the clear solution above the precipitate with a pipet or dropping tube — a process known as decanting. Precipitates are transferred using small spatulas (Figure 27-1). A brief description of some important semimicro techniques follows:

Transfer of Reagents. Since liquid reagents are added in drops, pipets and dropping tubes (Figure 27-2) must be both readily available and kept absolutely clean. A drop from a dropping tube has about twice the volume of one from a 15 μl pipet. In the procedures that follow, a drop refers to that from a dropping tube. If the 15 μl pipets are used, the number of drops should be doubled. Tubes should be *thoroughly rinsed immediately after each use.* Otherwise residues of previous uses may remain on the walls of the tubes causing contamination of the sample and misleading or incorrect results.

Great care should be taken to avoid adding a large excess of any reagent since this frequently displaces the chemical equilibria in solution so as to cause analytical difficulties. However, reagents must be added until precipitation appears to cease, that is, until an added drop of reagent fails to produce

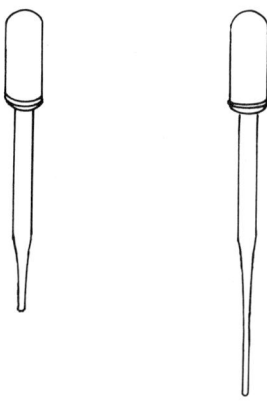

FIGURE 27-2

A. A dropping tube;   B. A micropipet.

A   B

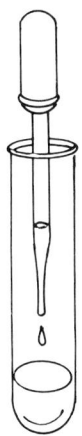

FIGURE 27-3

Transfer of a reagent using a dropping tube or a micropipet.

further precipitate as it diffuses through the reaction mixture. Figure 27-3 illustrates the transfer of a reagent.

Heating Solutions. Solutions can be heated by placing the test tubes in a beaker of hot or boiling water. If a flame is used, the test tube should be moved around the edge of the flame so a large area of the tube is being heated. Point the tube away from your neighbor's face, and take care to avoid over-heating the small test tubes to prevent the sudden ejection of liquid from the mouth of the tube.

Decantation. This consists of centrifuging followed by removal of the decantate (the clear liquid) from the precipitate.

Centrifugation is the most rapid method of separating a precipitate from a solution. Power driven centrifuges (Figure 27-4) are easy to operate but certain precautions must be observed in using them. The centrifuge must first be balanced. This is done by placing a second test tube containing an equivalent volume of water in the holder opposite the tube containing the mixture to be centrifuged. Care must be taken to stop the centrifuge slowly and without vibration. Otherwise the precipitate in the bottom of the tube will be disturbed. Care also must be taken to avoid catching ties, hair, apron strings, etc. in the centrifuge.

Most mixtures require no more than a minute of centrifugation to separate a precipitate from the decantate. After centrifugation, the clear liquid can be withdrawn from the precipitate using a pipet with a long stem.

Washing Precipitates. Following removal of the decantate, the precipitate must be washed to remove from it any ions adsorbed from the solution

FIGURE 27-4

A power driven centrifuge.

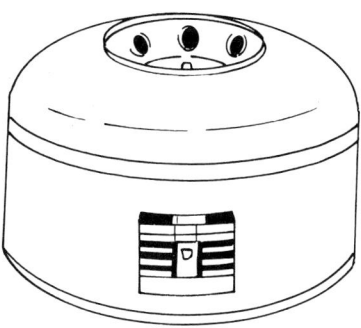

from which it was separated. Washing usually consists in adding 1 or 2 ml. of distilled water to the precipitate, agitating the mixture with a stirring rod, centrifuging it, and removing the wash water with a pipet.

Dissolving Precipitates. When possible precipitates are dissolved in the test tubes in which they are formed using the smallest possible volume of dissolving reagent. Large precipitates frequently are transferred to casseroles which can be manipulated easily.

Adjusting the pH of a Solution. To make acidic solutions basic or basic solutions acidic, the appropriate solution is added dropwise until, based on your knowledge of the situation, you feel the pH is close to that desired. Stir the mixture well with a stirring rod and then touch the end of the rod to a piece of pH paper. Add reagent to the mixture and repeat the test until the pH reaches the desired point.

NOTEBOOK AND REPORT

Good records are especially important in qualitative analysis. A highly useful approach is to record the procedure used and the observations made during each step of the analysis in the form of a flow sheet for each group (see page 341). When a confirmatory test is inconclusive or the results otherwise ambiguous, a careful examination of the observations made throughout the analysis may clarify matters. Moreover, your instructor will require a report on each sample analyzed, and you will be expected to cite evidence to justify your conclusions.

GENERAL SUGGESTIONS

Before analyzing an unknown you should work through the procedures with a sample that is known to contain all the ions included in the scheme. In this way you can familiarize yourself with procedures, techniques and the kinds of observations that can and must be made. As you work through the scheme try to relate what you see to the chemistry that is taking place in the reaction mixture.

In analyzing an unknown sample your main task is to identify the ions present. To do this you should have a good idea of how the ions will behave at each point. Hence you must have more than a cookbook comprehension of the scheme of analysis.

When your instructor issues an unknown, he may tell you that certain ions are not present in your sample. In such cases you may wish to omit or modify certain portions of the scheme. Considerable time and effort can be saved by planning each analysis in accordance with the special information provided with the unknown sample. If your unknown contains ions from only one specific group, omit the tests for ions in the other groups.

SCHEME FOR CATION ANALYSIS

The cations included in this scheme can be separated into three groups. The members of the groups and the conditions for precipitation are summarized below:

Group I Ions:	Pb^{2+}, Ag^+, Hg_2^{2+}	Precipitated as chlorides from acid solution
Group II Ions:	Hg^{2+}, Pb^{2+}, Bi^{3+}, Cu^{2+} Sb^{3+} or $Sb(OH)_6^-$, Sn^{2+} or Sn^{4+}	Precipitated as sulfides from 3 M acid solution.
Group III Ions:	Fe^{3+}, Al^{3+}, Cr^{3+}, Co^{2+}, Ni^{2+}, Zn^{2+}	Precipitated as hydroxides or sulfides from dilute base solutions

A. ANALYSIS OF GROUP I

Of the cations included here, only lead, Pb^{2+}, silver, Ag^+ and mercury(I), Hg_2^{2+}, form insoluble chlorides. Addition of cold dilute hydrochloric acid to a solution containing any or all of these ions brings about the following reactions:

$$Pb^{2+} + 2Cl^- \rightleftharpoons \underline{PbCl_2} \quad \text{(white precipitate)}$$

$$Ag^+ + Cl^- \rightleftharpoons \underline{AgCl} \quad \text{(white precipitate)}$$

$$Hg_2^{2+} + 2Cl^- \rightleftharpoons \underline{Hg_2Cl_2} \quad \text{(white precipitate)}$$

(A line under a formula in an equation will be used to indicate that this substance is present as a precipitate. The non-underlined formulas of other ions will indicate a soluble species.) Actually this group can be precipitated by any substance which provides chloride ions in solution (e.g., NaCl or KCl); HCl is used because the dilute acid environment prevents precipitation of Groups II and III ions as hydroxides (or as hydrolysis species such as BiOCl) and assures more complete precipitation of the Group I chlorides. The mixture is kept cool since the solubility of $PbCl_2$ increases rapidly with increasing temperature. Addition of excess HCl (or Cl^-) must be avoided because it causes the chlorides to dissolve, forming the complex ions $PbCl_4^{2-}$, $AgCl_2^-$ and $Hg_2Cl_3^-$.

Once the chlorides have been precipitated, the mixture can be centrifuged (see page 337) and the solution separated (or decanted) from the precipitate. The solution or decantate can be set aside to be analyzed later for Groups II and III. The precipitate can be analyzed to determine which members of Group I are present.

Separation of Lead. Lead ion can be separated from silver ion and mercurous ion by treating the precipitate of Group I chlorides with hot water. This will dissolve lead chloride but not silver chloride or mercury(I) chloride.

$$PbCl_2 \underset{\text{hot water}}{\overset{\text{cold water}}{\rightleftharpoons}} Pb^{2+} + 2Cl^-$$

The hot water containing Pb^{2+} can now be separated from the insoluble AgCl and Hg_2Cl_2, and this solution tested to confirm the presence of Pb^{2+}.

Confirmation of Lead. Potassium dichromate, $K_2Cr_2O_7$, is used to confirm the presence of Pb^{2+} ion. The reactions are:

$$Cr_2O_7{}^{2-} + 3H_2O \rightleftharpoons 2CrO_4{}^{2-} + 2H_3O^+$$

$$Pb^{2+} + CrO_4{}^{2-} \rightleftharpoons PbCrO_4 \quad \text{(yellow precipitate)}$$

In the first step the dichromate ion, $Cr_2O_7{}^{2-}$, reacts with water to establish an equilibrium in which chromate ion, $CrO_4{}^{2-}$, is produced. Lead ion reacts with this ion to give yellow lead chromate, $PbCrO_4$, which is less soluble than $PbCr_2O_7$.

Separation of Silver and Mercury(I) Chlorides. The precipitate remaining after lead chloride is dissolved with hot water can be treated with ammonia solution to separate Ag^+ from $Hg_2{}^{2+}$. The important reactions are:

$$AgCl + 2NH_3 \rightleftharpoons Ag(NH_3)_2{}^+ + Cl^-$$

$$Hg_2Cl_2 + 2NH_3 \rightleftharpoons \underset{\text{white}}{HgNH_2Cl} + \underset{\text{black}}{Hg} + NH_4{}^+ + Cl^-$$

Silver chloride dissolves with the formation of the soluble complex ion $Ag(NH_3)_2{}^+$. Mercury(I) chloride undergoes an unusual reaction in which the white mercury amido chloride, $HgNH_2Cl$, and black metallic mercury, Hg, are formed. The mixture of the two often has a salt and pepper appearance. Any remaining lead chloride reacts to form another white salt, PbOHCl, the appearance of which will not interfere with either of the other reactions.

Confirmation of Mercury(I) Ion. The salt and pepper appearing precipitate formed on treatment of Hg_2Cl_2 with NH_3 is usually sufficient to confirm $Hg_2{}^{2+}$. However, when silver is present, a small amount of the silver in AgCl may be reduced to metallic silver, Ag, which also appears black. To distinguish between silver and mercury, the black residue can be heated in aqua regia (a mixture of concentrated HCl (12 M) and concentrated HNO_3 (16 M)). The reactions are:

$$Ag + 2H_3O^+ + 2Cl^- + NO_3{}^- \rightleftharpoons NO_2 \uparrow + AgCl_2{}^- + 3H_2O$$

$$Hg + 4H_3O^+ + 3Cl^- + 2NO_3{}^- \rightleftharpoons 2NO_2 \uparrow + HgCl_3{}^- + 6H_2O$$

When this reaction mixture is cooled and added to water, a white precipitate of AgCl appears if some or all of the black residue was silver. Mercury, if present, can be confirmed by treating the resulting solution with tin(II) chloride. In this reaction Hg^{2+} is reduced to Hg_2^{2+} which precipitates with Cl^- as white Hg_2Cl_2.

$$Hg^{2+} + Sn^{2+} + 2Cl^- \rightleftharpoons \underset{\text{white}}{\underline{Hg_2Cl_2}} + Sn^{4+}$$

Hence, a white precipitate at this point confirms the presence of mercury.

Confirmation of Silver Ion. If silver is present in the sample, it will now be in the form of the complex ion $Ag(NH_3)_2{}^+$ in the decantate from the treatment of the $AgCl/Hg_2Cl_2$ precipitate with ammonia. To confirm its presence the complex ion must be decomposed and the silver ion allowed to reprecipitate as AgCl. This is accomplished by adding a small amount of nitric acid, HNO_3. The important reactions are:

$$Ag(NH_3)_2{}^+ + 2H_3O^+ \rightleftharpoons Ag^+ + 2NH_4{}^+ + 2H_2O$$

$$Ag^+ + Cl^- \rightleftharpoons \underline{AgCl}$$

The chloride ions for the second reaction above are present in the solution as a result of dissolving AgCl with NH_3.

FLOW SHEET GROUP I
A GRAPHICAL METHOD TO SHOW ION SEPARATION AND IDENTIFICATION.

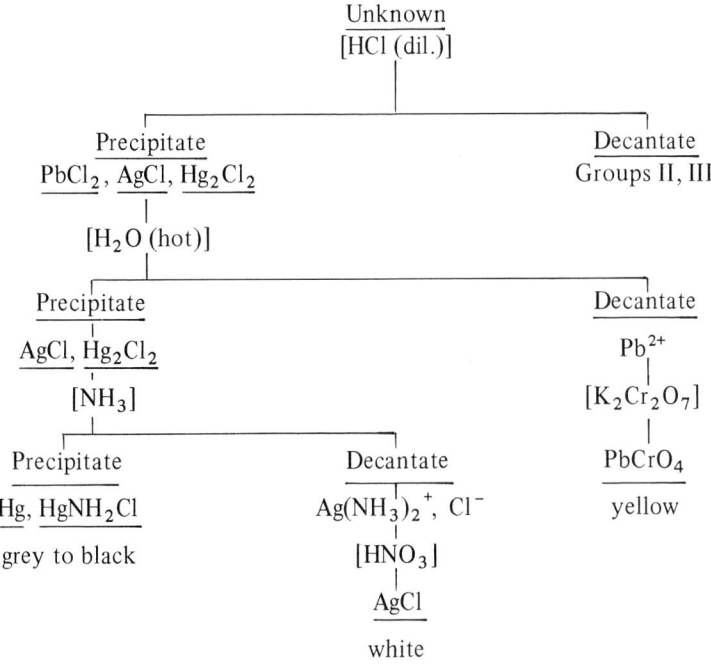

PROCEDURE

The solution to be analyzed may contain any or all of the cations of Groups I, II, III.* Add 5 drops of the cold solution to a 4-ml. test tube and then add 1 drop of 3 M HCl. Mix thoroughly and centrifuge the mixture. Add another drop of HCl solution and if precipitation is incomplete, add one more drop. Centrifuge the mixture and decant the supernatant liquid with a dropping tube. *This decantate is saved for the analysis of Groups II and III.*

Wash the precipitate remaining in the test tube with 1 ml. of cold water containing some HCl (1 drop of HCl in 4 ml. of water) and add the washings to the previous decantate.

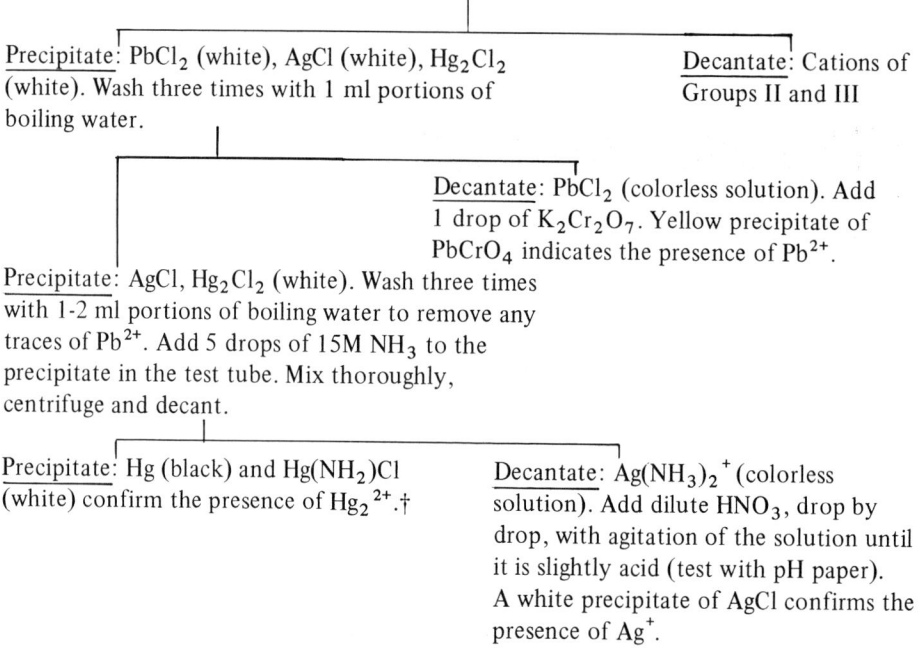

Precipitate: $PbCl_2$ (white), AgCl (white), Hg_2Cl_2 (white). Wash three times with 1 ml portions of boiling water.

Decantate: Cations of Groups II and III

Decantate: $PbCl_2$ (colorless solution). Add 1 drop of $K_2Cr_2O_7$. Yellow precipitate of $PbCrO_4$ indicates the presence of Pb^{2+}.

Precipitate: AgCl, Hg_2Cl_2 (white). Wash three times with 1-2 ml portions of boiling water to remove any traces of Pb^{2+}. Add 5 drops of 15M NH_3 to the precipitate in the test tube. Mix thoroughly, centrifuge and decant.

Precipitate: Hg (black) and $Hg(NH_2)Cl$ (white) confirm the presence of Hg_2^{2+}.†

Decantate: $Ag(NH_3)_2^+$ (colorless solution). Add dilute HNO_3, drop by drop, with agitation of the solution until it is slightly acid (test with pH paper). A white precipitate of AgCl confirms the presence of Ag^+.

---

*If your unknown is a solid sample, follow the directions given below for a Group I or a General Unknown.

Group I: Prepare 1 ml. of *cold* 12 M HCl to which has been added *1* drop (no more) of 3 M HNO_3. Place approximately 50 mg. of the unknown powder into this solution and stir vigorously for a *few* seconds (30 seconds *maximum*). If the precipitate is not completely white, add 1 more drop of 3 M HNO_3 and again stir vigorously for a *few* seconds (30 seconds maximum). Immediately dilute with an equal volume of water, stir, centrifuge and decant. The precipitate is a mixture of Group I chlorides and is treated according to the procedure starting at the asterisk above.

General   Treat 100 mg. of the unknown powder with 1 ml. of 12 M HCl. Immerse in boiling water
Unknown: and stir continuously until all has been dissolved except Group I. Add 1 ml. of water and stir. Cool thoroughly and centrifuge. The precipitate is a mixture of Group I chlorides and is treated as directed above at the asterisk. The decantate, which contains Groups II and III, is treated for Group II as directed on page 351.

†If silver is present and the test for Hg_2^{2+} is inconclusive, transfer the residue to a casserole. Add 3 drops of 12 M HCl and 1 drop of 16 M HNO_3 and heat gently. Dilute the solution with 2 ml. of water; any silver present will appear as white AgCl. Centrifuge if necessary and add 2 drops of $SnCl_2$ solution to the decantate. A white or gray precipitate confirms the presence of Hg_2^{2+}.

QUESTIONS, GROUP I ANALYSIS

1. Which of the separation procedures listed on pages 334 and 335 was involved in separation of the Group I cations from those of Groups II and III?

2. Which of the separation procedures listed on pages 334 and 335 was involved in the separation of lead chloride from the chlorides of mercury(I) and silver?

3. Which of the separation procedures listed on pages 334 and 335 was involved in the separation of silver ion from the mercury(I) ion?

4. Which of the identification items listed on page 335 was involved in identification of lead ion? Write the equation for the reaction that occurred on adding potassium dichromate to the lead ion solution.

5. Write the equation for the reaction which occurred in identification of silver ion. Does this identification correspond to any listed on page 335? If so, to which one?

6. Which of the identification items listed on page 335 was involved in identifying mercury(I) ion?

7. Using the information provided on Group I analysis, how would you distinguish between the substances in each of the following pairs (i) if both substances are in the same container? (ii) if each substance is in a separate container? Note that an answer to (i) requires a separation. An answer to (ii) requires only that you treat both substances in the same way or with the same reagent and show that each behaves in a different characteristic way.
   - a. Hg_2Cl_2 and $AgCl$
   - b. $PbCl_2$ and Hg_2Cl_2
   - c. $AgNO_3$ and $Hg_2(NO_3)_2$
   - d. $PbCrO_4$ and $AgCl$
   - e. KCl and Hg_2Cl_2
   - f. $Pb(NO_3)_2$ and K_2CrO_4

8. You are given an aqueous solution that may contain any, all or none of the ions given in each of the lettered items below. In each case cite a reagent, reagents or an operation that if added or performed would enable you to determine which, if any, of the ions are present. State the observations that would lead you to the various possible conclusions:
   - a. Ag^+ and Pb^{2+}
   - b. Ag^+ and Hg_2^{2+}
   - c. Cu^{2+} and Pb^{2+}
   - d. Sn^{2+} and Hg_2^{2+}
   - e. CrO_4^{2-} and Hg_2^{2+}
   - f. Hg^{2+} and Hg^{2+}

9. A solid sample is known to contain any or all of the following: $Pb(NO_3)_2$, $AgCl$, Hg_2Cl_2, K_2CrO_4, $CuCl_2$ and KCl. It gave a white precipitate and a colorless solution when treated with water. Approximately half of the precipitate dissolved when treated with concentrated ammonia, leaving a gray residue. On the basis of this information classify each of the following ions as: *definitely present, definitely absent* or *undetermined* in the original sample: Pb^{2+}, NO_3^-, Ag^+, Cl^-, Hg_2^{2+}, K^+, CrO_4^{2-}, Cu^{2+}.

REFERENCES

1. Garrett, A.B., Sisler, H.H., Bonk, H., and Stoufer, R.C.: *Semimicro Qualitative Analysis.* Blaisdell Publishing Company, Waltham, Massachusetts, 1966.
2. Layde, D.C., and Busch, D.H.: *Introduction to Qualitative Analysis,* 2nd edition. Allyn and Bacon, Inc., Boston, Massachusetts, 1971.

Experiment 27
REPORT SHEET FOR GROUP I

Name _____   Section _____

Results at Each Separation Part of the Flow Sheet

Ions Present in Unknown   _____

Observations That Differed from the Text Presentation

Attach Answers to the Assigned Questions

B. ANALYSIS OF GROUP II

Following the removal of the cations of Group I as chlorides by decantation, the cations of Group II may be separated from those of Group III by making use of the fact that the Group II cations form sulfides that are insoluble in dilute acid solution. The concentration of acid in this solution is critical: too high a concentration will cause incomplete precipitation of the Group II cations and too low a concentration will cause some of the Group III cations to precipitate as sulfides. The optimum concentration here is 0.3 M and the procedures have been developed so that if followed this concentration will be achieved.

The ions of Group II are: mercury(II), Hg^{2+}, lead, Pb^{2+} (which has not been completely removed by the Group I separation), bismuth, Bi^{3+}, copper, Cu^{2+}, antimony(III) and (V), Sb^{3+} and $Sb(OH)_6^-$, and tin(II) and (IV), Sn^{2+} and Sn^{4+}.

Before proceeding to precipitate Group II, it is necessary to be certain that all the tin present is in the IV oxidation state, since Sn(II) can reduce Hg^{2+} to Hg_2^{2+}. In addition, a separation to come later, namely that of separating tin and antimony from the other Group II elements, will not be effective if tin is present as Sn(II). Tin(II) is oxidized to tin(IV) using aqua regia. This oxidizing agent also oxidizes antimony(III) to antimony(V). This latter oxidation will cause no problem.

Precipitation of Group II. Two reagents are needed to precipitate Group II. The first is nitric acid which, as has been indicated, needs to be present in 0.3 M concentration. The second is thioacetamide, CH_3CSNH_2. The latter reagent is used as a convenient source of H_2S, which is formed by the following hydrolysis reaction:

$$CH_3CSNH_2 + 2H_2O \rightleftharpoons H_2S + NH_4^+ + CH_3CO_2^-$$

The H_2S produced reacts to establish the following equilibria and gives sulfide ions:

$$H_2S + H_2O \rightleftharpoons H_3O^+ + HS^-$$

$$HS^- + H_2O \rightleftharpoons H_3O^+ + S^{2-}$$

These equilibria indicate that the sulfide ion concentration may be controlled by adjustment of the hydronium ion concentration. Thus, if the hydronium ion concentration is increased, the sulfide ion concentration can be diminished and vice versa. By keeping the hydronium ion concentration at 0.3 M and using the amount of thioacetamide indicated in the procedures, the sulfide ion concentration can be controlled so as to precipitate the Group II cations effectively without precipitating those from Group III. However, if the hydronium ion concentration becomes too high, lead sulfide probably will not precipitate. Conversely, if the hydronium ion concentration is too low, the sulfide ion concentration in the H_2S equilibrium will be increased to the point that some sulfides of Group III, namely sulfides of zinc, cobalt and nickel, may precipitate.

While the equations for the precipitation reactions are complex in some cases, they can be represented simply as follows:

$$M^{2+} + S^{2-} \rightleftharpoons \underline{MS} \qquad\qquad \text{where } M^{2+} = Hg^{2+}, Pb^{2+}, Cu^{2+}$$

$$2Bi^{3+} + 3S^{2-} \rightleftharpoons \underline{Bi_2S_3}$$

$$SnCl_4 + 2H_2S + 4H_2O \rightleftharpoons \underline{SnS_2} + 4H_3O^+ + 4Cl^-$$

$$2Sb(OH)_6^- + 5H_2S + 2H_3O^+ \rightleftharpoons \underline{Sb_2S_5} + 14H_2O$$

It is important to know the colors of the sulfides of Group II. By observing the colors formed on the precipitation of this group, it frequently is possible to get a hint as to which cations are present in an unknown. The colors of these sulfides are:

HgS — most often black, sometimes red; PbS — black; Bi_2S_3 — dark brown; CuS — black; Sb_2S_5 — orange red; SnS_2 — yellow.

Separation of Group II into Subgroups. Because it is more convenient to work with smaller numbers of ions, Group II will be divided into subgroups known as the copper and antimony subgroups. The basis for this separation is the fact that antimony and tin are amphoteric elements and as a result their sulfides will dissolve in excess base, whereas those of the other members of Group II, the copper subgroup, will not dissolve under these conditions. A 3 M sodium hydroxide solution will effect this separation. The sulfides insoluble under these conditions are:

HgS, PbS, Bi_2S_2 and CuS — which constitute the copper subgroup. Equations for the dissolution of the antimony subgroup are:

$$4Sb_2S_5 + 24OH^- \rightleftharpoons 5SbS_4{}^{3-} + 3SbO_4{}^{3-} + 12H_2O$$

$$3SnS_2 + 6OH^- \rightleftharpoons 2SnS_3{}^{2-} + SnO_3{}^{2-} + 3H_2O$$

Copper Subgroup Analysis. A 4 M nitric acid solution will dissolve the sulfides of lead, bismuth and copper, but will not dissolve HgS. This is the basis for the separation of Hg^{2+} from the other cations of the copper subgroup. The reactions with HNO_3 are:

$$\underline{3MS} + 8H_3O^+ + 2NO_3^- \rightleftharpoons 3M^{2+} + 2NO + 12H_2O + 3S$$
$$\text{where MS} = PbS, CuS$$
$$\underline{Bi_2S_3} + 8H_3O^+ + 2NO_3^- \rightleftharpoons 2Bi^{3+} + 2NO + 12H_2O + 3S$$

Confirmation of Mercury. Mercury is confirmed by first dissolving the sulfide by oxidation using aqua regia. The key reaction here is:

$$3HgS + 8H_3O^+ + 2NO_3^- + 9Cl^- \rightleftharpoons 3HgCl_3^- + 3S + 2NO + 12H_2O$$

The solution is then evaporated carefully to dryness to get rid of excess HCl and HNO_3 and to convert $HgCl_3^-$ to $HgCl_2$. The residue is redissolved in water. This water solution is then treated with tin(II) chloride which reduces the mercury to the I and subsequently to the 0 oxidation state. Under these conditions, mercury(I) precipitates initially as white Hg_2Cl_2, which is then reduced to metallic mercury, which appears black. The confirmatory test for mercury then is observed as a white precipitate turning to gray as the black mercury is formed and mixes with the white Hg_2Cl_2. Ultimately the mixture turns black as all mercury present is reduced to the 0 oxidation state. The important equations here are:

$$2HgCl_2 + Sn^{2+} + 2Cl^- \rightleftharpoons SnCl_4 + \underline{Hg_2Cl_2} \text{ (white)}$$

$$Hg_2Cl_2 + Sn^{2+} + 2Cl^- \rightleftharpoons SnCl_4 + \underline{2Hg} \text{ (black)}$$

Separation and Confirmation of Lead. Pb^{2+}, Bi^{3+}, and Cu^{2+}, if present in the sample, are in the decantate resulting from the treatment of the copper subgroup precipitate with the nitric acid. Lead ion can be precipitated in the presence of the bismuth and copper ions by using sulfate ion, which forms white insoluble lead sulfate, $PbSO_4$. The reagent used to supply sulfate ions is ammonium sulfate, $(NH_4)_2SO_4$.

This reagent is added as a solid and stirred into the solution to provide a high concentration of sulfate ions. After stirring for a few minutes, the solution is cooled to minimize the possibility of lead sulfate forming supersaturated solutions. If no precipitate forms at this point, the solution should be allowed to stand for several minutes before it is concluded that lead is not present. Even small amounts of white precipitate at this point should be carried over to the confirmatory tests for lead. Since lead sulfate is highly granular and very efficiently packed, small amounts of this material often give voluminous precipitates of $PbCrO_4$ in the confirmatory test for lead. Lead ion is confirmed by separating the lead sulfate from its decantate and dissolving it in a solution of ammonium acetate, $CH_3CO_2NH_4$. Acetate ion forms a complex with lead which displaces the equilibrium

$$\underline{PbSO_4} + CH_3CO_2^- \rightleftharpoons PbCH_3CO_2^+ + SO_4^{2-}$$

to the right, causing the lead ion to dissolve. Once the lead is dissolved, it is reprecipitated as the brilliant yellow lead chromate, $PbCrO_4$, by adding potassium dichromate reagent, $K_2Cr_2O_7$, to the solution. The important reaction here is:

$$2PbCH_3CO_2^+ + Cr_2O_7^{2-} + H_2O \rightleftharpoons 2CH_3CO_2H + \underline{2PbCrO_4} \text{ (yellow)}$$

Separation and Confirmation of Bismuth. The decantate from the precipitation of lead sulfate may contain Bi^{3+} and Cu^{2+} ions. Bismuth ions are separated from copper ions by addition of concentrated ammonia, which leads to the formation of insoluble bismuth hydroxide, $Bi(OH)_3$, and the formation of the soluble complex ion, $Cu(NH_3)_4^{2+}$. Bismuth hydroxide is a white precipitate; the tetraammine copper(II) ion is a deep blue. Normally a white precipitate at this point indicates the presence of bismuth and a deep blue solution characteristic of the $Cu(NH_3)_4^{2+}$ ion is an indication of the presence of copper. The key reactions here are:

$$NH_3 + H_2O \rightleftharpoons NH_4^+ + OH^-$$

$$Bi^{3+} + 3OH^- \rightleftharpoons \underline{Bi(OH)_3}$$

$$Cu^{2+} + 4NH_3 \rightleftharpoons Cu(NH_3)_4^{2+} \text{ (blue)}$$

Bismuth ion is confirmed by adding a specific reducing agent which will reduce bismuth(III) ion to metallic bismuth. This will appear as a black precipitate in the reaction mixture. The specific reducing agent used for this is sodium stannite solution. The important reaction here is:

$$2Bi(OH)_3 + 3HSnO_2^- + 3OH^- \rightleftharpoons 2Bi(black) + 3SnO_3^{2-} + 6H_2O$$

Confirmation of Copper. This is accomplished by treating the decantate from the separation of bismuth with acetic acid to destroy the ammine complex. The reaction here is:

$$Cu(NH_3)_4^{2+} + 4CH_3CO_2H \rightleftharpoons Cu^{2+} + 4NH_4^+ + 4CH_3CO_2^-$$

The copper ion can now be precipitated as the reddish brown copper ferrocyanide, $Cu_2Fe(CN)_6$, by addition of potassium ferrocyanide reagent, $K_4Fe(CN)_6$. The key reaction here is:

$$2Cu^{2+} + Fe(CN)_6^{4-} \rightleftharpoons Cu_2Fe(CN)_6 \text{ (reddish brown)}$$

Antimony Subgroup Analysis. At this point in the analysis, the members of the antimony subgroup are found dissolved in a basic solution in the form of their anions, SbS_4^{3-}, SbO_4^{3-}, SnS_3^{2-}, or SnO_3^{2-}. Because the confirmatory tests for antimony and tin are much more easily performed on the chloro complex ions of these elements that on their oxo- or thio- anions, concentrated hydrochloric acid is added to the solution. This reagent functions first to neutralize the base and then to convert the anions of antimony and tin, if present, to the corresponding chloro anions of these elements, $SbCl_6^-$ and $SnCl_6^{2-}$. The reactions can be represented as:

$$SbS_4^{3-} + 8H_3O^+ + 6Cl^- \rightleftharpoons SbCl_6^- + 4H_2S + 8H_2O$$

$$SnS_3^{2-} + 6H_3O^+ + 6Cl^- \rightleftharpoons SnCl_6^{2-} + 3H_2S + 6H_2O$$

The solution formed as a result of the treatment with concentrated hydrochloric acid can now be divided into two parts, one part to be used to test for the presence of tin, the other for the presence of antimony.

To confirm tin, it is necessary to remove the antimony, if present, from solution and also to reduce the tin from the IV to the II oxidation state. A clean iron nail placed in the solution will serve as the reducing agent and will give rise to a large black precipitate if antimony is present. The reactions here are:

$$SnCl_6^{2-} + Fe \rightleftharpoons Fe^{2+} + 6Cl^- + Sn^{2+}$$

$$2SbCl_6^- + 5Fe \rightleftharpoons 5Fe^{2+} + 2Sb \text{ (black)} + 12Cl^-$$

A heavy black precipitate at this point is definite evidence that antimony is present. After separating the black residue of antimony and the iron nail, the solution can be tested for the presence of tin. If tin is present, it is now in its II oxidation state. In this state, it can reduce mercury from the II to the I and finally to the 0 oxidation state. This is accomplished by adding mercury(II) chloride, $HgCl_2$, and observing the white to gray to black color changes as the mercury is converted to mercury(I) and precipitated as Hg_2Cl_2, and this is further reduced to black metallic mercury according to the equations:

$$Sn^{2+} + 2HgCl_2 + 2Cl^- \rightleftharpoons SnCl_4 + \underline{Hg_2Cl_2} \text{ (white)}$$

$$Sn^{2+} + \underline{Hg_2Cl_2} + 2Cl^- \rightleftharpoons SnCl_4 + \underline{2Hg} \text{ (black)}$$

The second portion of the solution containing the antimony subgroup ions can be tested for the presence of antimony. To do this, the tin ions present must be complexed in such a way that they will not precipitate under the conditions created to precipitate the antimony. This is accomplished by adding oxalic acid, $H_2C_2O_4$. While both antimony and tin form complex ions with this substance, the tin complexes are somewhat more stable than those of antimony, hence, it is possible to add thioacetamide and to cause orange antimony sulfide, Sb_2S_5, to precipitate without having tin sulfide precipitate. The reactions are:

$$SbCl_6^- + 3C_2O_4^{2-} \rightleftharpoons Sb(C_2O_4)_3^- + 6Cl^-$$

$$SnCl_6^{2-} + 3C_2O_4^{2-} \rightleftharpoons Sn(C_2O_4)_3^{2-} + 6Cl^-$$

$$2Sb(C_2O_4)_3^- + 5S^{2-} \rightleftharpoons \underline{Sb_2S_5} + 6C_2O_4^{2-}$$

PROCEDURES FOR GROUP II ANALYSIS

Precipitation and Separation into Subgroups; Analysis of Copper Subgroup. The decantate from the separation of Group I may contain any or all the cations of Groups II and III.* Pour this decantate into a casserole and evaporate to a paste. Add 8 to 10 drops of 12 M HCl and 8 to 10 drops 16 M HNO_3 and evaporate again to a paste. Dissolve the residue in 8 drops of water and 6 drops of 6 M HNO_3. Add enough 5% thioacetamide solution to bring the volume of the mixture to 2 ml. Heat the mixture in a boiling water bath for 5 minutes, meanwhile stirring well. Centrifuge, remove the decantate, place half of it in a 4-ml. test tube and the other half in a second test tube. To each of these test tubes add 10 drops of 5% thioacetamide solution and 10 drops of H_2O. Heat 5 minutes more. To assure complete precipitation, again add 10 drops of 5% thioacetamide and 10 drops of water to each of the tubes and heat for 5 more minutes. Centrifuge, decant and combine the precipitates, save the decantate for analysis of Group III and use the precipitate in the analysis of Group II.

---

*If your unknown is a solid sample of Group II ions only, treat 50 mg. of the unknown powder with 1 ml. of 12 M HCl. Immerse the test tube in boiling water and stir continuously for a few seconds until the powder is dissolved entirely. The resultant solution is treated according to the procedure starting at the asterisk above.

Precipitate: HgS (black), PbS (black), Bi_2S_3 (brown), CuS (black), Sb_2S_5 (orange), SnS_2 (yellow). Wash the precipitate once with 2 ml. of distilled water and discard the washing. Now add 10 drops of 3 M NaOH and stir vigorously for 20-30 seconds. Centrifuge and decant the clear liquid.

Decantate: Stopper tightly and save for analysis of Group III.

Precipitate: HgS, PbS, Bi_2S_3, CuS. Wash the precipitate once with 2 ml. of water and discard the washing. Now add 6 drops of 6 M HNO_3 and heat on a boiling water bath for 2-3 minutes. Centrifuge and decant the liquid into a clean test tube. Repeat this procedure using another 6 drops of 6 M HNO_3 and add this decantate to the above decantate.

Decantate: Antimony subgroup. Stopper tightly and save for analysis of the antimony subgroup.

Precipitate: HgS (black or red) and S (yellow). Dissolve this precipitate in 3 drops of 12 M HCl and 1 drop of 16 M HNO_3, using moderate heat. Add 1 ml. of water and boil the solution just to dryness. Redissolve in 1 ml. of water, centrifuge to remove any sulfur, and add 2 drops of 0.5 M $SnCl_2$. A white to gray to black precipitate confirms the presence of mercury(II).

Decantate: Pb^{2+}, Bi^{3+}, Cu^{2+}. To this decantate add 0.4 grams of solid $(NH_4)_2SO_4$ and stir until this is dissolved. Allow this mixture to stand for 3-5 minutes. If no precipitate forms, cool the mixture for another 5 minutes. Centrifuge and decant.

Precipitate: $PbSO_4$ (white). Wash the residue with hot water three times and discard these washings. Then dissolve it in 4 drops of 3 M $CH_3CO_2NH_4$, heating if necessary. Add 1 drop of 6 M CH_3CO_2H and 2 drops of 0.5 M $K_2Cr_2O_7$. A voluminous yellow precipitate confirms the presence of Pb^{2+}.

Decantate: Bi^{3+} and Cu^{2+}. Add 15 M NH_3 dropwise until the solution is strongly basic. Centrifuge and decant.

Precipitate: $Bi(OH)_3$ (white). Wash this residue once with water and add to the precipitate 2 drops of freshly prepared sodium stannite solution.* A black precipitate of metallic bismuth confirms the presence of Bi^{3+}

Decantate: $Cu(NH_3)_4^{2+}$. If Cu^{2+} is present, this solution should now be a deep blue color. Make the solution acid with 6 M CH_3CO_2H, and add two drops of 0.05 M $K_4Fe(CN)_6$. A red precipitate of $Cu_2Fe(CN)_6$ confirms the presence of Cu^{2+}.

---

*The sodium stannite solution is prepared by adding 6 M NaOH drop by drop, with stirring, to two drops of 0.5 M $SnCl_2$ until the first formed precipitate just dissolves. $3\ OH^- + Sn^{2+} \rightarrow Sn(OH)_3^-$.

Antimony Subgroup Analysis. The decantate resulting from the addition of sodium hydroxide to the Group II precipitate should contain the oxo- and thio-anions of antimony and tin if either of these elements is present. To this solution, add 3 M HCl dropwise and with constant stirring until the solution is barely acidic. Centrifuge, discard the decantate, wash the precipitate with 1 ml. of water and discard the washing. To the precipitate, add 10 drops of 12 M HCl and place in a boiling water bath for at least 3 minutes. Centrifuge the mixture, decant, save the decantate and discard the residue which may consist of sulfur and traces of mercury sulfide if Hg^{2+} is present. *Now divide this decantate into equal parts.* Test one part for antimony, the other for tin.

Test for Antimony. Add 0.5 gram of oxalic acid and 5 ml. of water to the solution. Stir and heat until the oxalic acid is dissolved. Add 5 drops of thioacetamide and heat for 2 minutes. An orange-red precipitate of Sb_2S_5 indicates the presence of antimony. When tin is present, a slowly forming brown precipitate may also appear.

Test for Tin. Add a clean iron nail to the mixture in the test tube and heat for 5 minutes on the water bath. The appearance of a voluminous black precipitate at this point is a good indication of antimony. Centrifuge if necessary, and decant the clear liquid into a test tube containing 4 drops of 0.1 M $HgCl_2$. A white to gray to black precipitate confirms the presence of tin.

FLOW SHEET GROUP II

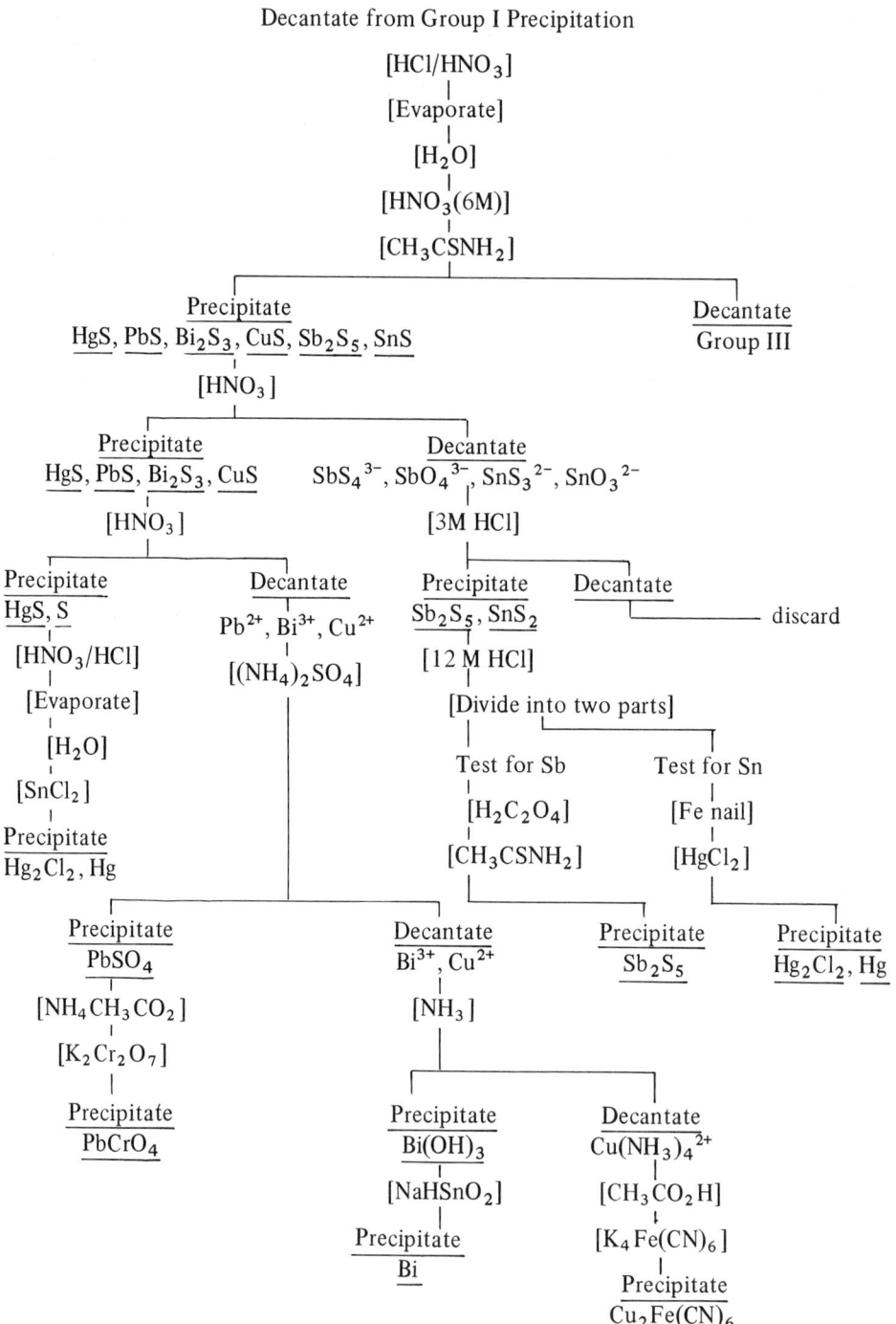

QUESTIONS—GROUPS I AND II

1. Why is 0.3 M acid used in the precipitation of the Group II cations? Calculate the equilibrium concentration of S^{2-} present in a 0.3 M H_3O^+ aqueous solution.

2. In analyzing Group II unknowns, several students in one class used sulfuric acid and several others used hydrochloric acid instead of the nitric

acid prescribed in the procedure for the precipitation of Group II. Some of these students obtained white precipitates upon adding the acid. Account for the white precipitates and for the fact that not all students obtained them.

3. What is the chemical basis for the separation of the sulfides of Group II cations into the copper and antimony subgroups?

4. Write equations based on procedures from Group II analysis to illustrate each of the following:
   a) control of the concentration of sulfide ion in solution
   b) the amphoteric behavior of a metal sulfide
   c) the formation of a complex ion
   d) oxidation by aqua regia
   e) removal of one cation in order to precipitate another
   f) reduction of a cation to its 0 oxidation state

5. How would you distinguish between the substances in each of the following pairs (i) if both members of the pair are in the same container, (ii) if each member of the pair is in a separate container.
   a) lead sulfide and tin(IV) sulfide
   b) bismuth sulfide and copper sulfide
   c) mercury(II) sulfide and lead sulfide
   d) antimony(V) sulfide and tin(IV) sulfide
   e) lead sulfate and bismuth sulfate
   f) lead chloride and lead sulfate

6. You are given an aqueous solution that may contain any, all or none of the ions given in each of the lettered items below. In each case cite a reagent, reagents or an operation that if added or performed would enable you to determine which, if any, of the ions are present. State the observations that would lead you to the various possible conclusions.
   a) Pb^{2+} and Cu^{2+} e) Hg^{2+} and Hg_2^{2+}
   b) Bi^{3+} and Cu^{2+} f) Bi^{3+} and Sb^{5+}
   c) Sn^{4+} and Sb^{5+} g) Sn^{2+} and Hg^{2+}
   d) Hg^{2+} and Sn^{4+} h) Pb^{2+} and Zn^{2+}

7. A solid sample is known to contain any or all of the following salts, but no others: $Hg_2(NO_3)_2$, $SbCl_3$, $Bi(OH)_3$, $CuCl_2$, and $HgCl_2$. It gave a white precipitate and a colorless solution A when treated with enough water to bring about solution. The precipitate (B) was separated from the decantate A. The precipitate was unchanged by the addition of concentrated NH_3, but it dissolved completely in dilute HCl. The decantate A gave a bright orange precipitate (C) when treated with thioacetamide. This precipitate dissolved completely when treated with NaOH. On the basis of these observations, indicate whether each of the following ions was *definitely present, definitely absent* or *undetermined:* Hg^{2+}, NO_3^-, Sb^{3+}, Cl^-, Bi^{3+}, OH^-, Cu^{2+}, Hg^{2+}.

8. A solid sample is known to contain any or all of the following salts, but no others: $Hg_2(NO_3)_2$, $CuCl_2$, $PbCl_2$, $Bi(NO_3)_3$, and $SnCl_4$. It gave a white precipitate and a blue solution A when treated with sufficient water to bring about solution. A major portion of the precipitate (B) dissolved in hot

water, with the remaining portion turning dark when treated with concentrated NH_3. The solution A gave a black precipitate (C) when treated with thioacetamide. A major portion of C dissolved when treated with NaOH. On the basis of these observations classify each of the following ions as being *definitely present, definitely absent* or *undetermined:* Hg_2^{2+}, NO_3^-, Cu^{2+}, Cl^-, Pb^{2+}, Bi^{3+}, Sn^{4+}.

REFERENCES

1. Garrett, A.B., Sisler, H.H., Bonk, H., and Stoufer, R.C.: *Semimicro Qualitative Analysis.* Blaisdell Publishing Company, Waltham, Massachusetts, 1966.
2. Layde, D.C. and Busch, D.H.: *Introduction to Qualitative Analysis.* 2nd edition. Allyn and Bacon, Inc., Boston, Massachusetts 1971.

Experiment 27
REPORT SHEET FOR GROUP II

Name _____ Section _____

Results at Each Separation Part of the Flow Sheet

Ions Present in Unknown   _____

Observations That Differed from the Text Presentation

Attach Answers to the Assigned Questions

C. ANALYSIS OF GROUP III CATIONS

The Group III cations include: iron(II) and (III), Fe^{2+} and Fe^{3+}, aluminum, Al^{3+}, chromium, Cr^{3+}, cobalt, Co^{2+}, nickel, Ni^{2+} and zinc, Zn^{2+}. Any or all of these ions present in the original sample will be found in the decantate of the precipitation of Group II. They are precipitated from this decantate in two groups of three. Iron, aluminum and chromium hydroxides are initially precipitated by adding a concentrated ammonia solution to the decantate. These hydroxides are separated and analyzed as the aluminum subgroup. The remaining Group III cations are precipitated as sulfides using ammonium sulfide $(NH_4)_2 S$. They are analyzed as the nickel subgroup.

Before attempting to precipitate iron, aluminum and chromium hydroxides, it is necessary to remove hydrogen sulfide from the solution; otherwise, the sulfides of the nickel subgroup will precipitate. This is accomplished by evaporating the decantate just to dryness. Addition of concentrated ammonia then precipitates the aluminum subgroup hydroxides. The separation of the subgroups is accomplished in this step, since ammonia also forms soluble complex ions with cobalt, nickel and zinc. The important reactions here are:

$$Al^{3+} + 3NH_3 + 3H_2O \rightleftharpoons 3NH_4^+ + \underline{Al(OH)_3} \quad \text{(white)}$$

$$Fe^{3+} + 3NH_3 + 3H_2O \rightleftharpoons 3NH_4^+ + \underline{Fe(OH)_3} \quad \text{(rust)}$$

$$Cr^{3+} + 3NH_3 + 3H_2O \rightleftharpoons 3NH_4^+ + \underline{Cr(OH)_3} \quad \text{(blue-gray)}$$

$$Ni^{2+} + 6NH_3 \rightleftharpoons Ni(NH_3)_6{}^{2+} \quad \text{(blue solution)}$$

$$Co^{2+} + 6NH_3 \rightleftharpoons Co(NH_3)_6{}^{2+} \quad \text{(blue solution)}$$

$$Zn^{2+} + 4NH_3 \rightleftharpoons Zn(NH_3)_4{}^{2+} \quad \text{(colorless solution)}$$

The hydroxides of nickel subgroup members are prevented from precipitating here not only through the formation of the complex ammine ions, but also because the hydroxide ion concentration is carefully controlled by an ammonium ion–ammonia buffer system. This equilibrium is represented by the equation

$$NH_3 + H_2O \rightleftharpoons NH_4^+ + OH^-$$

This means the hydroxide ion concentration can be maintained at a level high enough to precipitate the aluminum subgroup hydroxide, but too low (because of the added ammonium ion) to permit precipitation of the nickel subgroup.

Aluminum Subgroup Analysis. Aluminum and chromium hydroxides are separated from ferric hydroxide by making use of their amphoteric properties. Thus, the addition of sodium hydroxide and hydrogen peroxide solutions and subsequent heating of the mixture result in the dissolution of aluminum hydroxide through formation of hydroxo complexes. Ferric hydroxide does not form such complexes. The equation is:

$$Al(OH)_3 + OH^- \rightleftharpoons Al(OH)_4{}^-$$

At the same time chromium hydroxide is dissolved through oxidation of chromium from the III to the VI oxidation state, using hydrogen peroxide in a basic solution. Ferric hydroxide is not oxidized under these conditions. The key reaction is:

$$2Cr(OH)_3 + 3H_2O_2 + 4OH^- \rightleftharpoons 2CrO_4{}^{2-} \text{ (yellow)} + 8H_2O$$

Confirmation of Aluminum. Aluminum ion is confirmed by allowing it to precipitate as the hydroxide in the presence of a red organic dye called aluminon. As the aluminum hydroxide precipitate forms in this solution, the dye is absorbed on it, thereby enhancing and concentrating the color of the dye and giving a specific test for aluminum.

Because the conditions for the adsorption of the dye must be carefully controlled, the decantate containing the hydroxo complex of aluminum is treated with a series of reagents prior to bringing the aluminum ions in contact with the dye. Thus, the decantate is first neutralized to convert the hydroxo complex to Al^{3+}, according to the reaction:

$$Al(OH)_4{}^- + 4H_3O^+ \rightleftharpoons Al^{3+} + 8H_2O$$

Under these conditions the yellow solution, because of the chromate ion, $CrO_4{}^{2-}$, will turn orange as the dichromate ion, $Cr_2O_7{}^{2-}$, is produced.

$$2CrO_4{}^{2-} + 2H_3O^+ \rightleftharpoons Cr_2O_7{}^{2-} + 3H_2O$$

Then aluminum hydroxide is carefully precipitated again by using ammonia and ammonium chloride. Under these conditions the $CrO_4{}^{2-}$ ion will be regenerated, but no chromium species should precipitate. The purpose of these reagents is to control the hydroxide ion concentration in the solution so as to get a maximum volume of aluminum hydroxide precipitate. The use of ammonium ion avoids dissolving some of the aluminum as $[Al(OH)_4]^-$ by controlling the amount of excess hydroxide ion.

Aluminum hydroxide is then dissolved in a mixture of acetic acid and ammonium acetate, $CH_3CO_2H–CH_3CO_2NH_4$, and the aluminon dye is added to this solution. Finally the solution is made basic by adding ammonia. This provides hydroxide ions for still another precipitation of aluminum hydroxide. This time as aluminum hydroxide precipitates, it carries the dye with it, giving the enhanced and concentrated color.

Chromium is confirmed by reconverting chromate ion, $CrO_4{}^{2-}$, to dichromate ion, $Cr_2O_7{}^{2-}$, by making the solution acidic. Treatment of the dichromate ion with hydrogen peroxide in acid solution produces a peroxy acid of chromium having the formula, H_2CrO_5. This acid is blue in color and is reasonably stable in ether. Thus, a few drops of ether are added to the solution following the oxidation of chromium to chromate and its conversion to dichromate. The presence of a blue color congregating in the ether layer is indicative of the presence of chromium. The important reactions in these procedures are:

$$2CrO_4{}^{2-} + 2H_3O^+ \rightleftharpoons Cr_2O_7{}^{2-} + 3H_2O$$

$$Cr_2O_7{}^{2-} + 2H_3O^+ + 2H_2O_2 \rightleftharpoons 2H_2CrO_5 \text{ (blue in ether)} + 3H_2O$$

Confirmation of Iron. $Fe(OH)_3$ is dissolved in HCl and the solution treated with ammonium thiocyanate, NH_4SCN: This forms a blood red complex having the formula: $Fe(NCS)^{2+}$. The important reactions here are:

$$\underline{Fe(OH)_3} + 3H_3O^+ \rightleftharpoons Fe^{3+} + 6H_2O$$

$$Fe^{3+} + SCN^- \rightleftharpoons Fe(NCS)^{2+} \quad \text{(blood red)}$$

Nickel Subgroup Analysis. The cations of the nickel subgroup are found as their ammine complexes, $Ni(NH_3)_6{}^{2+}$, $Co(NH_3)_6{}^{2+}$, and $Zn(NH_3)_4{}^{2+}$ in the decantate from the precipitation of the aluminum subgroup hydroxides at the beginning of the analysis of Group III. The nickel subgroup members are first precipitated as sulfides by adding ammonium sulfide, $(NH_4)_2S$ to this solution and heating. The key reactions are:

$$Ni(NH_3)_6{}^{2+} + S^{2-} \rightleftharpoons \underline{NiS} \text{ (black)} + 6NH_3$$

$$Co(NH_3)_6{}^{2+} + S^{2-} \rightleftharpoons \underline{CoS} \text{ (black)} + 6NH_3$$

$$Zn(NH_3)_4{}^{2+} + S^{2-} \rightleftharpoons \underline{ZnS} \text{ (white)} + 4NH_3$$

Ammonium sulfide, rather than thioacetamide, is used in this procedure since the sulfides of cobalt, nickel and zinc are more soluble than are those of the Group II ions. As a consequence, a higher concentration of sulfide is needed for these precipitations than in the case of Group II. Ammonium sulfide provides sulfide ions, and a basic environment to maintain the concentration of these ions at a reasonably high level.

Separation of Zinc from Cobalt and Nickel. This is accomplished by first heating the sulfides so as to convert those of nickel and cobalt to less soluble crystalline modifications that dissolve very slowly in acid. Then the mixture is treated with 1 M HCl which permits the zinc sulfide to dissolve without affecting the NiS or CoS. The important reaction here is:

$$ZnS + 2H_3O^+ \rightleftharpoons Zn^{2+} + H_2S + 2H_2O$$

The zinc ion, having been dissolved and separated can now be reprecipitated as white ZnS using ammonium sulfide solution, $(NH_4)_2S$.

Analysis of Nickel and Cobalt. At this point in the analysis, cobalt and nickel are present as a precipitate containing NiS and CoS. These sulfides are readily soluble in aqua regia. The important reactions are:

$$3\underline{NiS} + 8H_3O^+ + 2NO_3^- \rightleftharpoons 3Ni^{2+} + 2NO + 3S + 12H_2O$$

$$3\underline{CoS} + 8H_3O^+ + 2NO_3^- \rightleftharpoons 3Co^{2+} + 2NO + 3S + 12H_2O$$

Following the treatment with aqua regia, the solution must be evaporated just to dryness to remove this strong oxidizing agent and any excess sulfur that may be present. The dry residue is then treated with dilute hydrochloric acid and the solution divided into equal parts. One is tested for nickel, the other for cobalt.

The test for nickel involves making the solution basic with ammonia and then adding a small amount of dimethylglyoxime, a specific and sensitive test reagent for nickel. This substance on contact with nickel ion gives a brilliant red color. The reaction is:

$$Ni^{2+} + 2NH_3 + 2(CH_3)_2C_2(NOH)_2 \rightleftharpoons 2NH_4^+ + \underline{Ni(C_4H_6N_2O_2)_2} \text{ (red)}$$
(green)

Cobalt(II) forms a brownish, soluble complex, $[Co(DMG)_3]^-$. Although this complex does not interfere with identification of Ni^{2+}, its color is noted more quickly, and excess dimethylglyoxime reagent may have to be added.

The test for cobalt ions involves their reaction with thiocyanate ion, SCN^-, which gives a blue complex ion having the formula $Co(NCS)_4^{2-}$. The color of this ion is more pronounced if a small amount of acetone is added to the solution. The reaction is:

$$Co^{2+} + 4SCN^- \rightleftharpoons Co(NCS)_4^{2-} \quad \text{(blue in acetone)}$$

PROCEDURE FOR GROUP III ANALYSIS

We are now ready to analyze the decantate from the precipitation of Group II which may contain any or all of the ions, Al^{3+}, Cr^{3+}, Fe^{2+} or Fe^{3+}, Co^{2+}, Ni^{2+}, Zn^{2+}.* These ions are in an acid solution containing H_2S and S^{2-} ions. Pour the decantate from the precipitation of Group II into a casserole and evaporate just to dryness. Add 2 drops of 6 M HCl and 25 drops of 3% H_2O_2, and again evaporate just to dryness. Dissolve the residue in 10 drops of water to which has been added 2 drops of 6 M HCl. To this, add 3 drops of 3 M NH_4Cl and enough 15 M NH_3 to make the solution distinctly basic. Do not add a large excess. Stir well, centrifuge quickly, and decant the clear liquid. Add 1 drop of 6 M NH_3 to the decantate to test for completeness of precipitation.

---

*If your unknown is a solid sample of Group III ions only, treat 50 mg. of the unknown powder with 1 ml. of 12 M HCl. Immerse the test tube in boiling water and stir continuously for a few seconds. When the powder is dissolved entirely, treat the solution according to the procedure starting at the asterisk above.

Precipitate: Aluminum subgroup $\underline{Al(OH)_3}$ (white), $\underline{Cr(OH)_3}$ (blue-gray), $\underline{Fe(OH)_3}$ (rust). Wash the precipitate once with 0.5 ml. of 1.5 M ammonia and proceed as described under the aluminum subgroup.

Decantate: $Ni(NH_3)_6{}^{2+}$, $Co(NH_3)_6{}^{3+}$, $Zn(NH_3)_4{}^{2+}$. Add 1 ml. of $(NH_4)_2S$ solution to the decantate and heat on a boiling water bath for 10 minutes. Centrifuge and decant the clear liquid. Save the precipitate. Add 1 drop of $(NH_4)_2S$ to the decantate following separation to test for completeness of precipitation. Discard the decantate.

Precipitate: $\underline{NiS}$ (black), $\underline{CoS}$ (black), $\underline{ZnS}$ (white). Wash this precipitate once with water and discard the washings. To the precipitate, add 10 drops of cold 1 M HCl, and stir constantly for 1 minute. Centrifuge and decant the clear liquid.

Precipitate: $\underline{NiS}$, $\underline{CoS}$. Wash the precipitate once with water and discard the washings. Now treat the precipitate with 6 drops of 12 M HCl and 2 drops of 16 M HNO_3. Heat this mixture until the black residue dissolves, discarding any white or yellow residue of sulfur which remains. Transfer the solution to a casserole and evaporate just to dryness. Add 5 drops of water and 1 drop of 6 M HCl. Divide the solution into 2 equal parts. Test one part for nickel, the other for cobalt.

Test for Nickel: Drop 6 M NH_3 into the solution until it is basic. Do not add an excess of NH_3. Then add 2 drops of dimethylglyoxime solution and mix. A red precipitate of nickel dimethylglyoxime indicates the presence of Ni^{2+}. If a brownish solution forms initially, cobalt(II) is quite likely present. Add several more drops of dimethylglyoxime to ensure complete precipitation of the red nickel compound. Centrifuge and decant, if necessary, to observe the red precipitate.

Test for Cobalt. Add 5 drops of 1 M NH_4SCN and 20 drops of acetone to the solution and mix well. The appearance of a blue solution due to $Co(NCS)_4{}^{2-}$ confirms the presence of Co^{2+}.

Decantate: Zn^{2+}. Neutralize with 6 M NH_3 and add 1 drop excess. Add 2 drops of $(NH_4)_2S$ solution to this decantate. Heat in a boiling water bath for 2 minutes. A white precipitate of ZnS confirms the presence of Zn^{2+}.

Aluminum Subgroup Analysis. We find the aluminum subgroup as a precipitate containing any or all of the hydroxides of iron, aluminum and chromium. To this precipitate add 6 drops of 6 M NaOH and 10 drops of 3% H_2O_2 and stir vigorously. When the evolution of gases has ceased, add 3 more drops of H_2O_2 and boil gently until the evolution of gases again ceases. Dilute the solution to 2 ml. with thorough mixing, centrifuge and remove the decantate.

Precipitate: $\underline{Fe(OH)_3}$. Wash the precipitate once with distilled water and discard the washings. Dissolve the precipitate in 6 M HCl and add 2 drops of NH_4SCN with stirring. A deep red color due to $FeNCS^{2+}$ verifies the presence of Fe^{3+}.

Decantate: $\underline{Al(OH)_4^-} . CrO_4^{2-}$. Acidify this decantate with 6 M HNO_3, add 0.4 gram of NH_4Cl, and mix well. Now add 6 M NH_3 until the solution is distinctly alkaline. Approximately 2-3 ml. is needed. Centrifuge and remove the decantate.

Precipitate: $\underline{Al(OH)_3}$. Add 10 drops of 6 M CH_3CO_2H and heat the mixture for 2-3 minutes. Discard any residue which remains. Now add to the solution 10 drops of 3 M $CH_3CO_2NH_4$ and 5 drops of aluminon test solution. While stirring the mixture, make it basic by adding 6 M NH_3. Formation of a red precipitate confirms the presence of Al^{3+}.

Decantate: $\underline{CrO_4^{2-}}$ (yellow solution). Pour the decantate into a casserole and boil down to approximately 0.5 ml.; acidify this solution with 6 M HNO_3 and cool to room temperature. Add 15 drops of ether and 5 drops of 3% H_2O_2 to a test tube. Then quickly pour the above decantate into the test tube containing the ether and H_2O_2. A sky blue coloration in the ether layer which disappears on standing a few seconds confirms the presence of Cr^{3+}.

ANION ANALYSIS

Procedures for identifying the anions NO_3^-, CO_3^{2-}, SO_4^{2-}, F^-, Cl^-, Br^- and I^- are given in Experiment 3. A fresh sample of the unknown should be used for each of the anion identification tests.

FLOW SHEET: GROUP III

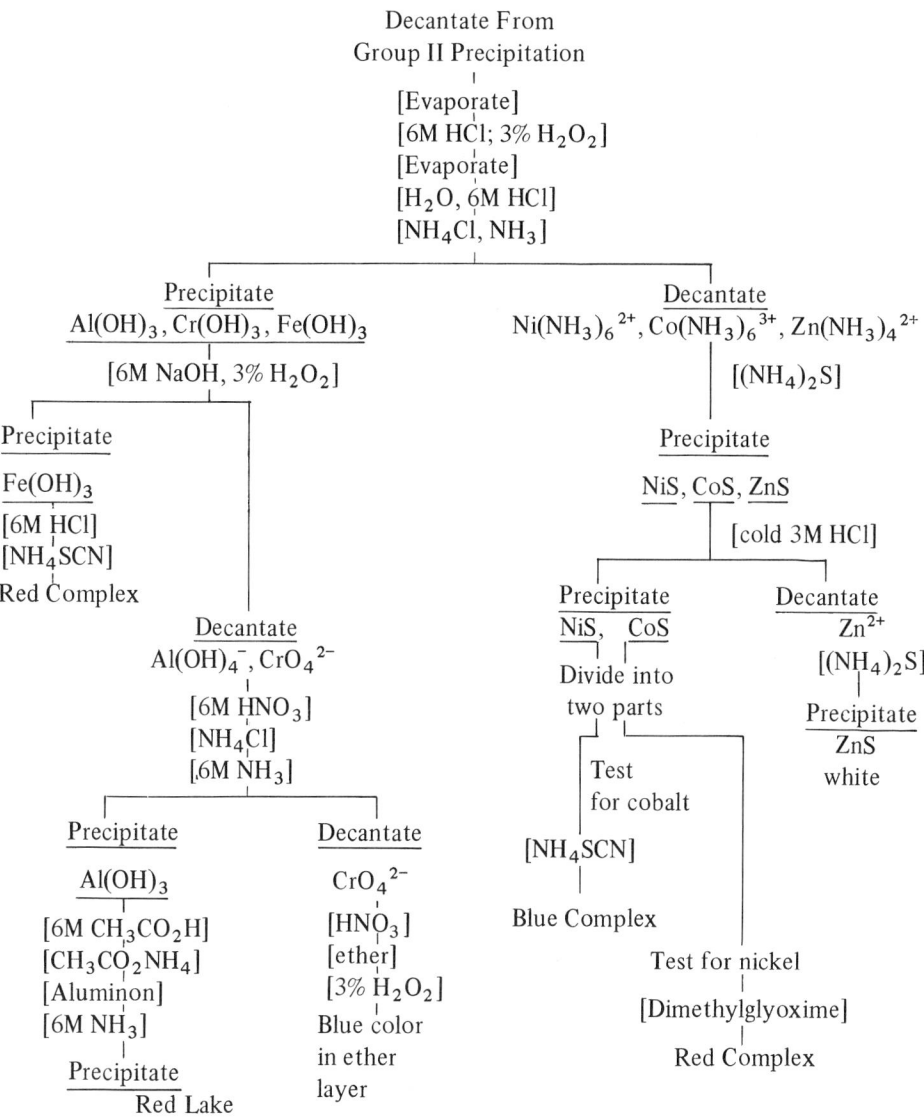

QUESTIONS—GROUPS I, II AND III

1. What property of the ions of cobalt, nickel and zinc is used to separate them from the ions of iron, aluminum and chromium?

2. If the sulfides of Group II were not all removed in the Group II precipitation procedure, would they precipitate with the Group III ions in the precipitation of Group III? Why or why not?

3. Write equations based on Group III procedures to illustrate each of the following:
   a) the separation of iron(III) ions from nickel(II) ions
   b) the separation of aluminum ions from chromium(III) ions
   c) the separation of zinc ions from cobalt(III) ions
   d) the oxidation of chromium from the III to the VI state

e)   the separation of iron ions from chromium ions

f)   the formation of three complex ions, each of which has different ligands (groups bonded to the metal ion).

4. How would you distinguish between the substances in each of the following pairs (i) if both are in the same container; (ii) if each is in a separate container?

a) $Al(NO_3)_3$ and $Zn(NO_3)_2$
b) $ZnCl_2$ and $NiCl_2$
c) $Cr(OH)_3$ and $Fe(OH)_3$
d) CoS and ZnS

e) $Ni(NO_3)_2$ and $Co(NO_3)_2$
f) $Al(OH)_3$ and $Cr(OH)_3$
g) $NiCl_2$ and $CuSO_4$
h) $AlCl_3$ and $PbCl_2$

5. You are given an aqueous solution that may contain any, all or none of the ions given in each of the lettered items below. In each case cite a reagent, reagents or an operation that if added or performed would enable you to determine which, if any, of the ions are present. State the observations that would lead you to the various possible conclusions.

a) Cr^{3+} and Zn^{2+}
b) Cr^{3+} and Fe^{3+}
c) Ni^{2+} and Co^{2+}
d) Al^{3+} and Cr^{3+}

e) Ni^{2+} and Bi^{3+}
f) Ag^+ and Al^{3+}
g) Zn^{2+} and Hg^{2+}
h) Sn^{2+} and Zn^{2+}

6. A solid sample known to contain any or all of the following salts, but no others: $Hg_2(NO_3)$, $HgCl_2$, SnS, $NiCl_2$, and $ZnCl_2$. It gave a brown precipitate and a colorless solution A when treated with water. The precipitate was completely soluble in NaOH with a little NH_3 added. Solution A gave a white precipitate when treated with thioacetamide in basic solution. On the basis of these observations, classify each of the following ions as definitely absent, definitely present or undetermined: Hg_2^{2+}, Hg^{2+}, Sn^{2+}, Ni^{2+}, Zn^{2+}, NO_3^-, Cl^-, S^{2-}.

7. A solid sample is known to contain any or all of the following salts, but no others: $PbCl_2$, $AgNO_3$, $BiCl_3$, $CoCl_2$, and $Fe(OH)_3$. It gave a white precipitate and a colored solution when treated with water. The white precipitate (A) was insoluble in hot water but soluble in NH_3. The colored solution (B) gave a white precipitate (C) when treated with concentrated NH_3. On the basis of these observations, classify the following ions as being definitely absent, definitely present or undetermined: Pb^{2+}, Cl^-, Ag^+, NO_3^-, Co^{2+}, Fe^{3+}, OH^-.

8. Give the name or formula of a reagent that if added to each of the following pairs would bring about a separation of the cations

a) HgCl and $PbCl_2$
b) $Hg_2(NO_3)_2$ and $Hg(NO_3)_2$
c) $Pb(NO_3)_2$ and $Bi(NO_3)_3$
d) $Cu(NO_3)_2$ and $Bi(NO_3)_3$
e) $Hg(NO_3)_2$ and $Fe(NO_3)_3$

f) $SbCl_3$ and $Bi(OH)_3$
g) $NiCl_2$ and NaCl
h) $Cu(NO_3)_2$ and $SnCl_2$
i) $AlCl_3$ and $HgCl_2$
j) $AlCl_3$ and $ZnCl_2$

9. A solid sample is known to contain two or more of the following salts in equivalent amounts, but no others: $Al(NO_3)_3$, $CrCl_3$, $Cu(NO_3)_2$, $SnCl_2$, and $CoCl_2$. It was completely soluble in concentrated NH_3 giving a colored solution A. Solution A was neutralized and the NH_3 boiled out. It

was then made 0.30M in HNO_3 and thioacetamide was added, whereupon a black precipitate (B) and a colored solution C appeared. A significant portion of precipitate B dissolved when treated with NaOH. On the basis of these observations, classify each of the following ions as: *definitely absent, definitely present* or *undetermined:* Al^{3+}, NO_3^-, Cr^{3+}, Cl^-, Cu^{2+}, Sn^{2+}, Co^{2+}.

REFERENCES

1. Garrett, A.B., Sisler, H.H., Bonk, H., and Stoufer, R.C.: *Semimicro Qualitative Analysis.* Blaisdell Publishing Company, Waltham, Massachusetts, 1966.
2. Layde, D.C. and Busch, D.H.: *Introduction to Qualitative Analysis.* 2nd edition. Allyn and Bacon, Inc., Boston, Massachusetts, 1971.

Experiment 27
REPORT SHEET FOR GROUP III

Name _____ Section _____

Results at Each Separation Part of the Flow Sheet

Ions Present in Unknown _____

Observations That Differed from the Text Presentation

Attach Answers to the Assigned Questions

QUANTITATIVE ANALYSIS
OF SOLUTIONS

QUANTITATIVE ANALYSIS OF A SOLUTION CONTAINING TWO ACIDS

OBJECTIVE

To illustrate how a proper choice of indicator permits the experimenter to determine quantitatively by titration with a base the concentrations of two weak acids that have different ionization constants.

EQUIPMENT NEEDED

25-ml. buret; 10-ml. pipet; 5-ml. pipet; three 125-ml. Erlenmeyer flasks; 10-ml. graduated cylinder; pipet rubber bulb.

REAGENTS NEEDED

Standardized 0.05 M NaOH solution; 0.1 M NaH_2PO_4; solutions of bromcresol green, o-cresolphthalein, phenolphthalein, thymol blue; 125 ml. of unknowns, which are mixtures of phosphoric acid and sodium dihydrogen phosphate; 0.1 M Na_2HPO_4.

INTRODUCTION

Phosphoric acid, H_3PO_4, is a tribasic acid. Ionization occurs in three stages:

$$H_3PO_4 + H_2O \rightleftharpoons H_3O^+ + H_2PO_4^-$$

$$H_2PO_4^- + H_2O \rightleftharpoons H_3O^+ + HPO_4^{2-}$$

$$HPO_4^{2-} + H_2O \rightleftharpoons H_3O^+ + PO_4^{3-}$$

The purpose of this experiment is to analyze solutions containing the first two of these acids, using titration procedures. The problems to be surmounted in carrying out this analysis with precision leads to a discussion of some important aspects of acid-base behavior and of titration of weak acids.

DISCUSSION

In the titration of a weak acid, HA, ionization constant K_{ion}, with a strong base, a plot of pH against milliliters of base added will look like that in Figure 28-1.

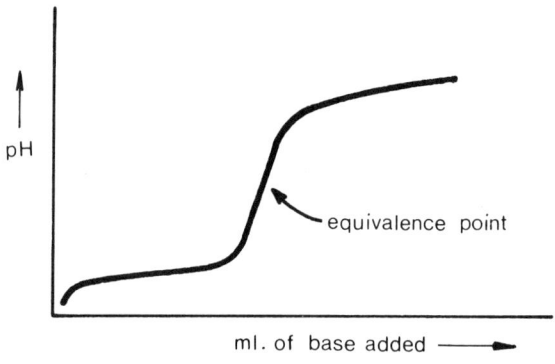

FIGURE 28-1

Titration curve for the titration of a weak monoprotic acid with a strong base.

That the graph has this appearance is readily shown by the following calculation. Since

$$K_{ion} = \frac{[H_3O^+]_{eq}\,[A^-]_{eq}}{[HA]_{eq}}$$

then

$$[H_3O^+]_{eq} = K_{ion}\,\frac{[HA]_{eq}}{[A^-]_{eq}}\;.$$

Taking the logarithm of both sides of the expression, and multiplying by −1, we get

$$-\log[H_3O^+]_{eq} = -\log K_{ion} + \log\frac{[A^-]_{eq}}{[HA]_{eq}}$$

or in terms of pH and pK_{ion}

$$pH = pK_{ion} + \log\frac{[A^-]_{eq}}{[HA]_{eq}} \tag{1}$$

This equation gives the pH at any $\dfrac{[A^-]}{[HA]}$ ratio. This ratio changes as base is added to the acid, because the reaction

$$HA + OH^- \longrightarrow H_2O + A^- \tag{2}$$

is pushed farther toward the right with each addition of base. At the beginning of the titration only the acid HA and the small concentrations of H_3O^+ and A^- from its ionization are present. As base is added the acid is neutralized, thus decreasing [HA], and salt is formed, increasing [A⁻]. It is easy to calculate these two concentrations, and from equation (1) the pH, of the solution at each stage of the titration. Allowance must be made for the change in the total volume of the solution as base is added. The results of such a calculation are given in Table 28-1.

TABLE 28-1 pH CHANGES DURING THE TITRATION OF 50 ml. OF
0.1M HA (K_{ion} = 1 $\times$ 10^{-5}) WITH 0.1M NaOH

NaOH added (ml.)	Total Volume (ml.)	pH	NaOH added (ml.)	Total Volume (ml.)	pH
0.00	50.0	3.00	49.95	99.95	8.00
5.00	55.0	4.05	50.00*	100.0*	8.85*
10.00	60.0	4.40	50.05	100.05	9.70
20.00	70.0	4.82	50.10	100.1	10.00
30.00	80.0	5.18	50.50	100.5	10.70
40.00	90.0	5.60	51.00	101.0	11.00
45.00	95.0	5.95	55.00	105.0	11.68
49.00	99.0	6.69	60.00	110.0	11.96
49.50	99.5	7.00	70.00	120.0	12.23
49.90	99.9	7.70			

*Equivalence point.

Note that the first addition of base produces a noticeable change in pH
followed by a region where the solution is buffered by the presence of both
acid and salt. This buffer region is followed by a region near the equivalence
point (50 ml. of base added) where a very rapid change occurs, and that in
turn by another region where the change is not so rapid. It is this rapid
change near the equivalence point that makes a quantitative titration of acid
by base a feasible experiment. Note that 2 drops of base solution near the
equivalence point (from 49.95 ml. of base to 50.05 ml.) causes a change in pH
of 1.70 units in this solution. Such a change in pH is sufficient to change the
color of an indicator from its acid color to its basic color, so that we can
easily measure to 0.05 ml. in a total volume of 50 ml. added, a precision of 1
part in 1000.

Note, too, in Table 28-1, that the pH at the equivalence point (8.85) is
not the value for the pH of pure water (7.00). We can see why this should be
if we remember that at the equivalence point the reaction

$$HA + NaOH \longrightarrow NaA + H_2O \tag{3}$$

has just gone to completion: that is, it is as if we had a solution of the salt
NaA, at a concentration, C, determined by the initial concentration of the
acid and the dilution during the titration. The pH of the solution is then
determined by the position of equilibrium in the reaction of the acid H_2O
reacting with the base A^-

$$\underset{\text{acid}}{H_2O} + \underset{\text{base}}{A^-} \longrightarrow \underset{\text{acid}}{HA} + \underset{\text{base}}{OH^-} . \tag{4}$$

This equilibrium is governed by the hydrolysis constant of A^-

$$K_{hyd} = \frac{[HA]_{eq}\,[OH^-]_{eq}}{[A^-]_{eq}} = \frac{[HA]_{eq}}{[A^-]_{eq}\,[H_3O^+]_{eq}} \times [H_3O^+]_{eq}\,[OH^-]_{eq} = \frac{K_w}{K_{ion}} \tag{5}$$

Since, according to equation (4), $[HA]_{eq} = [OH^-]_{eq}$

$$[OH^-]_{eq} = \sqrt{\frac{K_w}{K_{ion}}[A^-]_{eq}} \simeq \sqrt{\frac{K_w}{K_{ion}}C} \qquad (6)$$

where, in the right-most term, we have made the approximation

$$[A^-]_{eq} = \left(C - [OH^-]_{eq}\right) \simeq C.$$

From (6), we calculate the hydronium ion concentration

$$[H_3O^+]_{eq} = \frac{K_w}{[OH^-]_{eq}} = K_w\sqrt{\frac{K_{ion}}{K_w\,C}} = \sqrt{\frac{K_w\,K_{ion}}{C}} \qquad (7)$$

Thus the pH at the equivalence point is given by

$$pH = \tfrac{1}{2}\,(pK_w + pK_{ion} + \log C) \qquad (8)$$

For the particular situation of Table 28-1, the concentration of NaA at the equivalence point is $\dfrac{50 \times 0.1}{100} = 0.05$ M, and

$$pH = \tfrac{1}{2}\,(14 + 5 - 1.3) = 8.85$$

as given in the table.

It is evident from equation (8) that the value of pH at the equivalence point depends upon the magnitude of K_{ion} and thus that the *indicator chosen for the titration of a weak acid should be matched to the acid,* since the color change must come at the equivalence point.

The matching is accomplished by noting that an indicator is itself a weak acid for which the acid and base forms have different colors

$$H\,Ind + H_2O \rightleftharpoons H_3O^+ + Ind^- \qquad (9)$$
$$\text{Color A} \qquad\qquad\qquad \text{Color B}$$

The indicator thus has an ionization constant

$$K_{ion,\,Ind} = \frac{[H_3O^+]_{eq}\,[Ind^-]_{eq}}{[HInd]_{eq}}$$

and the concentration of hydrogen ion in the solution determines the ratio of concentration of the two colored forms

$$\frac{[HInd]_{eq}}{[Ind^-]_{eq}} = \frac{[H_3O^+]_{eq}}{K_{ion,\,Ind}} \qquad (10)$$

The human eye can detect color difference if this ratio is between 10:1 and 1:10 for most indicators, that is,

$$\frac{1}{10} < \frac{[HInd]_{eq}}{[Ind^-]_{eq}} < 10$$

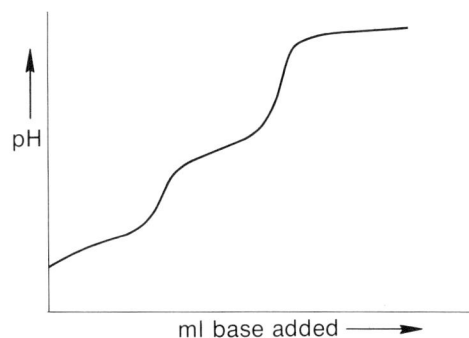

FIGURE 28-2

Titration curve for a diprotic acid with widely spaced ionization constants, using a strong base.

and will be most sensitive to color changes when the ratio is about 1:1. Substituting the extreme values in equation (10) and taking logarithms, we see that the useful "range" of the indicator will be about 2 pH units, and that the best match will be when the pH at the equivalence point is near the mid-point of that range. These conclusions must be modified for some indicators because of the eye's different sensitivity to some colors. To match an indicator to the titration of Table 28-1, we would want one which will have a range from about 7.85 to 9.85; phenolphthalein (colorless at pH = 8.2 to red at pH = 10.0) would be suitable if the end-point is chosen as a barely pink color.

Polyprotic acids. For a diprotic acid, H_2A, with a ratio between the first and second ionization constants of 10^4 or more, neutralization of the first of the ionizable protons is essentially complete before neutralization of the second begins, and addition of base causes the pH to change as in Figure 28-2.

For the triprotic acid, phosphoric acid, the curve looks like that of Figure 28-3. Such behavior is characteristic of titration of acids with ionization constants less than about 1×10^{-8}. Thus any attempt to titrate to the equivalence-point of the third stage would be grossly imprecise.

Titration for the first two steps, however, will be valid if we can calculate the pH at the equivalence points and choose the proper indicators. At the equivalence point for the neutralization of the first ionizable proton, the reaction

$$H_3PO_4 + NaOH \longrightarrow NaH_2PO_4 + H_2O \tag{11}$$

FIGURE 28-3

Change of pH with added base for phosphoric acid. While the first two steps are sharp, the acid HPO_4^{2-} is so weak that the pH does not change rapidly enough to allow the experimenter to pinpoint the exact drop or two which represents the end-point, as indicated in the gentle slope.

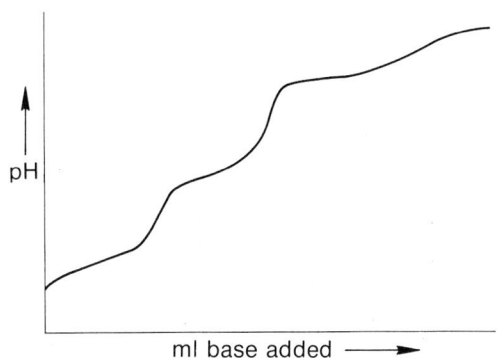

has gone just to completion, and at the equivalence point for the second step, the reaction

$$NaH_2PO_4 + NaOH \longrightarrow Na_2HPO_4 + H_2O \qquad (12)$$

has just gone to completion. In order to choose indicators which will change color at completion of these reactions, we must calculate the pH of solutions of the salts NaH_2PO_4 and Na_2HPO_4, just as we did for NaA. Let us do this for NaH_2PO_4.

The ion $H_2PO_4^-$ differs from the ion A^- of the earlier case in that it is not only a base as A^- was (since it can acquire a proton to form H_3PO_4) but is also an acid (since it can lose a proton to form HPO_4^{2-}). Being an acid and a base, it can undergo reaction with itself

$$\underset{\text{acid}}{H_2PO_4^-} + \underset{\text{base}}{H_2PO_4^-} \rightleftharpoons \underset{\text{acid}}{H_3PO_4} + \underset{\text{base}}{HPO_4^{2-}} \qquad (13)$$

The equilibrium constant for this reaction is related to the first and second ionization constants of phosphoric acid

$$K_{eq} = \frac{[H_3PO_4]_{eq}[HPO_4^{2-}]_{eq}}{[H_2PO_4^-]_{eq}^2} = \frac{[H_3PO_4]_{eq}}{[H_2PO_4^-]_{eq}[H_3O^+]_{eq}} \times \frac{[H_3O^+]_{eq}[HPO_4^{2-}]_{eq}}{[H_2PO_4^-]_{eq}} = \frac{K_2}{K_1} \qquad (14)$$

If equation (13) represents the only source of H_3PO_4 and HPO_4^{2-}, then the equilibrium concentrations of these two are equal, and from (14)

$$\frac{[H_3PO_4]_{eq}^2}{[H_2PO_4^-]_{eq}^2} = \frac{K_2}{K_1}.$$

But H_3PO_4 is also in ionization equilibrium with hydronium ion

$$H_3PO_4 + H_2O \rightleftharpoons H_3O^+ + H_2PO_4^-$$

$$K_1 = \frac{[H_3O^+]_{eq}[H_2PO_4^-]_{eq}}{[H_3PO_4]_{eq}}$$

$$[H_3O^+]_{eq} = K_1\frac{[H_3PO_4]_{eq}}{[H_2PO_4^-]_{eq}} = K_1\sqrt{\frac{K_2}{K_1}} = \sqrt{K_1 K_2}$$

Thus at the equivalence point, where it is as if only NaH_2PO_4 was in the solution

$$pH_{1st\,equiv.\,pt.} = \frac{1}{2}(pK_1 + pK_2)$$

From the values on p. 379 for phosphoric acid in dilute salt solutions similar to those involved in this experiment, $pH_{1st\,equiv.\,pt.} = 4.41$. Thus an indicator with the mid-point of its color range near this value will be suitable for this titration.

IONIZATION OF PHOSPHORIC ACID IN DILUTE SALT SOLUTIONS*

(0.1 Molar in NaClO$_4$)

$$pK_1 \quad 2.10$$
$$pK_2 \quad 6.71$$
$$pK_3 \quad 11.8$$

*Schwarzenbach, G., and Geier, G., Helv. Chim. Acta., *46*, 906 (1963).

A similar calculation will show that titration to the second step of Figure 28-3 (solution as if only HPO$_4{}^{2-}$ was present) demands an indicator in the pH range 9-10. Choose one from the list below.

pH COLOR RANGE OF INDICATORS

Bromcresol green	3.8 (yellow) to 5.4 (blue)
o-Cresolphthalein	8.2 (colorless) to 9.8 (red)
Thymol blue	8.0 (colorless) to 9.6 (blue)
Phenolphthalein	8.2 (colorless) to 10.0 (red)

We thus have the information we need in order to determine the amounts of H$_3$PO$_4$ and NaH$_2$PO$_4$ in a solution. Titration of a sample to the bromcresol green end-point will determine the quantity of H$_3$PO$_4$ present, changing it to H$_2$PO$_4{}^-$. Titration of a second sample to the proper end-point will convert both the H$_3$PO$_4$ and the H$_2$PO$_4{}^-$ to HPO$_4{}^{2-}$. Knowing, from the first titration, how much H$_3$PO$_4$ is present, we can calculate how much of the added base was needed to change it to HPO$_4{}^{2-}$; any amount of base greater than this must have been used to change the H$_2$PO$_4{}^-$, originally present, to HPO$_4{}^{2-}$.

PROCEDURE

Your unknown may contain only phosphoric acid (various concentrations are possible), only sodium dihydrogen phosphate (various concentrations are possible), or a mixture of the two in various proportions. On the basis of your results, you are to report what is present, and the molar concentration of the acid or acids found.

Titration to the First Stage.   Using a dry transfer pipet, pipet 10 ml. samples of the unknown to be analyzed into two 125-ml. Erlenmeyer flasks, and add 10 ml. of H$_2$O and 5 drops of bromcresol green solution to each. In a third flask place 30 ml. of NaH$_2$PO$_4$ solution and 5 drops of indicator to give a comparison color. Titrate each of the samples with standardized sodium hydroxide solution until the color matches that of the reference solution. If the results of the two titrations are not concordant, repeat until you have confidence that your results are correct. Estimate the volume of base used to the nearest 0.02 ml.

Accurately pipet 10 ml. aliquots of the unknown solution and titrate.

Repeat for consistent results.

Titration to the Second Stage. Since titration to the second stage will require at least twice as much base as needed for titration to the first stage, it may be desirable to use a smaller quantity of sample, lest you exceed the capacity of your buret. On the basis of your first titration, decide whether you can safely use 10 ml. of sample or whether 5 ml. would be better.

(Remember that if your pipets are accurate to 0.02 ml., a 10-ml. pipet will deliver with a precision of 1 part in 500, but a 5-ml. pipet with a precision of only 1 part in 250; hence, other things being equal, greater precision is possible with the larger volume.) Titrate two samples (more if the results are not concordant) of the unknown, adding water to make the initial volume about 20 ml., and using 5 drops of indicator. Use a comparison solution of Na_2HPO_4 containing the same indicator concentration.

QUESTIONS–PROBLEMS

1. Show how to calculate the pH at the equivalence point for the second stage of the titration of phosphoric acid.

2. The pK_1 and pK_2 values of oxalic acid, $H_2C_2O_4$, are 1.25 and 4.28. Methyl orange has a color range from pH 3.1 to 4.5. Would this be a suitable indicator for either stage of an oxalic acid titration? Why?

3. Calculate the pH at the equivalence point in the titration of a 0.1 M NH_3 solution with 0.1 M HCl. The basic ionization constant of ammonia is 1.47×10^{-5} at 25°C.

4. In view of the need to know the ionization constant of an acid in order to choose an indicator, suggest experiments for determining the three ionization constants of phosphoric acid. Would application of equations analogous to (1) be helpful?

5. Prepare a table similar to Table 28-1, to represent the change of pH with volume of base added for the titration of 50 ml. of 0.1 M solution of an acid of ionization constant 1×10^{-9} using 0.1 M NaOH as titrant, over the range 45 ml. of base added to 55 ml. using the same 11 volumes in that range as appear in Table 28-1. Plot these data and those for the same range from Table 28-1 on the same graph, and comment upon the statement that an acid of ionization constant 10^{-9} or less cannot be reliably titrated with a strong base.

6. A 5 g. sample of dried NaH_2PO_4 was dissolved in 100 ml. of water and a 10-ml. portion of that solution titrated with 0.1 M NaOH to the $HPO_4{}^{2-}$ equivalence-point; 40 ml. was required. What was the purity of the NaH_2PO_4?

REFERENCE

Pecsok, R.L. and Shields, L.D.: *Modern Methods of Chemical Analysis.* John Wiley and Sons, New York, 1968, Chapter 15.

Experiment 28
REPORT SHEET

Name _____   **Section No.** _____

	Trial 1	Trial 2	Trial 3
Concentration of standardized NaOH used _____			
Volume of NaOH to titrate 10 ml. of unknown to first equivalence-point	_____	_____	_____
Volume of NaOH to titrate _____ ml. of unknown to second equivalence-point	_____	_____	_____
Indicator used	_____	_____	_____
Concentration of H_3PO_4 in unknown (molarity)	_____	_____	_____
Concentration of NaH_2PO_4 in unknown (molarity)	_____	_____	_____

Attach sheets with sample calculations.

Attach sheets with answers to questions.

Analysis of a
Mixture of Salts

The purpose of this series of three experiments is to illustrate a method for quantitatively determining the concentrations of components in a mixture and to provide experience with the important analytical methods known as ion exchange chromatography, precipitation titrations, and complexometric titrations. More·specifically, you will be asked to analyze a mixture of sodium nitrate, sodium chloride, and magnesium chloride to determine the concentration of each ion in the mixture. This will be accomplished by determining the total concentration of cations by titrating the protons liberated from a cation exchange resin, using standard sodium hydroxide; the chloride will be determined by a modified Volhard precipitation titration; the magnesium ion concentration will be determined by a complexometric titration. From these determinations, the concentration of sodium ion can be estimated.

ION EXCHANGE CHROMATOGRAPHY
Analysis for Total Cationic Charge

EQUIPMENT NEEDED

25-ml. buret; 10-ml. pipet; aspirator bulb; cation exchange column; two 125-ml. Erlenmeyer flasks.

REAGENTS NEEDED

3 M HCl solution; standardized solution of sodium hydroxide ($\sim$0.1 M); phenolphthalein; cation exchange resin Dowex 50W-4X

INTRODUCTION

Chromatography (Exp. 6) is a method of separation in which the components to be separated are distributed between two phases. In ion exchange chromatography charged ions are exchanged between a mobile solution phase and a stationary solid phase. Such exchange processes have been known for a long time. They are important in the interaction of soils with fertilizers and plant roots and are used with naturally occurring mineral or man-made solid phases in various industrial operations, such as water softening.

Synthetic *ion exchange resins* are very insoluble polymers of high molecular weight and an open netlike molecular structure into which water and ions in solution can readily penetrate. Commercial resins may be divided into two main groups, *acidic* or *basic;* the names indicate their chemical nature. The acidic cation exchange resins are composed of organic polymers with acidic groups, such as the strongly acidic sulfonic acid groups ($-SO_3H$) or the more weakly acidic carboxyl group ($-CO_2H$). The negative part of each sulfonate radical remains fixed to the resin molecule, but the hydrogen ions are free to move and, thus, can be replaced by cations in the aqueous solution that permeates the pores of the resin. For example:

$$\text{Resin} \cdot SO_3H \cdot H_2O + Na^+ \longrightarrow \text{resin} \cdot SO_3^- Na^+ + H_3O^+$$

The extent to which a cation displaces the hydronium ion of such a resin depends upon several factors, among them the temperature and pH value of the solution. Of greater effect, however, is the relative concentration of the cation of the electrolyte, because the cation exchange is a reversible one and

Le Chatelier's principle applies here just as it does to other chemical equilibria. The foregoing equilibrium can, therefore, be displaced to the left and the original resin compound regenerated if acid of moderately high concentration is introduced into the system.

For many ion exchange equilibria, equilibrium can be most conveniently and efficiently established by allowing the solution to percolate slowly through a tube containing a sufficiently long column of resin. As the solution percolates down the column, it continually contacts fresh resin and the exchange occurs before the solution emerges from the end of the column. Thus the concentration of a neutral salt in solution can be quantitatively determined by exchanging hydronium ions for the cation and then determining the acidity of the eluted solution by titration. Usually washing the column with pure solvent is necessary to assure that all ions that should be displaced from the column have actually been removed. Thus, after the solution has been percolated, follow it with pure solvent until the solvent comes through unchanged. You can then assume that the exchanger has been returned to its stoichiometrically neutral state, i.e., when the effluent is neutral to pH paper.

In this experiment, we shall use a cation exchange resin to determine the total concentration of positive charge in an unknown solution; the solution may contain any, or all, of the compounds sodium nitrate, sodium chloride, and magnesium chloride. The resin absorbs all the cations from an aliquot of the solution and liberates H_3O^+, which one can titrate with standardized sodium hydroxide.

PROCEDURE

The equipment required for ion exchange chromatography is in general very simple. Obtain a small glass column, filled with cation resin and equipped with a rubber tube and screw clamp at the bottom, from the instructor and treat the column with 30 ml. of 3 M HCl solution. Pass the acid solution through the column at a rate not exceeding three ml. per minute. *Never treat ion exchange resins with nitric acid or other oxidizing acids.* Then wash the resin column with distilled water until the effluent from the column tests neutral to pH paper. *Never allow the water level to fall below the level of the top of the resin.* The rate of flow of the wash solution can be 6 to 7 ml. per minute.

Regenerate the column to the H-form using 3 M HCl.

If a large number of air bubbles appear in the column, repack the column. If necessary, the ion exchange column, filled with distilled water and tightly stoppered, can be kept in the laboratory locker.

Pipet 10 ml. (accurate to 0.02 ml.) of the unknown solution onto the resin column and collect the effluent in a clean 125-ml. Erlenmeyer flask. Wash the column with distilled water, which has been boiled (to remove carbon dioxide) and cooled, until the effluent is neutral, i.e., until you have washed all the liberated H_3O^+ from the column.

Carefully pipet 10 ml. of unknown solution onto the resin column.

Pipet a second aliquot onto the column and begin collecting the effluent in a second clean flask. If you are careful to watch both the column and your titration, you can begin titrating the first solution while collecting the second. Titrate with standardized sodium hydroxide (~0.1 M), using four to

Repeat with a second aliquot.

Titrate with standard base.

five drops of phenolphthalein as an indicator. The results of the two titrations should agree to within 0.1 ml.; if they do not, regenerate the column with hydrochloric acid and make additional trials.

All data for this experiment should be recorded in the notebook. Calculate and report the moles of hydronium ion liberated per liter of sample taken (or what can be called the molarity of positive charge in your unknown solution) using the equation

$$M_{PC} = \frac{\text{moles of positive charge}}{\text{volume of sample in liters}}$$

QUESTIONS–PROBLEMS

1. If a solution was known to contain 0.1 M magnesium ion, Mg^{2+}, and 0.15 M sodium ion, Na^+, what would be the molarity of positive charge in this solution?

2. What error will be introduced in M_{PC} if the ion exchange resin absorbs all the magnesium ion but only 90 percent of the sodium ion from the solution described in Question 1?

3. Why is phenolphthalein a satisfactory indicator for the titration in this experiment?

4. Can ion exchange resin columns be used more than once without regeneration of the resin? Explain.

5. Could an aqueous mixture of hydrochloric acid and barium chloride be analyzed by ion exchange chromatography? Why or why not?

REFERENCE

1. Day, R.A., and Underwood, A.L.: *Quantitative Analysis.* 2nd edition. Prentice-Hall, Englewood Cliffs, N.J., 1967, Chapter 17.

Experiment 29
REPORT SHEET

Name _____   Section No. _____

Data

	Trial 1	Trial 2	Trial 3
Volume of sample used	_____	_____	_____
Molarity of standardized NaOH used	_____	_____	_____
Volume of NaOH used	_____	_____	_____
Molarity of positive charge in the sample	_____	_____	_____
Averaged value of molarity	_____		

Example of Calculations

Attach Answers to Questions

Analysis of a Mixture
of Salts (Continued)

A PRECIPITATION TITRATION

Analysis for Chloride Ion by the Volhard Method

OBJECTIVE

To determine the concentration of chloride ion in a mixture of salts using a precipitation titration.

EQUIPMENT NEEDED

25-ml. buret; 10-ml. pipet; aspirator bulb; three 125-ml. Erlenmeyer flasks.

REAGENTS NEEDED

Standardized solutions of sodium chloride (~0.14 M), potassium thiocyanate (~0.01 M), and silver nitrate (~0.1 M); 2 M ferric nitrate; concentrated nitric acid.

INTRODUCTION

Gravimetric methods of analysis can be used to determine a large number of metal ions by quantitative precipitation, drying, and weighing the solid. These procedures are usually tedious and time-consuming. If, however, it were possible to determine by titration the point at which a stoichiometrically equivalent quantity of precipitant had been added to a solution of the substance being determined, information necessary for the analysis could be obtained without collecting, drying, and weighing the precipitate. This is the basis of precipitation titrations.

Chemical reactions suitable for precipitation titrations must yield a very slightly soluble product that forms rapidly, quantitatively, and according to a definite stoichiometric relationship. The reaction also must lend itself easily to end-point detection techniques. Not many reactions meet all these criteria but an important reaction that can be used in this way is the formation of silver chloride, represented by the reaction

$$Ag^+ + Cl^- \longrightarrow \underline{AgCl}$$

The concentration of chloride ion in the sample can be determined by a precipitation titration with standard silver nitrate solution.

Volhard Method

Several volumetric procedures have been developed for the determination of silver or chloride (and other halides). However, the method developed by Volhard is especially sensitive. In this method the chloride to be determined is precipitated as silver chloride by titrating the sample with standard silver nitrate. The equivalence point in the titration is detected by the disappearance of the soluble pink-colored complex $FeNCS^{2+}$ in the following way:

Iron(III) nitrate, $Fe(NO_3)_3$, and potassium thiocyanate are added to the unknown sample just prior to the titration. They react according to the equation

$$Fe^{3+} + SCN^- \rightleftharpoons FeNCS^{2+}$$

to give the pink iron(III) isothiocyanato complex ion. This ion remains in solution until enough silver nitrate has been added so that the silver ion has reacted with all the chloride ion in the sample. At this equivalence point the following reaction occurs:

$$FeNCS^{2+} + Ag^+ \longrightarrow \underline{AgSCN + Fe^{3+}}$$
$$\text{(pink)} \qquad\qquad \text{(colorless)}$$

In calculating the amount of chloride ion in the sample, a correction must be made for the silver ion that reacts with the iron thiocyanato complex ion to bring about the color change at the end-point.

The equilibrium considerations that show that this procedure is feasible are as follows:

$$\text{Total moles } Ag^+ \text{ added} = \text{moles } Cl^- + \text{moles } SCN^- \tag{1}$$

At the equivalence point these substances are distributed as follows:

$$\text{Total moles } Ag^+ = \text{moles } AgCl + \text{moles } AgSCN + \text{moles } Ag^+ \text{ present at equilibrium} \tag{2}$$

$$\text{Total moles } Cl^- = \text{moles } AgCl + \text{moles } Cl^- \text{ at equilibrium} \tag{3}$$

$$\text{Total moles } SCN^- = \text{moles } AgSCN + \left\{\begin{matrix}\text{moles } SCN^- \\ \text{at equilibrium}\end{matrix}\right\} + \left\{\begin{matrix}\text{moles } FeNCS^{2+} \\ \text{at equilibrium}\end{matrix}\right\} \tag{4}$$

Substituting in the first equation, one obtains

$$\left\{\begin{matrix}\text{Moles } Ag^+ \\ \text{at equilibrium}\end{matrix}\right\} = \left\{\begin{matrix}\text{moles } Cl^- \\ \text{at equilibrium}\end{matrix}\right\} + \left\{\begin{matrix}\text{moles } SCN^- \\ \text{at equilibrium}\end{matrix}\right\} + \left\{\begin{matrix}\text{moles } FeNCS^{2+} \\ \text{at equilibrium}\end{matrix}\right\} \tag{5}$$

Experimentally, it has been found that the color of the $FeNCS^{2+}$ complex is just observable when 10^{-6} moles of KSCN are added to 0.1 liter of a solution which is 0.013M in Fe^{3+}. For the reaction

$$FeNCS^{2+} \rightleftharpoons Fe^{3+} + SCN^-$$

K is 7.5×10^{-3}. Therefore,

$$[FeNCS^{2+}] + [SCN^-] = \frac{10^{-6} \text{ moles}}{0.1 \text{ liter}} = 10^{-5} \text{ moles/liter}$$

$$\frac{[Fe^{3+}] \ [SCN^-]}{[FeNCS^{2+}]} = 7.5 \times 10^{-3}$$

$$[Fe^{3+}] = 0.013$$

$$\frac{0.013 \times (10^{-5} - [FeNCS^{2+}])}{[FeNCS^{2+}]} = 7.5 \times 10^{-3}$$

$$1.3 \times 10^{-7} - 0.013 \ [FeNCS^{2+}] = 0.0075 \ [FeNCS^{2+}]$$

$$[FeNCS^{2+}] = \frac{1.3 \times 10^{-7}}{0.0205} = 6.4 \times 10^{-6} M$$

At this concentration $FeNCS^{2+}$ is just barely perceptible to the human eye, and it is assumed that this is the concentration of $FeNCS^{2+}$ at the equivalence point of the titration. The SCN^- concentration at the equivalence point can now be calculated by use of equation (5) and the following relations:

$$[Ag^+]_{eq} \ [Cl^-]_{eq} = K_{sp} = 1 \times 10^{-10} \tag{6}$$

$$[Ag^+]_{eq} \ [SCN^-]_{eq} = K_{sp} = 1 \times 10^{-12} \tag{7}$$

Dividing equation (5) by the volume of the solution at the equivalence point:

$$[Ag^+]_{eq} = [Cl^-]_{eq} + [SCN^-]_{eq} + [FeNCS^{2+}]_{eq} \tag{8}$$

but since

$$[Ag^+]_{eq} = \frac{K_{spAgSCN}}{[SCN^-]_{eq}}$$

and

$$[Cl^-]_{eq} = \frac{K_{spAgCl}}{[Ag^+]_{eq}} = \frac{K_{spAgCl}}{K_{spAgSCN}} [SCN^-]_{eq}$$

$$\frac{K_{spAgSCN}}{[SCN^-]_{eq}} = \left(\frac{K_{spAgCl}}{K_{spAgSCN}} + 1 \right) [SCN^-]_{eq} + [FeNCS^{2+}]_{eq} \tag{9}$$

Substitution of known quantities and multiplication of both sides by $[SCN^-]_{eq}$ gives

$$10^{-12} = (100 + 1) \ [SCN^-]^2_{eq} + 6.4 \times 10^{-6} \ [SCN^-]_{eq} \tag{10}$$

Solution of equation (10) by means of the quadratic formula yields

$$[SCN^-]_{eq} = \frac{-6.4 \times 10^{-6} + \sqrt{(6.4 \times 10^{-6})^2 + 4.04 \times 10^{-10}}}{202}$$

$$[SCN^-]_{eq} = 7.3 \times 10^{-8} M$$

We have calculated the concentrations of $FeNCS^{2+}$ and SCN^- at the equivalence point of a perfect titration in which solid silver chloride and solid silver thiocyanate are both present. All that remains is to calculate the concentration of Fe^{3+} that should be used to obtain these desired concentrations. This is determined by use of the equilibrium constant for the dissociation of $FeNCS^{2+}$:

$$\frac{[Fe^{3+}]_{eq}\,[SCN^-]_{eq}}{[FeNCS^{2+}]_{eq}} = 7.5 \times 10^{-3}$$

$$[Fe^{3+}]_{eq} = 7.5 \times 10^{-3} \times \frac{[FeNCS^{2+}]_{eq}}{[SCN^-]_{eq}}$$

$$[Fe^{3+}]_{eq} = 7.5 \times 10^{-3} \times \frac{6.4 \times 10^{-6}}{7.3 \times 10^{-8}} = 0.66M$$

This result indicates the concentration of Fe^{3+} needed to get a perceptible indicator color at the true equivalence point of the titration. Note that solid silver chloride and silver thiocyanate are both present and in equilibrium with the solution at the equivalence point. The SCN^- concentration can be so low and still give an indicator color because of the higher than usual Fe^{3+} concentration. Swift and co-workers used this calculation to approximate the concentration of Fe^{3+} to be used. In a series of experiments with Fe^{3+} concentrations near this value, they found that the Fe^{3+} concentration that resulted in zero titration error was 0.2M. They found, further, that a concentration of nitric acid of 1M was necessary in the solution to remove the dark orange color of partially hydrolyzed Fe^{3+}. The nitric acid functions by reversing equilibria such as

$$Fe(H_2O)_6{}^{3+} + H_2O \rightleftharpoons Fe(H_2O)_5OH^{2+} + H_3O^+$$

PROCEDURE[1]

Standardization and Practice Titrations. Obtain approximately 200 ml. of $AgNO_3$ solution from the container in the hood. *Be careful to keep this solution away from your skin and clothes.* Silver salts are readily reduced to metallic silver, which causes stains that are difficult to remove. Swirl the silver nitrate solution and then rinse the clean buret twice with 5- to 6-ml.

Carefully pipet NaCl and KSCN solutions. Add Fe(NO$_3$)$_3$ and HNO$_3$.

portions of the solution. Transfer three 10.0 ml. portions of the standard sodium chloride with a pipet to three clean 125-ml. Erlenmeyer flasks containing 10 ml. H_2O each. Add 3 ml. of concentrated HNO_3, 4 ml. of 2 M $Fe(NO_3)_3$, and 1 ml. of 0.01 M KSCN. (The potassium thiocyanate should be added from a 1-ml. pipet, which may be obtained from the instructor and then returned immediately after use.) Immediately titrate the solution with silver nitrate solution under artificial light until the pick color disappears.

Standardize AgNO$_3$ solution.

Swirl the flask vigorously during the entire titration. Time is saved by using the first sample for the approximate location of the end-point. Silver nitrate solution should be added 1 ml. at a time in the approximate titration until the pink color disappears. Read the buret and then add a few drops in excess.

Observe the color of the suspension from above. In the second titration it is safe to add silver nitrate without hesitation to within 1 ml. of the buret

---

[1] The entire procedure usually takes a full three hours to complete. However, the silver nitrate solution can be standardized during the latter half of one laboratory period and used in the next provided it is placed in a tightly stoppered flask, wrapped in a towel to deter photochemical decomposition, and stored in the laboratory locker. Approximately 100 ml. of standard $AgNO_3$ should suffice for the titration of the unknown.

reading just preceding the disappearance of the pink color in the approximate titration. When the titration is within 1 ml. of the end-point, rinse down the inside walls of the flask with chloride-free distilled water. Continue the titration slowly, while observing the color of the suspension carefully. When the end-point is obviously near, add only part of a drop at a time. Allow a droplet of solution to collect on the tip of the buret, touch it against the wall of the flask, rinse the wall, swirl the suspension, and observe its color. When the end-point appears to have been reached, read the buret to the nearest 0.02 ml. and add another drop of silver nitrate solution. The color should now be completely absent. Compare the end-point color with that of the approximate titration, which now contains a few drops of silver nitrate in excess. This solution will definitely be overtitrated, and comparison of a titration near the end-point with it will give a good indication when the end-point has been passed. The first reading is taken as the end-point. Record the observed volume. Calculate the molarity of the silver nitrate solution to the fourth decimal place. Checks of molarity should agree within 0.001.

Repeat titration with a second sample.

Do a third sample.

Titration of Chloride in Unknown Sample. Transfer with the pipet three 10-ml. portions of your unknown solution to separate 125-ml. Erlenmeyer flasks as in the preceding standardization. Add 10 ml. of H_2O, 3 ml. of concentrated HNO_3, 4 ml. of 2 M $Fe(NO_3)_3$, and pipet 1 ml. of 0.01 M KSCN and titrate to the end-point as before. Calculate the chloride concentration. Obtain at least two results that agree within 0.3 percent.

Now use your standardized $AgNO_3$ solution to titrate the unknown Cl^- solution.

CALCULATIONS

The molarity of the standard sodium chloride and potassium thiocyanate solutions should be given on the bottles. The molarity of the silver nitrate solution, based on the titration with standard sodium chloride solution, is calculated from equation (1):

$$\text{Moles } AgNO_3 = \text{moles } Cl^- + \text{moles } SCN^-$$

$$\frac{V_{AgNO_3} \times M_{AgNO_3}}{1000} = \frac{V_{NaCl} \times M_{NaCl}}{1000} + \frac{V_{KSCN} \times M_{KSCN}}{1000}$$

$$M_{AgNO_3} = \frac{V_{NaCl} M_{NaCl} + V_{KSCN} M_{KSCN}}{V_{AgNO_3}}$$

where M_{NaCl} is the molarity of the sodium chloride solution, and V_{NaCl} is the volume of standard sodium chloride solution in milliliters.

Develop an expression for calculating the chloride concentration in the unknown.

QUESTIONS–PROBLEMS

1. Describe as briefly as possible the mechanism by which the thiocyanate and ferric ions serve as indicators in the Volhard titration.

2. How is it possible for silver chloride to redissolve after it has been precipitated in this experiment?

3. What error in the molarity of chloride ion is introduced by redissolving of some silver chloride in this determination?

4. Why is it necessary to standardize silver nitrate solutions shortly before using them? What error is introduced if these solutions are standardized and then allowed to stand for some time before being used?

5. Discuss several advantages of determining chloride content by a volumetric procedure rather than by a gravimetric procedure.

REFERENCES

1. Day, R.A., and Underwood, A.L.: *Quantitative Analysis.* 2nd edition. Prentice-Hall, Englewood Cliffs, N.J., 1967, Chapter 6.
2. Swift, E.H.: *Introductory Quantitative Analyses.* Prentice-Hall, Englewood Cliffs, N.J., 1950, p. 79.

Experiment 30
REPORT SHEET

Name _____   Section No. _____

I. Standardization of Silver Nitrate

	Titration		
	1	2	3
Volume of NaCl solution taken	_____	_____	_____
Molarity of standardized NaCl solution	_____	_____	_____
Initial buret reading	_____	_____	_____
Final buret reading	_____	_____	_____
Volume of $AgNO_3$ used	_____	_____	_____
Molarity of $AgNO_3$	_____	_____	_____
Average molarity of $AgNO_3$		_____	

II. Titration of Chloride in Unknown Solution

	Titration		
	1	2	3
Volume of sample taken	_____	_____	_____
Initial buret reading	_____	_____	_____
Final buret reading	_____	_____	_____
Volume of $AgNO_3$ used	_____	_____	_____
Molarity of Cl^- in unknown	_____	_____	_____
Average molarity of Cl^- in unknown		_____	

Analysis of a Mixture
of Salts (Concluded)

A COMPLEXOMETRIC TITRATION

Ethylenediaminetetraacetic Acid (EDTA) Determination of Magnesium Ion

OBJECTIVE

To determine magnesium ion in the presence of sodium ion using a complexometric titration.

EQUIPMENT NEEDED

25-ml. buret; 10-ml. pipet; 1-ml. pipet (instructor); aspirator bulb; two 125-ml. Erlenmeyer flasks; 10-ml. graduated cylinder.

REAGENTS NEEDED

Standardized solutions of EDTA (~0.05 M); buffer solution 0.5 M ammonium chloride plus 0.5 M ammonium hydroxide buffered at pH ~ 10); Eriochrome Black T-potassium chloride mixture.

INTRODUCTION

In this experiment the concentration of magnesium ion in the sample will be determined by titration of a portion of the mixture with a solution containing ethylenediaminetetraacetate.

EDTA·Metal Complexes

Titration of a solution containing a metal ion with a suitable standard solution of a complexing agent or ligand is an important method of analysis for metal ions. To be suitable for such a titration the complex-formation reaction must proceed rapidly and according to well-defined stoichiometry. It also must lend itself to a sensitive end-point detection.

Ethylenediaminetetraacetic acid, EDTA, is an important and well known tetraprotic acid.

$$HOOC-CH_2 \qquad CH_2-COOH$$
$$:N-CH_2CH_2-N:$$
$$HOOC-CH_2 \qquad CH_2-COOH$$

403

Its four pK_a values are 1.99, 2.67, 6.16, and 10.26, respectively. In the form of its anion ethylenediaminetetraacetate ion, $EDTA^{4-}$, or simply Y^{4-}, it acts as a hexadentate ligand, forming very stable one-to-one complexes with nearly every metal ion. The reaction may be represented as

$$M^{n+} + Y^{4-} \rightleftharpoons MY^{n-4}$$

and the equilibrium constant for this reaction is given by

$$K_{MY} = \frac{[MY^{n-4}]_{eq}}{[M^{n+}]_{eq}\,[Y^{4-}]_{eq}}$$

Values of this equilibrium constant for several metal salts at 25°C are:

METAL ION	K_{MY}	METAL ION	K_{MY}
Ca^{2+}	$10^{10.70}$	Fe^{2+}	$10^{14.33}$
Co^{2+}	$10^{16.31}$	Mg^{2+}	$10^{8.69}$
Cu^{2+}	$10^{18.80}$	Zn^{2+}	$10^{16.50}$

The very large values for K_{MY} indicate that the complexes are very stable. The high stability of the complex ion formed and the fact that the reaction has a one to one stoichiometry are the reasons for sharp breaks at the end-points in the titration of EDTA with metal ions, as illustrated in Figure 31-3. The value of K_{MY} for the sodium salt is very small, indicating that the sodium complex is unstable. Hence, magnesium ion can be titrated successfully in the presence of sodium ion.

Infrared and x-ray studies have indicated that the EDTA complex with cobalt(III) ion has a structure that may be represented as:

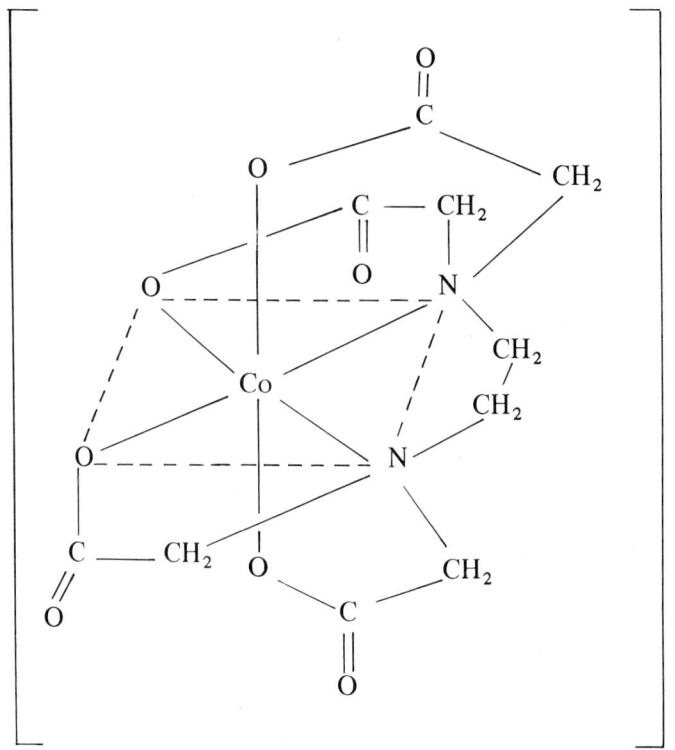

The magnesium complex ion presumably has a similar structure, and a charge of 2^-.

Metal Ion Indicators

Some organic compounds react with metal ions to form colored complexes in reactions analogous to those undergone by acid-base indicators. Some of these compounds are useful for metal ion indicators in EDTA titrations. Eriochrome Black T (ErioT) is a dye molecule that has suitable properties to serve as an indicator for Mg^{2+} ions. $ErioT^-$ has the structure given in Figure 31-1.

Metal chelates are formed with $ErioT^-$ by loss of hydrogen ions from the phenolic (−OH) groups and the formation of bonds between the metal ions and the oxygen atoms as well as the azo group (Fig. 31-2). The molecule is usually represented in abbreviated form as a tribasic acid, H_3In. The sulfonic acid group is shown in Figure 31-1 and 31-2 as being ionized; this is a strong acid group, which is dissociated in aqueous solution regardless of pH, and, thus, the structure shown is that of the ion H_2In^-. This form of the indicator is red. The pK_a value for the dissociation of H_2In^- to form HIn^{2-} is 6.3. The latter species is blue. The pK_a value for the ionization of HIn^{2-} to form In^{3-} is 11.6; the latter ion is yellowish orange.

The color of the uncomplexed dye depends upon its degree of ionization; for the discussion below assume In = ErioT.

$$H_2In^- \rightleftharpoons HIn^{2-} \rightleftharpoons In^{3-}$$

(red)   (blue)   (yellow-orange)

The color of the magnesium complex $MgIn^-$ is red.

FIGURE 31-1

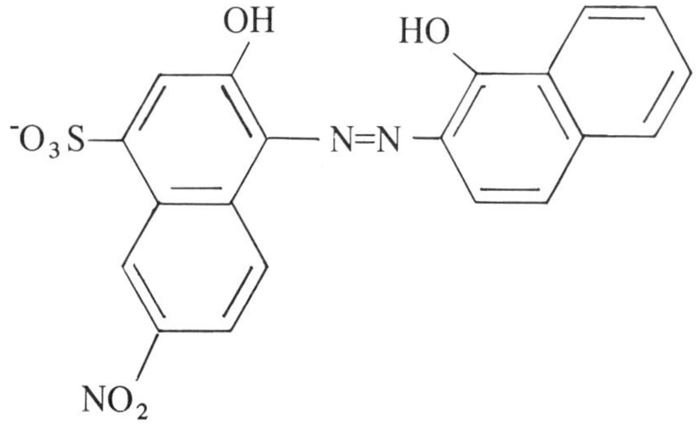

FIGURE 31-2

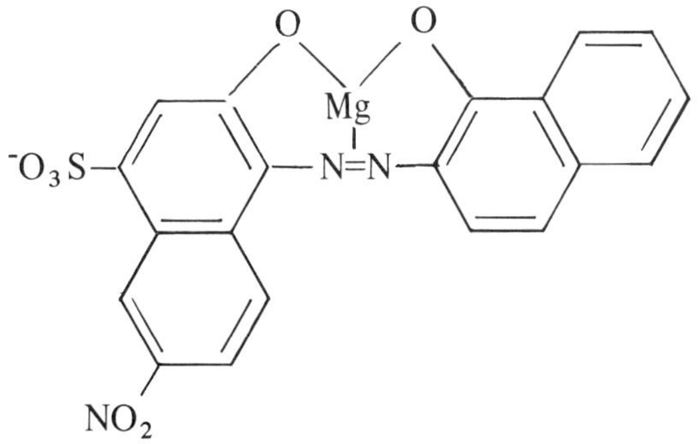

Upon addition of EDTA to a solution containing Mg^{2+} and the indicator (Eriochrome Black T) (ErioT) a sharp color change of wine-red to blue occurs when the magnesium ions have been removed by EDTA, with which magnesium forms a more stable complex than it does with $ErioT^-$.

$$MgIn^- + H_2Y^{2-} \rightleftharpoons MgY^{2-} + HIn^{2-} + H^+$$
(wine-red)   (colorless)   pH 10   (colorless)   (blue)

The hydrogen ion on the right side of the equilibrium is neutralized in the basic buffer solution, and the equilibrium shifts to the right. The color reaction for magnesium and ErioT is sharpest in solutions buffered to pH 10 to 11 with ammonium hydroxide and ammonium chloride. Because the stabilities of both the magnesium EDTA complex and the magnesium indicator complex are affected by pH changes, the titrations of magnesium with EDTA are performed in solutions of controlled pH.

Difficulties in the Titration

The equilibrium constant for the reaction $Mg^{2+} + Y^{4-} \rightleftharpoons MgY^{2-}$ is $10^{8.69}$ at 25°C. Under many conditions, however, this value is not directly applicable because the principal equilibria involve protonated EDTA species, e.g.,

$$Mg^{2+} + HY^{3-} \rightleftharpoons MgY^{2-} + H^+$$

$$Mg^{2+} + H_2Y^{2-} \rightleftharpoons MgY^{2-} + 2H^+, \text{ and so forth.}$$

The protons in these reactions can be considered to compete with magnesium ions for Y^{4-}. This competition effectively reduces the stability of the magnesium complex. Furthermore, in acidic solution, hydrogen ions may also associate with the complex to form less stable complex species; these effects will not be considered here. To take into account such effects it is convenient to use equilibrium constants for the complex formation that are adapted to the particular conditions, such as the pH, that prevail in the experiments. These are called *effective* or *conditional formation constants* and give the effective values of the equilibrium constants under a given set of conditions.

The effects of pH in the Mg^{2+} equilibrium can be taken into account by describing the equilibrium

$$Mg^{2+} + Y_S \rightleftharpoons MgY^{2-} \tag{1}$$

where

$$[Y_S] = [H_4Y] + [H_3Y^-] + [H_2Y^{2-}] + [HY^{3-}] + [Y^{4-}] \tag{2}$$

The conditional constant is defined as

$$K_{cond} = \frac{[MgY^{2-}]}{[Mg^{2+}][Y_S]} \tag{3}$$

Using the acid dissociation constants

$$K_1 = \frac{[H^+]\,[H_3Y^-]}{[H_4Y]} \; , \; K_2 = \frac{[H^+]\,[H_2Y^{2-}]}{[H_3Y^-]} \; , \text{ and so forth}$$

and substituting into equation (2)

$$Y_S = \left(\frac{[H^+]^4}{K_1K_2K_3K_4} + \frac{[H^+]^3}{K_2K_3K_4} + \frac{[H^+]^2}{K_3K_4} + \frac{[H^+]}{K_4} + 1 \right) [Y^{-4}]$$

Let A equal the sum on the right side,

$$A = \left(\frac{[H^+]^4}{K_1K_2K_3K_4} + \frac{[H^+]^3}{K_2K_3K_4} + \frac{[H^+]^2}{K_3K_4} + \frac{[H^+]}{K_4} + 1 \right)$$

and solving for $[Y^{4-}]$

$$[Y^{-4}] = \frac{[Y_S]}{A} \qquad\qquad (4)$$

$$\text{By definition } K_{MgY^{2-}} = \frac{[MgY^{2-}]}{[Mg^{2+}]\,[Y^{4-}]} \qquad\qquad (5)$$

Substituting equation (4) into equation (5), rearranging and employing equation (3)

$$K_{cond} = \frac{K_{MgY^{2-}}}{A} = \frac{[MgY^{2-}]}{[Mg^{2+}]\,[Y_S]}$$

As the pH decreases, A increases, causing K_{cond} to decrease.

At pH 10 the value of K_{cond} for MgY^{2-} is calculated to be about 2×10^9. The curve for the titration of 0.01 M Mg^{2+} with EDTA in solutions buffered at pH 10 is presented in Figure 31-3.

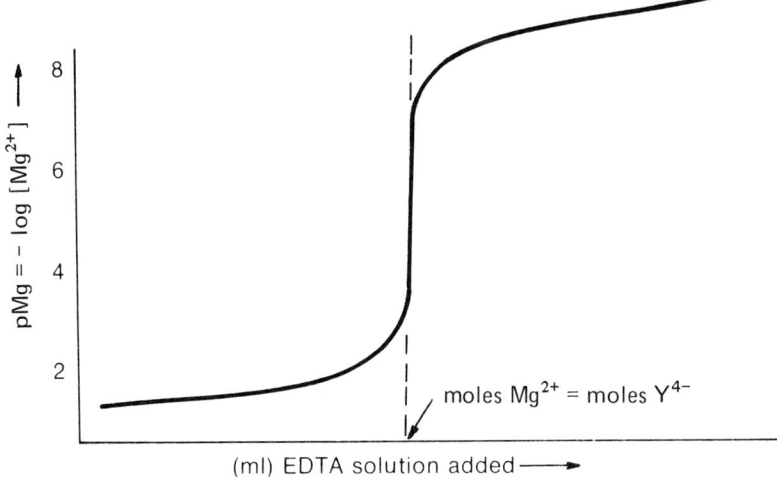

FIGURE 31-3

Titration curve for EDTA titration of magnesium ion (pH 10, 0.5 M NH_4^+ + 0.5 M NH_3).

PROCEDURE

Determine the magnesium ion concentration of your unknown solution directly, i.e., without treating it with the ion exchange resin.

Pipet 10 ml. aliquots of unknown solution.

Test with pH paper.

Titrate solution with EDTA.

Carefully pipet 10 ml. aliquots of your unknown solution into two clean 125-ml. Erlenmeyer flasks and dilute each with 20 ml. of distilled water. Add enough buffer solution to give a pH of approximately 10. Ten milliliters of buffer probably will suffice; test with pH paper. Obtain the Eriochrome Black T mixture at the instructor's desk and add 20 mg. or more (enough to produce a slight color) using a micro spatula. Return the indicator mixture to the desk immediately so that other students may use it. Titrate with standardized EDTA solution, stirring constantly, until there is a color change from wine-red to grey to pure blue. The equivalence-point occurs with the disappearance of the last trace of red. Neither red nor purple should be visible at the end-point. If, during the titration, the end-point is not sharp or the ErioT indicator appears to be fading, this is because the pH has dropped below 10. In this case add 2 to 3 ml. of the NH_3-NH_4Cl buffer and continue the titration.

CALCULATIONS

From the data obtained, calculate and report the molar concentration of magnesium ions in the sample. Using this information and the results of Experiments 29 and 30 calculate the concentrations of the three salts, sodium chloride, sodium nitrate, and magnesium chloride, in the original sample, and include the results in your report.

QUESTIONS—PROBLEMS

1. Why is it possible in this experiment to titrate magnesium ions accurately even though sodium ions also are present in the solution?

2. Is it possible to use this procedure to titrate magnesium ions in the presence of zinc ions? Why or why not?

3. Briefly discuss the chemistry of the indicator used in this experiment.

4. What is the function of the ammonium chloride-ammonium hydroxide buffer used in this experiment?

5. How would you prepare the ammonium chloride-ammonium hydroxide buffer so that it would be at pH 10?

REFERENCE

1. Day, R. A., and Underwood, A. L.: *Quantitative Analysis.* 2nd edition. Prentice-Hall, Englewood Cliffs, N.J., 1967, Chapter 7.

Experiment 31
REPORT SHEET

Name _____ Section No. _____

Data

	Titration		
	1	2	3
Molarity of EDTA solution used	———	———	———
Volume of sample taken	———	———	———
Volume of EDTA used	———	———	———
Molarity of Mg^{2+} in sample	———	———	———
Averaged molarity of Mg^{2+}	———		

Sample Calculations

Observations and/or Comments Concerning any Unusual Aspects of the Titration

Attach Answers to Questions

ORGANIC AND
BIOCHEMISTRY

IDENTIFICATION OF ORGANIC COMPOUNDS

OBJECTIVE

To identify specific organic compounds and to study briefly the methods used in making such identifications.

EQUIPMENT NEEDED

16 X 150-mm. test tubes; 100-, 250-ml. beakers; 100-ml. round-bottomed flask; reflux condenser; Büchner funnel; suction filter flask; melting point tubes; infrared spectrophotometer (optional); rubber bands; high-temperature burner; 260°C or 400°C thermometer.†

CHEMICALS NEEDED

0.5 M NaHCO$_3$ solution; 95% ethanol; 2,4-dinitrophenylhydrazine solution*; 0.3 M AgNO$_3$ solution; 3 M NaOH solution; 15 M NH$_3$ solution; acetyl chloride; thionyl chloride; 3,5-dinitrobenzoyl chloride (solid); pyridine; unknown samples of carboxylic acids, aldehydes, ketones, and alcohols; mineral oil or dibutylphthalate for m.p. baths (according to the temperature range of the unknowns chosen; see Appendix VIII-B).†

INTRODUCTION

When confronted with the qualitative analysis of an organic mixture, the chemist seeks to identify each of the organic materials present. Since more than two million different organic compounds have been characterized and identified, the methods used for determining just what substances are present in an organic mixture involve somewhat different strategies and techniques than those employed in many analyses of inorganic mixtures such as the cation analysis of Experiment 27. In this experiment we shall have time to develop only a few of the strategies and techniques an organic analyst would need to have at his fingertips in order to identify properly the components of an organic mixture.

One of the first steps in such an analysis would be the separation of the mixture into its individual components. We have indicated in earlier

---

*Prepared by dissolving 1 g. of 2,4-dinitrophenylhydrazine in 10 ml. 18 M H$_2$SO$_4$ and adding dropwise with stirring 12 ml. H$_2$O and 40 ml. of 95% ethanol. This reagent deteriorates on standing and must be prepared fresh weekly.

†Use mineral oil up to ~180°C; use dibutylphthalate up to ~300°C.

experiments how this is possible by crystallization, distillation, and chromatographic techniques. Once the individual substances have been separated and isolated, it is possible to proceed with their identification. While there is no single approach to identifying an organic substance, the chemist usually starts out by seeking four kinds of information. He is first interested in certain physical properties of the material, particularly its melting or boiling point, its solubility in various types of solvents, its color, odor, etc. Second, he seeks an elemental analysis to determine just what elements are present in addition to carbon and hydrogen. Third, he seeks to determine which functional groups are present in the substance. This frequently is done by a series of chemical tests and by careful examination of the infrared spectrum of the substance. When he knows which functional groups are present, he then is able to prepare derivatives of the substance by causing it to react through its characteristic functional group with other substances that will result in the formation of a solid product having a definite melting point. In many cases this combination of information will enable the chemist to identify the compound unequivocally.

To illustrate how this information could lead to an unequivocal identification of a substance, let us consider the following. An unknown solid organic substance was found to have a melting point of $121 \pm 1°$ and to contain only carbon, hydrogen, and oxygen. Its infrared spectrum showed the presence of hydroxyl, carbonyl, and aromatic functional groups. When this substance was placed in a 5% aqueous sodium bicarbonate solution, carbon dioxide was liberated. The elemental analysis, the presence in the infrared of a carbonyl and an OH group, when added to the evolution of CO_2 from the bicarbonate solution, strongly suggests the presence of a carboxylic acid, RCO_2H. The number of solid organic carboxylic acids with melting points in the vicinity of $121°$ is relatively small and probably less than 100. We, therefore, have narrowed the possibilities for our substance from somewhat over two million compounds to somewhat less than 100. We can now prepare two derivatives of this acid, making use of the reactivity of the acid functional group. The first derivative is prepared by allowing the acid to react with ammonia to form an amide. The amide is purified and its melting point is found to be $128 \pm 1°C$. A second derivative of the acid is made by causing it to react with aniline. This results in the formation of a solid product known as an anilide, melting point $160 \pm 1°C$. Since the amides and anilides of a very high percentage of the known organic acids are also known and their melting points available, it is easy to show that only one organic acid having a melting point near $120°$ gives an amide that melts at $128°$ and an anilide that melts at $160°$. This is benzoic acid, $C_6H_5CO_2H$. Hence it is reasonably certain that the original substance is benzoic acid. To make an unequivocal identification, however, one could take a small sample of benzoic acid and mix it with the unknown substance and determine the melting point. If the melting point of this mixture is $121 \pm 1°$ one can be reasonably certain that there is only one molecular species present in the mixture. Otherwise, the melting range would be much wider and probably the actual melting point much lower than this due to the fact that more than one molecular species is present.

In this experiment you will have the opportunity to identify one or more organic substances. You will be provided with a sample of the substance to be identified and with its elemental analysis and its infrared spectrum. You will be expected to conduct tests to determine which functional groups are present, to prepare at least one derivative of the substance and to obtain the melting point of this derivative. Using this information you should be able to make an identification of the substance.

TECHNIQUES

Melting and Boiling Points. Procedures for taking the melting or boiling point of a substance have been outlined in Experiment 3 (melting point) and in Experiment 13 (boiling point).

Infrared Spectra. Infrared spectroscopy is one of the most useful tools for determining the structure of organic molecules. Photons of infrared radiation can be absorbed by molecules to increase their vibrational energy. In simple molecules absorption of a given infrared wavelength often is associated with excitation of a certain chemical bond. For example, molecules containing the OH bond absorb infrared photons having wavelengths in the 2.8 – 3.2 μ range, and those containing the carbonyl, C=O, group absorb in the 5.4 – 6.1 μ region. A given chemical bond can be vibrationally excited in more than one way, as is illustrated for water in Figure 32-1 below. The vibrations a and b correspond to the stretching of oxygen-hydrogen bonds; a illustrates the symmetric vibration where hydrogen atoms stretch in and out together along the bond axis; b represents the asymmetric vibration in which one hydrogen atom moves out, the other moves in, while the oxygen atom, moving slowly due to its larger mass, shifts slightly to one side. Diagram c involves a bending of the OH bonds. Each of these vibrational excitations can be associated with absorption of a photon of definite wavelength. Studies of infrared spectra of organic compounds have advanced to the point that we now can identify absorption due to symmetric or asymmetric stretching and to various types of bending vibrations including scissoring, rocking, and twisting. Stretching vibrations appear at characteristically higher frequencies than do bending vibrations. When organic molecules become more complex, that is, contain more than ten atoms, the infrared spectrum will become too complex to associate every absorption with a specific type of vibrational excitation. However, in nearly all organic compounds, it is possible to associate specific absorptions with certain functional groups. Usually the chemist seeks to identify the major absorptions from 2.5 to 10 μ, associating each of these with a functional group. Figure 32-2 below indicates the regions of strong absorption for various functional groups in the range 2.5 to 10 μ.

Infrared spectrophotometers are instruments in which a beam of infrared radiation of known wavelength and intensity is allowed to fall on a cell containing a sample being studied, and the intensity of the radiation after it

(apart) (together)

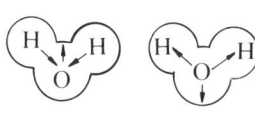 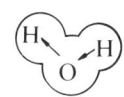 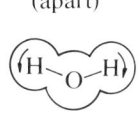 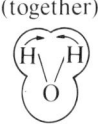

a. symmetric vibrations   b. asymmetric vibrations   c. bending vibrations

FIGURE 32-1.

Vibrations in water molecules.

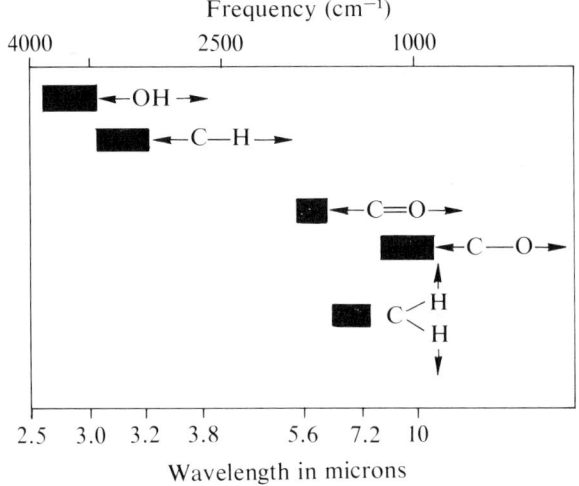

FIGURE 32-2.

Regions of strong infrared absorption by various functional groups.

passes through the cell is measured. The difference between the incident and the transmitted intensity represents the light absorbed, and this magnitude is displayed upon a recorder. The wavelength of the incident light is now slowly varied so that the per cent of light absorbed can be automatically traced out by the recorder as a function of wavelength.

Such a record or graph, known as an infrared spectrum of a substance, is shown in Figure 32-3 below. As discussed above, each peak is associated with the presence in the molecule of a certain structure or bond. The peak height or the amount of absorption of a certain wavelength depends on certain characteristics of the bond being excited and on the amount of material in the light path.

Appendix VIII contains six infrared spectra in which each of the major peaks is identified with a certain functional group or structure. When you receive the infrared spectrum for your substance, you might first attempt to associate each of the major peaks with a functional group, using the information in Figure 32-3, or you may refer directly to the infrared spectra in the Appendix VIII-A to make this association.

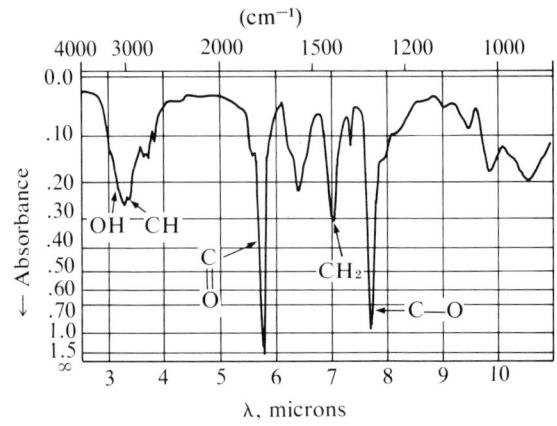

FIGURE 32-3.

The infrared spectrum of acetic acid.

Chemical Characterization of Functional Groups

To keep this experiment manageable, we shall limit the unknowns to four types of organic compounds. These are carboxylic acids, aldehydes, ketones, and alcohols. Following are tests that can be used to characterize the substance as belonging to one of these several classes of compounds.

Test for Carboxylic Acids. Add about 0.2 g. of the unknown to 1 ml. of 0.5 M aqueous sodium bicarbonate. Evolution of carbon dioxide indicates the presence of a carboxylic acid group, since neither aldehydes, ketones, nor alcohols liberate carbon dioxide under these conditions.

Test for Aldehyde or Ketone. If the test above is negative, dissolve about 0.5 g. of the compound in 20 ml. of 95% ethanol. Add this to 15 ml. of 2,4-dinitrophenylhydrazine test solution. Allow this mixture to stand at room temperature. The formation of a colored precipitate in ten minutes or less indicates the presence of an aldehyde or ketone.

Differentiation between Aldehydes and Ketones. Tollen's test. Place 2 ml. of a 0.3 M silver nitrate solution in a clean 16 × 150-mm. test tube and add one drop of 3 M sodium hydroxide, and just enough ammonium hydroxide to dissolve the precipitate that forms. Be careful to avoid an excess. Now add about 0.3 g. of the substance to be tested and place the test tube in a beaker of hot water. The presence of a silver mirror on the walls of the test tube or a precipitate of black metallic silver is a positive test for an aldehyde (Footnote: all ammoniacal solutions of silver salts should be washed down the drain with lots of water as soon as possible after use, since prolonged standing of these salts produces explosive precipitates.)

A negative test with Tollen's reagent following a positive test with 2,4-dinitrophenylhydrazine indicates the presence of a ketone.

Test for Alcohols. If the tests for acid, aldehyde, and ketone are all negative, the unknown can be tested for the presence of alcohol as follows. To 0.5 g. of the unknown in a 16 × 150 mm. test tube, add acetyl chloride dropwise and with great care while shaking the contents of the tube. If the unknown is an alcohol, heat will be liberated and hydrogen chloride gas will be evolved and can be identified by the fog which forms if one breathes gently across the mouth of the test tube. A sweet smelling product, namely the acetate ester of the alcohol, also should be produced.

Preparation of Derivatives. Having identified the functional group present, further identification requires the preparation of a derivative of that group.

Acids. *Preparation of Amides.* Place 2 g. of the acid and 3 ml. of thionyl chloride ($SOCl_2$) *(Care: this reagent reacts vigorously with water and skin!)* into a small, round-bottomed flask, and attach a reflux condenser. (See Figure 33-2.) Heat the mixture for 30 minutes under reflux. During this time the acid will be converted to the acid halide.

Prepare the amide by adding the solution of the acid halide dropwise with vigorous stirring to approximately 30 ml. of ice-cold concentrated ammonium hydroxide. (*Caution:* if the addition is too rapid, spattering will take place and the reaction may get out of hand.) Separate the amide which will precipitate from solution by pouring it on a Büchner funnel. Wash the precipitate once with water and discard this washing and the solution from which the amide was separated. Dissolve the precipitate in a small amount of hot water. If it will not dissolve in 10 ml. of hot water, add a few ml. of ethanol. Allow the mixture to cool and once again separate the solid product. Allow the solid to dry in the air and determine its melting point.

Derivatives of Aldehydes and Ketones. *2,4-Dinitrophenylhydrazines.* The procedure described for characterizing aldehydes or ketones above can be used to prepare the derivative of the aldehyde or ketone. The colored solid that forms when the unknown is treated with 2,4-dinitrophenylhydrazine is separated from the reaction mixture and washed with water. The reaction mixture and the water washings are discarded. The precipitate is dissolved in about 30 ml. of hot ethanol and this solution allowed to cool. The resultant solid is again filtered and allowed to dry, after which the melting point is obtained.

Derivatives of Alcohols. *Preparation of 3,5-dinitrobenzoates.* Mix about 0.5 g. of the alcohol, 0.5 g. of 3,5-dinitrobenzoyl chloride, and 1 ml. of pyridine in a 16 × 150 mm. test tube and heat the mixture on a boiling water bath in a hood for 30 minutes. Cool it and add about 10 ml. of 0.5 M sodium bicarbonate solution. If the product has not crystallized by this time, place the container in an ice bath and scratch the interior walls with a glass stirrer until crystallization occurs. Separate the solid by pouring the reaction mixture onto a Büchner funnel, wash it with 0.5 M sodium bicarbonate, and dry in air on filter paper. Recrystallize it by dissolving it in a small amount of hot ethanol, and allowing the mixture to cool. Again separate the solid. Allow it to dry and take a melting point.

Appendix VIII-B contains tables of acids, aldehydes, ketones, and alcohols, including those that will be used for your unknown. These tables include not only the melting point of the compound itself, but also the melting point of its appropriate derivative. Once you have obtained the melting or boiling point of your unknown and have characterized it as to functional group, you may be able to narrow it to 2 or 3 substances from these tables. Preparation of a derivative should enable you to make a definite identification. A mixed melting point (Experiment 3) should make the identification certain.

NOTEBOOK AND REPORT

Reports on this experiment should include all important observations needed to justify the identification made and equations for all reactions you carried out on the substance.

QUESTIONS

1. Using the infrared spectra in Appendix VIII as a guide, is it possible to distinguish an aldehyde from a ketone solely on the basis of its infrared spectrum? Why or why not?

Identification of Organic Compounds

419

2. What is the basis in terms of chemical structure and reactivity for the Tollens test that distinguishes an aldehyde from a ketone? Could comparable tests be designed to distinguish among primary, secondary and tertiary alcohols? Explain.

3. What advantages does 3,5-dinitrobenzoyl chloride have over other acyl chlorides such as acetyl or benzoyl chlorides in preparing derivatives of alcohols?

4. If two substances have the same melting point, will a mixture of the two also melt at the same temperature? Why or why not?

5. Why are carboxylic acids converted to acyl chlorides before reaction with ammonia? What products are formed when acyl halides react with primary, secondary and tertiary amines?

Experiment 32
REPORT SHEET

Name _____   **Section No.** _____

Unknown No: _____

Elemental Analysis   C _____ H _____ N _____ O _____ others _____

Major infrared peaks μ _____   _____   _____   _____   _____

Possible functional groups present based on infrared

_____   _____   _____   _____

Melting point _____ Boiling point _____

Other physical properties: Color _____ : Odor _____

_____   _____   _____   _____

Results of tests

   $NaHCO_3$ test   _____

   2,4-dinitrophenylhydrazine test _____

   Tollen's test   _____

   Acetyl chloride test   _____

Derivatives: Name   m.p.

   _____   _____

   _____   _____

Tentative identification   _____

Mixed melting point   _____

Final identification   _____

ORGANIC SYNTHESIS
Preparation of Medicinally Useful Compounds

OBJECTIVE

To illustrate several aspects of synthetic organic chemistry by preparing several medicinally useful compounds from a given starting material.

EQUIPMENT NEEDED

Two 250-ml. Erlenmeyer flasks; 100-, 250-, 800-ml. beakers; boiling chips; condenser; Büchner funnel; suction filter flask; 125-ml. separatory funnel; hot plate; long scoopula or spatula; Meker burner; melting point tubes, 25 × 200 mm. test tubes, rubber bands, plastic bag; cork stopper; 400°C thermometer.

CHEMICALS NEEDED

Salicylic acid, thionyl chloride, concentrated ammonia, methanol, ether, acetic anhydride, concentrated H_2SO_4, 95% ethanol, phthalic anhydride, dilute sodium hydroxide solution, 1.0 M $NaHCO_3$ solution, 12 M HCl, ice; dibutylphthalate for m.p. bath.

INTRODUCTION

One reason organic chemistry has made considerable contributions to our health and welfare is the very highly developed skill of the organic chemist in synthesizing complex and uniquely structured compounds in high purity and often in large quantities from plentiful and readily available starting materials. Among the important synthetic compounds prepared by organic chemists are many of the medicinal products we use regularly.

In this experiment we shall illustrate several aspects of synthetic organic chemistry by preparing compounds that have known medicinal applications. The synthetic principles illustrated here include: 1) the selection of appropriate organic reactions, such as substitution, condensation, acetylation, decarboxylation, and esterification; 2) the concept of multistep syntheses whereby the chemist, using a given starting material, must convert it by appropriate reactions into several intermediate substances before synthesizing the final product; 3) the choice of suitable procedures for working up reaction mixtures, and isolating and purifying products of a reaction.

Starting material for the reactions in this experiment is salicylic acid.

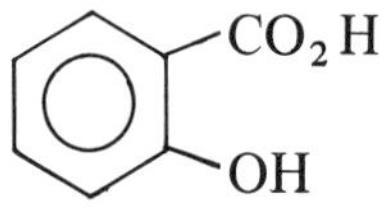

salicylic acid

This is a solid melting at 157-159°C. Its esters occur naturally in certain plants, notably in the bark of sweet birch and in wintergreen.

In one part of the experiment the acid will be converted to the acyl chloride by reaction with thionyl chloride.

Substitution

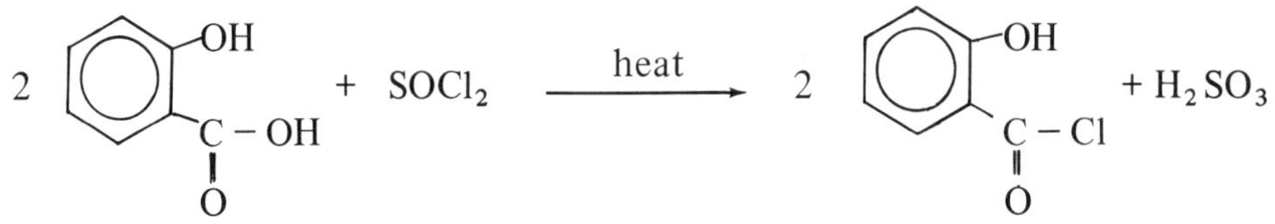

$$2 \quad \text{(salicylic acid)} \quad + \text{ SOCl}_2 \quad \xrightarrow{\text{heat}} \quad 2 \quad \text{(acyl chloride)} \quad + \text{ H}_2\text{SO}_3$$

The acid chloride can then be treated with ammonia to give the amide or with methanol to give the methyl ester, commonly known as oil of wintergreen.

Acylation of ammonia to form an amide

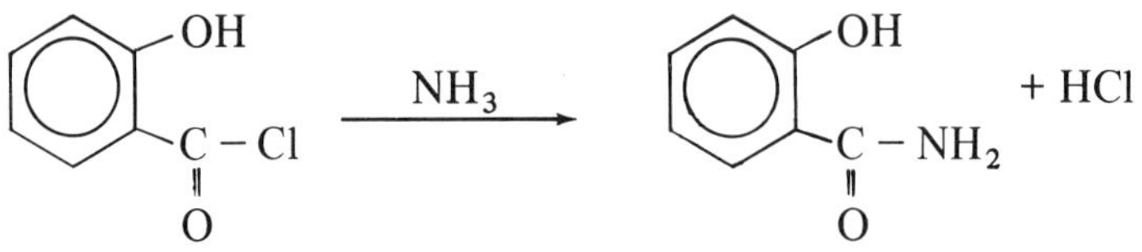

$$\text{(acid chloride)} \quad \xrightarrow{\text{NH}_3} \quad \text{(amide)} \quad + \text{ HCl}$$

Salicylamide (m.p. 140°)

Esterification by acylation of an alcohol

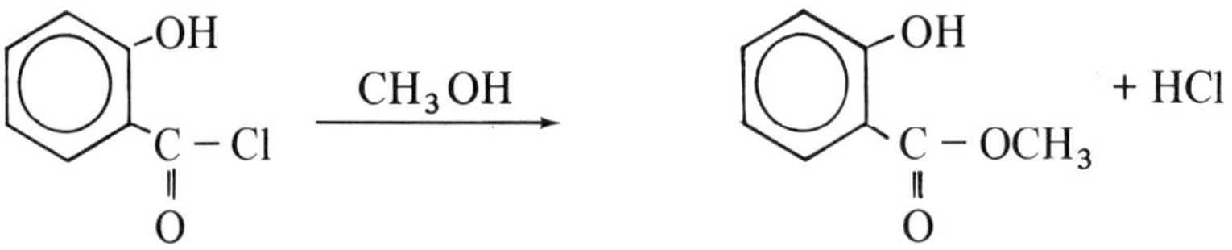

$$\text{(acid chloride)} \quad \xrightarrow{\text{CH}_3\text{OH}} \quad \text{(methyl ester)} \quad + \text{ HCl}$$

methyl salicylate (b.p. 220-224°)

In another part of the experiment, salicylic acid will be converted to acetylsalicylic acid, commonly known as aspirin, by treatment with acetic anhydride.

Acetylation

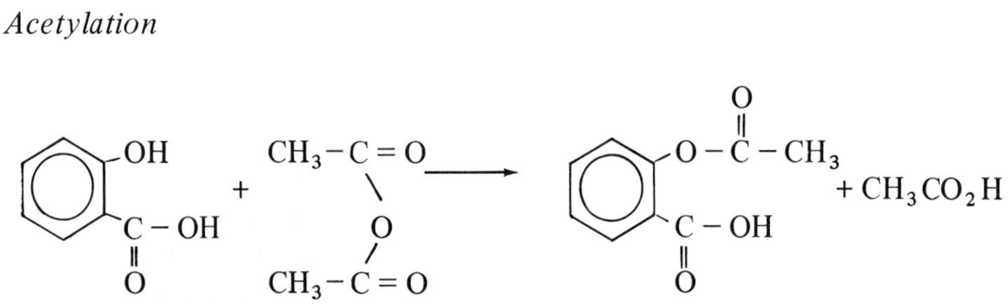

$$\text{(salicylic acid)} \quad + \quad \text{(acetic anhydride)} \quad \longrightarrow \quad \text{(acetyl salicylate)} \quad + \text{ CH}_3\text{CO}_2\text{H}$$

acetyl salicylate (m.p. 129°)

In a third part of the experiment salicylic acid will be converted to phenol and the phenol allowed to react with phthalic anhydride to give phenolphthalein. The first of these reactions is a decarboxylation; the second is a condensation.

Decarboxylation

phenol (m.p. $43°$)

Condensation

phthalic anhydride   phenolphthalein (m.p. 258-$262°$)

The products of each of these reactions have important uses. Salicylamide is used as an analgesic, an antipyretic and an antirheumatic. Methyl salicylate is used medicinally as a counterirritant. It also is used for flavoring candies and in perfumery. Aspirin, of course, is the most commonly used antipyretic. Phenol is used topically as an anesthetic for pruritic lesions in 1% solutions or ointments. In full strength it is very escharotic. Phenolphthalein is used as a cathartic; it also is used as an indicator in acid-base titrations.

Salicylic acid is used in ointments and alcohol solutions for fungus and ringworm infections.

TECHNIQUES

Some of the reactions selected for this exercise proceed readily at room temperature; others require only heating. The equipment used is much simpler than that which might be employed in many organic syntheses. The reactions that occur readily at room temperature require that certain safety precautions be strictly and constantly adhered to. These include the addition of small amounts of material at reasonable intervals so as to prevent the reactions from getting out of hand and, in some cases, the ready availability of ice baths in which the reaction flask can be placed should the reaction become too violent. Two reagents, thionyl chloride and acetic anhydride, must be kept away from moisture and the skin at all times. Open flames in the vicinity of organic vapors should be avoided, and safety glasses must be worn at all times.

The techniques used in separating the product from the other components of the reaction mixture and the purification of the product may

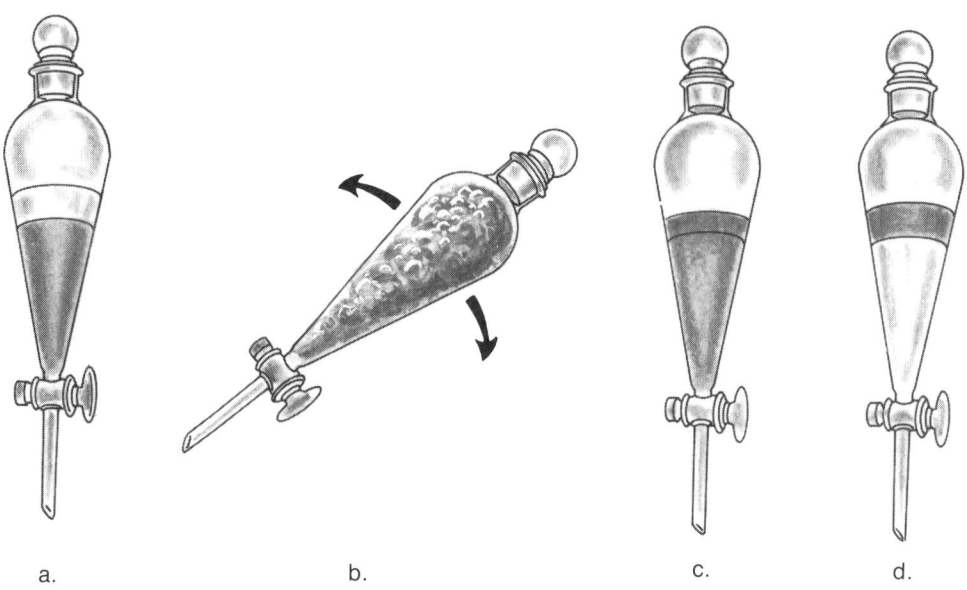

a.   b.   c.   d.

FIGURE   33-1.

Extractions using a separatory funnel.

a. The solution containing the material to be extracted is added to the separatory funnel; the immiscible extracting solvent is carefully poured into the funnel.

b. The funnel and its contents are shaken to bring about contact among solvents and the material to be extracted. Pressure created by evaporating solvent or released gas must be relieved frequently during shaking.

c. and d. After shaking, the liquids again separate, but the material to be extracted may now be concentrated in the extracting solvent.

include extraction, followed by recrystallization or distillation. In extraction, use is made of the fact that the organic product ordinarily will be more soluble in a nonpolar solvent such as ether than in water, whereas many of the inorganic products of the reaction will be more soluble in water than in ether. Hence, at the completion of the reaction, water may be added and this followed by the addition of small volumes of ether. After each addition the ether and water are shaken to allow the organic product to dissolve in the ether. The mixture is then allowed to stand so that the ether and water layers can separate, after which the ether layer is removed and saved. A second small volume of fresh ether is then added to the water layer, the mixture shaken, the layers allowed to separate, and the ether layer again removed and combined with the previous ether extract. Using two or three small volumes of ether in separate extraction operations will remove the product more efficiently than a single extraction with a large volume of ether. This is illustrated in Figure 33-1.

The product can be removed from the ether solution by distillation if it is a liquid or by crystallization if it is a solid.

In part A of this experiment methyl salicylate will be extracted from the reaction mixture by using ether. In parts B and C the products will be removed by crystallizing them from ice water or from ethanol. Here, the product that is formed is separated from the solvent and further purified by dissolving it in a small amount of a solvent in which the impurities present in the crude solid are soluble, and again either evaporating some of the solvent or reducing its temperature to the point where the more pure crystals will form. These are then separated, dried and their melting point determined. As indicated in earlier experiments, the melting range of a solid is an indication of its purity. Samples that melt over a 0.1 degree range are usually quite pure. Samples that melt over a 1-1.5 degree range can be considered acceptable for most work. Procedures for determining the melting points of substances are given in Experiment 3.

PROCEDURES

Part A: Preparation of Salicylamide and Methyl Salicylate.

Support a 125-ml. Erlenmeyer flask in an 800-ml. beaker in such a way that the beaker can be used as a water bath to heat the contents of the flask. Put a few boiling chips in the water in the beaker. Using a cork stopper, attach a condenser to the top of the flask, and arrange for the flow of water through the condenser (Figure 33-2). Remove the condenser and add about 4 grams of salicylic acid (weighed to the nearest 0.1 g.) to the flask. *Carefully and using precautions to avoid contact of the reagent with the skin, nasal passages, clothing or with moisture*,* add 6 ml. of thionyl chloride to the

Assemble apparatus.

Add salicylic acid.

Add thionyl chloride.

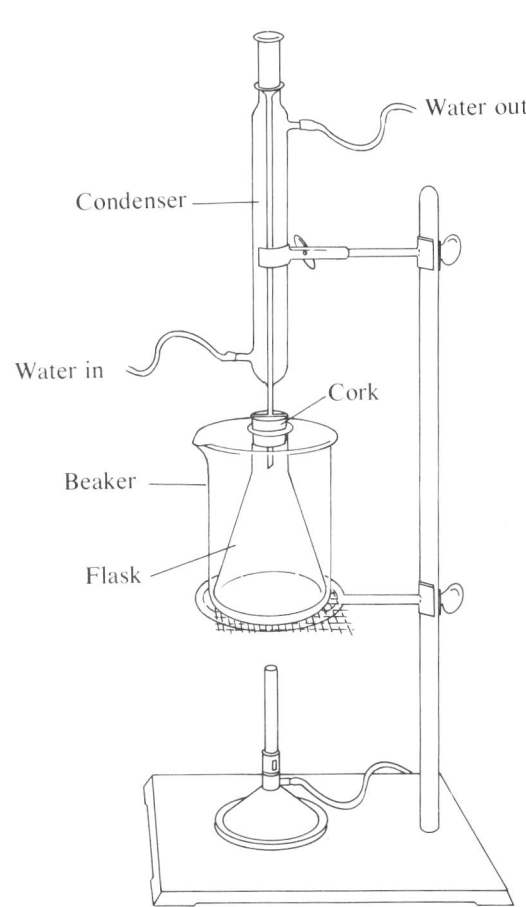

FIGURE 33-2.

Diagram of apparatus for reaction of salicylic acid with thionyl chloride.

---

*If $SOCl_2$ makes contact with the skin or clothing, wash the affected area immediately with large quantities of water.

flask. Replace the condenser and heat the water in the bath. When the liquid in the flask begins to boil, reduce the heat slightly but keep the reaction liquid boiling and condensing back into the flask (refluxing) for 30 minutes. Remove the condenser from the flask and the flask from the beaker. Allow the reaction mixture to cool. (Note 1.)

Reflux.

Meanwhile add about 25 ml. of concentrated ammonia to a beaker or an Erlenmeyer flask and cool it by placing the flask in a beaker of ice. Add 10 ml. of methanol to a second flask.

Cool the ammonia.

Carefully, and using precautions to avoid contact of the reaction mixture with the skin, transfer half of it – with the aid of a long spatula or scoopula – to the flask containing the methanol. Allow the mixture to stand while the remaining half of the reaction mixture is treated with ammonia and the product isolated, as in the following paragraphs.

Carefully transfer ½ of reaction mixture.

Carefully, and with cautious swirling, transfer the second half of the reaction mixture to the ice-cold ammonia solution. [*Caution: If added too rapidly the reaction will get out of hand and considerable splattering of material may occur.*] Stir this mixture with a stirring rod for two minutes; allow the solid to settle and decant the liquid from it. Discard the solid. Heat the liquid in a hood, maintaining a gentle boil for five minutes. Cool the solution and carefully neutralize it by adding, a little at a time, about 30 ml. of concentrated hydrochloric acid. Cool the mixture by placing it in a beaker of ice. Separate the solid that forms by suction filtration. That is, pour the contents of the flask into a Büchner funnel attached to a filter flask, and using an aspirator attached by suction tubing to the filter flask pull the liquid through the filter. (See Figure 2-1 for set-up.)

Transfer second half of mixture in hood.

Discard solid.

Add HCl.

Filter.

Wash the precipitate on the filter once with water and discard this washing and the solution from which the solid was separated. Dissolve the precipitate in 10 ml. or less of hot water. Allow the mixture to cool and once again separate the solid product. Set it aside to dry in the air and, when dry, determine its mass and its melting point. The pure amide melts at $140°$.

Wash the precipitate and dissolve in hot water.

While the amide is drying, return to the flask containing methanol and the crude salicylyl chloride. Gently heat this on a hot plate in a hood for 5 minutes. DO NOT use an open flame because methanol may burn on contact with it.

Allow the contents of the flask to cool and pour them carefully into 50 ml. of a 1 M $NaHCO_3$ solution in a separatory funnel. Extract the methyl salicylate from this mixture, using three 10-ml. samples of ether as follows. (Also see Figure 33-1.) Add 10 ml. of ether to the separatory funnel. Stopper the funnel, shake gently and relieve the pressure. Repeat with more vigorous shaking, relieving pressure periodically. Allow the mixture to separate and remove and save the ether layer. Return the water layer to the separatory funnel and repeat the extraction with a second, and then a third, 10-ml. portion of ether. Combine the three ether extracts. Allow the ether to evaporate, using a procedure to be suggested by your instructor. Determine the mass, and note other physical properties such as odor, color, and viscosity of the resulting product.

Extract with ether.

The crude methyl salicylate obtained here can be further purified by distillation or chromatography. Because these require more elaborate equipment than is available to us, we shall not purify the product further.

Submit both products to your instructor in labeled vials when you turn in your report.

Part B: Acetylation of Salicylic Acid –
Preparation of Aspirin.

Arrange a 250-ml. Erlenmeyer flask in an 800-ml. beaker in such a way that the beaker can be used as a water bath to heat the contents of the flask as in Part A (Figure 33-2. [The reflux condenser shown in Figure 33-2 is not needed for this part of the experiment, and should be removed.]) Put a few boiling chips in the water in the beaker. Remove the flask from the water bath and place a known amount (approximately 2 g.) of salicylic acid in it. Obtain 8 ml. of acetic anhydride and add this to the flask. [*Caution:* acetic anhydride is irritating to the skin and eyes. It should be used only with adequate ventilation. Should this get on your skin, the affected area should be washed immediately with large amounts of water.]

In the apparatus, combine salicylic acid and acetic anhydride.

To the mixture, add carefully and slowly 5 drops of concentrated sulfuric acid, swirling the flask gently after each addition. [*Caution:* concentrated sulfuric acid is extremely irritating to the skin and eyes. It is destructive to clothing and it reacts violently with water. It should be handled with extreme care. Affected areas of the skin should be washed immediately and with large amounts of water.] After the sulfuric acid has been added, place the flask in the beaker of boiling water for ten minutes. Remove the flask from the water bath, cool it under the cold-water tap, and carefully add 25 ml. of ice water to the flask. This will destroy any unreacted acetic anhydride and dilute the sulfuric acid present. Set the flask in a beaker of ice until crystallization appears to be complete. Use a Büchner funnel and suction filtration to separate the crystals from the liquid. Recrystallize the aspirin by dissolving the crystals in 20 ml. of ethanol in a 100-ml. beaker, warming the mixture on a water bath if necessary to effect solution. Now pour 40-50 ml. of warm water into this solution. Place the beaker in a large beaker of ice to facilitate crystallization. Again separate the crystals from the liquid using a Büchner funnel and suction filtration. Allow them to dry, determine their melting point, and weight. Record this and calculate the yield of aspirin based on the amount of salicylic acid you used as starting material.

Cool the reaction mixture.

Recrystallize the aspirin.

Submit the product to your instructor in a labeled vial with your report.

Part C: Decarboxylation of Salicylic Acid;
Preparation of Phenolphthalein.

To a large (25 × 200 mm.) test tube add about 2 grams of salicylic acid. Heat this rapidly with a burner flame until it melts. Continue heating the liquid acid for five minutes, moving it in and out of the flame to prevent splattering or boiling over. Cool the test tube and its contents and note the odor of the product after cooling. This is the characteristic odor of phenol.

In a large test tube, heat the salicylic acid.

Add about 2 grams of phthalic anhydride and two drops of concentrated sulfuric acid [*CAUTION*] to the test tube and once again heat the mixture gently until it melts. Continue heating the liquid and moving it in and out of the flame until the liquid becomes dark red but not black.

Add phthalic anhydride and sulfuric acid.

Cool the test tube and contents, add 20 ml. of water and, using a stirring rod, break up the lumps of solid that have formed in the test tube. Carefully decant and discard the water solution (which should contain the sulfuric acid and unreacted phenol). Dissolve the remaining solid in about 10 ml. of warm

Cool the test tube and add water.

Dissolve solid in ethanol.

ethanol. Pour the ethanol solution into a small beaker and cool it in ice. If a good quantity of crystals forms, separate them from the ethanol using suction filtration. If the yield of crystals is not large add 1 or 2 ml. of water to the ethanol solution and collect the crystals. Save the ethanol solution.

Dry crystals and obtain melting point.

Allow the crystals to dry in air and obtain the melting point and the mass of the phenolphthalein and calculate its yield based on the moles of salicylic acid used. Submit the product to your instructor in a labeled vial.

Add a few drops of the ethanol solution to a dilute solution of sodium hydroxide and explain your observations.

NOTEBOOK AND REPORT

Record the amounts of all starting materials, any changes in procedure, unusual observations and the masses, melting points and yields of all isolated products.

In the report include (balanced) equations for all reactions and describe the method used in calculating yields of products.

SUGGESTIONS FOR FURTHER WORK

Phenol reacts at room temperature with salicylyl chloride, acetic anhydride and phthalic anhydride. One of the products from these reactions is phenyl salicylate. This is used in medicine as an intestinal antiseptic. It passes through the stomach unchanged and is slowly hydrolyzed in the intestinal tract. It also is used as a coating for pills that are intended to pass through the stomach before dissolving. As an additional project you may wish to carry out some of the reactions described above and isolate the products. You should plan your work carefully and discuss it with your instructor before proceeding.

QUESTIONS

1. Why is the reaction of salicylic acid with thionyl chloride not carried out in the presence of a solvent such as water or ethanol? Would any solvent be suitable for this reaction? Explain.

2. What is the role of sulfuric acid in the acetylation of salicylic acid? Suggest a "mechanism" for its involvement in this reaction.

3. Why does the acetylated product in Part B not contain acetyl groups at both the $-CO_2H$ and $-OH$ functions on salicylic acid?

4. Why is the extraction with several portions of ether more efficient than a single extraction with a larger volume?

5. Write equations for: a) the reaction of base with each of the following: salicylic acid, methyl salicylate, and aspirin; b) the reaction of water

with each of the following: thionyl chloride, acetic anhydride and salicylyl chloride

;c) the reaction of ethanol with each of the following: phthalic anhydride, acetic anhydride and salicylic acid; the reaction of phenol with each of the above.

Note 1. The top of the reflux condenser may fume badly. If so, arrange a tube leading from the condenser to within 1 cm of the surface of 3M NaOH contained in an open flask, or to an open soda lime tube.

Experiment 33
REPORT SHEET

Name _____   Section _____

PART A

Mass of salicylic acid taken _____

Volume of thionyl chloride taken _____

Mass of salicylamide obtained _____

Melting point of salicylamide _____

Mass of methyl salicylate obtained _____

Yield of products: salicylamide _____ methyl salicylate _____

PART B

Mass of salicylic acid used _____

Volume of acetic anhydride used _____

Mass of aspirin obtained _____

Melting point of aspirin _____

Yield of aspirin obtained _____

PART C

Mass of salicylic acid used   _____

Mass of phthalic anhydride used   _____

Melting point of phenolphthalein   _____

Yield of phenolphthalein   _____

AMINO ACIDS ON HUMAN SKIN

OBJECTIVE

To separate and identify some amino acids from human skin. To illustrate the sensitivity and effectiveness of partition chromatography.

EQUIPMENT NEEDED

Wattman 3MM chromatography paper 8 × 6 in. sheets; small casserole; 800-ml. beaker; 15-μl. pipet; staples and staple gun; spray bottle; hotplate; 12-in. ruler; oven (optional).

CHEMICALS NEEDED

6 M acetic acid; acetonitrile developing solvent (60 parts acetonitrile, 40 parts 0.1 M ammonium acetate, V/V); n-butanol developing solvent (40 parts n-butanol, 10 parts glacial acetic acid, 10 parts water, V/V/V); ninhydrin spray solution (0.5% ninhydrin in 95 ml. n-butanol and 5 ml. 60% aqueous acetic acid); samples of reference amino acids (0.5% weight solutions of each of the following: leucine, tyrosine, valine, alanine, glutamic acid, glycine, histidine, asparagine, arginine, serine and phenylalanine); soap; *Saran* film or equivalent. Optional: casein, wheat gluten, soybean meal and other protein material.

INTRODUCTION

The amino acids in the proteins of human skin include alanine, arginine, asparagine, glutamic acid, glycine, histidine, leucine, phenylalanine, proline, serine, tyrosine, and valine. In this experiment you will obtain some amino acids from the palm of your hand, separate them using paper chromatography, and identify them by comparing their rate of movement across the paper with that of known amino acids used as references.

Techniques. The principal technique used is known as ascending paper chromatography. It is similar to that described for thin layer chromatography in Experiment 6 except that the stationary phase here is water on a cellulose surface. As with ascending thin layer chromatography, the sample of the mixture to be separated is placed near the bottom of the strip of chromatography paper and this strip dipped into a solution of a developing solvent. The developing solvent is then allowed to rise by capillary action across the paper strip where it eventually hits the sample and moves the components with it. These move at varying rates depending upon their relative solubilities in the water of the stationary phase and in the components of the developing mixture. Either of two developing mixtures can be used here: 60 parts acetonitrile to 40 parts 0.1 M ammonium acetate, or 40 parts *n*-butanol to 10 parts glacial acetic acid and 10 parts water. The acetonitrile solvent is faster; the *n*-butanol mixture gives a better separation.

To aid in identification of the amino acids after they are separated, samples of known amino acids will be chromatographed on the same paper and simultaneously with the mixture to be separated. For example, a single sheet of chromtography paper 8″ wide will be prepared by spotting 7 separate samples at one inch intervals along a line 1″ from the bottom edge of the paper. Two of these samples will contain the mixture to be separated. Each of the others will be a different amino acid to be used as references. When the samples have been spotted and dried, the paper is bent to form a cylinder and placed in the development tank.

The chromatogram is allowed to develop until the solvent front has moved approximately 6″. It is then removed, allowed to dry, sprayed carefully with a ninhydrin solution, and heated at 60° for about 10 minutes. The colored spots on each of the samples are circled with a pencil since the color may change upon standing.

The amino acids in the original samples can be identified by associating a given spot in the unknown sample with that of a known amino acid that has moved the same distance during development or by comparing its R_F value with those in Table 34-1 below. Spraying the sample with a ninhydrin solution and heating will make the locations of the amino acids visible.

TABLE 34-1 R_F VALUES FOR AMINO ACIDS USING ACETONITRILE AND AMMONIUM ACETATE

Amino Acid	R_F Value
leucine	0.76
valine	0.64
proline	0.57
alanine	0.52
glutamic acid	0.44
glycine	0.43
histidine	0.40
asparagine	0.31

PROCEDURE

Obtaining Amino Acids. Wash your hands thoroughly with soap and water, rinse them for one minute with tap water and for about 30 seconds with distilled water. Then hold approximately 10-15 ml. of distilled water in the palm of your hand for no less than one minute. Carefully pour this into a clean casserole and evaporate it just to dryness on a hotplate. Dissolve the residue in 6 drops of 6 M acetic acid, being sure to allow the acetic acid to wash the sides and bottom of the casserole thoroughly.

Obtain sample of amino acid from your hand.

Evaporate solution just to dryness.

Obtain an 8″ × 6″ piece of Wattman 3MM Chromatography Paper, and carefully but lightly draw a pencil line across the entire sheet approximately 1″ from the bottom edge of the longer side. Using a 15 microliter pipet, place about 5 microliters of the sample of amino acids at a spot on this line approximately 1″ from the left-hand side of the paper. Place a second 5-microliter spot on the line approximately 1″ from the right-hand edge of the paper. Using clean pipets for each sample, place 5-microliter spots of 5 different amino acids at 1″ intervals along this line between the original sample spots. Record the position on the paper of each of the reference amino acids. Allow the samples to dry.

Add samples to Chromatography Paper.

Carefully grasp the paper at the edge farthest from the spots, turn it into a cylinder, and then staple it at the upper edge with the smallest amount of overlap between the edges necessary to hold the staple.

Make a cylinder of the Chromatography Paper.

Prepare the developing tank by cleaning and carefully drying an 800-ml. beaker. Place approximately one-half inch of one of the developing solvents in the bottom of the beaker. Cover the beaker with a piece of *Saran* film and shake it gently to saturate the air inside the beaker with the developing solvent. Remove the film and carefully place the cylinder of paper in the flask with the edge closest to the sample spots down. Cover the beaker with the *Saran* film and allow the chromatogram to develop until the solvent front has moved approximately 6″ across the paper. Remove the chromatogram and mark the position of the solvent front with a pencil. Allow it to dry and spray lightly with the ninhydrin solution. Heat the sprayed chromatogram at approximately 60-80° for about 10 minutes. Circle the spots in all samples with a pencil. Identify the amino acids in your sample by comparing them with the reference samples or by determining their R_F factors and comparing these with those in the table. Compare the amino acids obtained in largest amounts from your hand with those obtained by other students.

Prepare developing tank and saturate the air with solvent vapor.

Remove chromatogram, mark, and dry.

Compare the unknown with known amino acids.

You may also want to repeat this experiment using a small sample of casein, wheat gluten, soybean meal, or other protein material. Small amounts of amino acids should be extractable by washing these samples with distilled water and following the procedures indicated earlier. Comparisons of the kind and relative amounts of amino acids from the various sources should prove interesting.

NOTEBOOK AND REPORT

Record your observations in the notebook and try to explain any unusual observations in the report. The report should also include the Report Sheet and answers to the questions that are assigned by your instructor.

QUESTIONS

1. What assumption is made in identifying the amino acids in the sample by associating their positions on the developed chromatogram with that of reference amino acids? How could the validity of this assumption be tested?

2. Write structural formulas for three of the amino acids obtained from your skin and offer an explanation in terms of their structure and the

composition of the developing solvent for their relative rates of movement across the paper, i.e., their R_F values.

3. Compare the structural formulas of the amino acids in your sample with others known to be in the proteins of the skin but not found on the chromatogram. After studying the bonding and structure of proteins, offer an explanation for the observation that certain amino acids are more easily obtainable from skin than are others.

4. Write the structural formula for ninhydrin and postulate a reaction between ninhydrin and amino acids. Offer an explanation for the observation that ninhydrin does not give the same color with all amino acids.

5. Compare the R_F values of several amino acids in the two developing solvents suggested in this experiment (one containing butanol, the other acetonitrile) and offer an explanation for observed differences and similarities.

6. Development using the acetonitrile solvent is considerably faster than that using the butanol solvent. Offer an explanation for this.

7. The effectiveness of the acetonitrile developing solvent can be improved for certain combinations of amino acids if it is buffered at a lower or higher pH. For example, by adding acetic acid it can be buffered at pH 4, or by adding ammonia it can be buffered at pH 9.2. What structural features of the amino acids of skin proteins are expected to be important in bringing about better separations at the lower pH? at the higher pH?

REFERENCES

Brinkman, U. A. Th. and DeVries, G.: *J. Chem. Educ., 49*:545(1972).
Heimer, E.P., *J. Chem. Educ., 49:*547(1972).

Experiment 34
REPORT SHEET

Name _____ **Section No.** _____

Data

Developing Solvent _____

Spot No.	1	2	3	4	5	6
Color with ninhydrin						
Reference acid at comparable position						
R_F value						
Identification						

Unusual Observations

EVALUATION OF EXPERIMENTAL DATA

Now that you have gained experience in calibrating and using instruments that measure temperature, mass, and spectra, we will examine the way in which one scientifically evaluates experimental data. In all of the experiments you should be aware of the precision required, the accuracy of the data, and the sources of systematic and random errors. (These topics are treated in the following sections.) Then you will be asked to evaluate the data obtained in some of the previous experiments.

MEASUREMENT

Most scientific discoveries begin as a personal experience for the investigator. In order to study this experience objectively and to communicate it accurately to others, the investigator usually finds it necessary to use instruments to measure the factors involved. For example, centuries ago man first realized that he had a heartbeat. Perhaps he realized that the rate of heartbeat seemed to change at different times of the day. Upon comparing his heartbeat with that of another person, man undoubtedly discovered differences in rates between the two. Then came the question: How can we communicate to others what we have found? One way to do this was to design an instrument to measure heartbeat rates objectively so that differences between persons or differences in one person at different times of the day could be measured precisely. Of course, this instrument is the clock! Perhaps this legend of its inception is convincing justification for the importance scientists give to measurement.

Measurement makes it possible to obtain more exact information about the properties of matter, such as the size, shape, mass, temperature, or composition. Measuring instruments are merely extensions or refinements of our senses. For example, the balance makes it possible to determine the mass of an object more accurately than we could by lifting it, and the clock allows us to measure time more accurately than we could by observing with the naked eye the position of the sun in the sky.

Although great importance is given to measurement, equal importance should be given to realizing and reporting the limitations of the measuring instrument and the reliability of the data obtained from an instrument. Viewed broadly, the limitations of instruments are summarized in two questions:

1. *Is the instrument suitable for making the measurement under consideration?* For example, is an ordinary meter stick, which can be read to the nearest millimeter, suitable for measuring the thickness of

a human hair, which is approximately 0.1 mm. thick? The distance between the smallest interval ruled on the meter stick is 1 mm. If the meter stick were used for this purpose, the result would not be very reliable. It is unlikely that one could detect a difference in thickness between one hair and another with the meter stick. Therefore, the meter stick is not a suitable instrument for this purpose. Such measurements are much more valid when they are made with the aid of a microscope and a rule calibrated to 0.01 mm.

2. *Given a suitable measuring instrument, can the measurements with it be made with the required precision?* As an illustration, consider the results of measuring the length of this laboratory manual by using a meter stick. We can read without difficulty to the nearest millimeter, and with care we can estimate the length to the nearest 0.5 mm. With difficulty, much squinting, and proportionate uncertainty, we might be able to estimate the length to the nearest 0.1 mm. But we cannot go further than this because of the limitation inherent in our instrument and in our vision. Therefore, measurement of length with the meter stick is reliable at best to ±0.1 mm. If we need to know the length more precisely than this, a different measuring instrument is required.

Accuracy and Precision. There is some degree of uncertainty in every measurement, which may come from either limitations of accuracy or limitations of precision. One should note carefully the difference between the terms accuracy and precision. *Accuracy* involves a comparison of the average result found for a measurement with that of a true or accepted value. *Precision,* on the other hand, involves comparison of a series of measurements, made in the same way, to one another (Fig. 35-1). We can always obtain an exact value for the precision on a given set of measurements, but a true or accepted value must be known in order for the accuracy to be determined. Otherwise the accuracy can only be estimated.

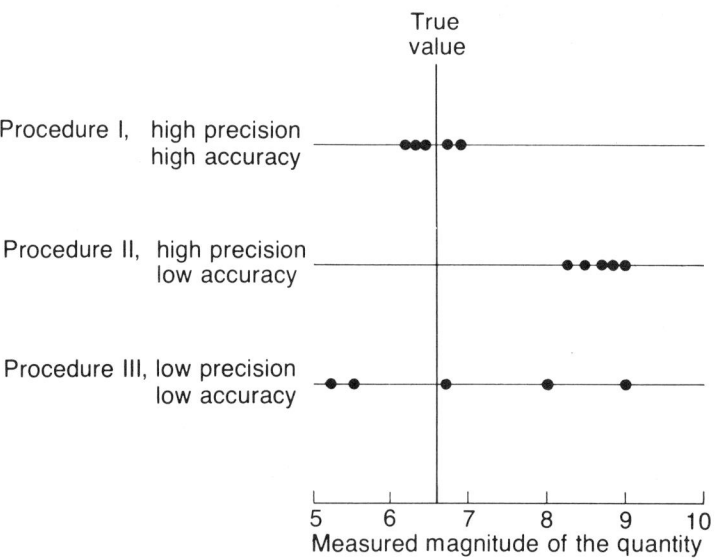

FIGURE 35-1

Accuracy and precision illustrated.

Most of the experiments in this laboratory program have a quantitative basis; they rely on measurements of factors such as temperature, mass, and volume. In order to evaluate quantitatively the accuracy and precision of a set of measurements, one needs to consider a set of N replicate measurements, X_1, X_2, $X_3 \ldots X_N$. The mean (sometimes called the average or arithmetic mean) is defined as

$$\overline{X} = \frac{X_1 + X_2 + X_3 + \ldots + X_N}{N} \tag{1}$$

For each measurement (and also for the mean value) there is a difference between the observed value (e.g., X_i) and the true value, T. This difference is the *error* of the measurement.

$$\text{Error} = \text{observed value} - \text{true value} \tag{2}$$

One can refer to the difference between the true value and a single measurement as the absolute error. A better estimate of the error can be made if the mean error (the difference between the mean value of the measurements and the true value) is calculated.

Often it is even more meaningful to express error as a value *relative* to the magnitude of the quantity being measured, i.e.,

$$\text{Relative error} = \frac{\text{error}}{\text{true value}} \tag{3}$$

This ratio is often multiplied by 100, 1000, or 1,000,000 and expressed as percentage (%), parts per thousand (p.p.t.), or parts per million (p.p.m.), respectively.

A quantitative representation of the precision of a given measurement is the *deviation,* defined as the difference between the mean of a set of measurements and the individual measurement. Thus

$$\text{Deviation, } \delta = \overline{X} - X_i \tag{4}$$

In any experimental science we are more often concerned with the precision of a set of measurements than with the deviation of a single measurement. We are interested in the *average deviation,* $\overline{\delta}$, which is the average value of all the individual deviations, δ,

$$\overline{\delta} = \frac{|\delta_1| + |\delta_2| + |\delta_3| \ldots + |\delta_N|}{N} \tag{5}$$

where the vertical bars, | |, enclosing a quantity signify that the absolute value of the quantity is used. The average deviation, $\overline{\delta}$, is easy to calculate, but two other measures of deviation, the *range* and the *standard deviation,* are also useful. The *range,* W, is merely the difference between the largest measured value and the smallest measure value, i.e.,

$$W = X_{\text{largest}} - X_{\text{smallest}}$$

The *standard deviation, σ,* is defined for a finite set of data as

$$\sigma = \sqrt{\frac{\delta_1{}^2 + \delta_2{}^2 + \delta_3{}^3 \ldots + \delta_N{}^2}{N - 1}} \tag{6}$$

Comparisons of the actual deviation of a measurement with the standard deviation of the set of measurements provides a criterion for discarding a given measurement or evaluating the precision of a set of measurements.

ERRORS

The numerical results obtained in an experiment are never completely accurate, but are always in error by variable amounts. The magnitude of the error depends upon the skill of the experimenter and the precision of his measuring instruments. Errors are divided for convenience into two broad categories, determinate (systematic) and indeterminate (random) errors. *Determinate (systematic) errors* are those that are introduced as a result of some inherent error in the method of measurement or in the calibration or manufacturing specifications of the measuring device. Determinate errors affect every measurement of a set to the same extent. They do not alter the precision, but do, of course, affect the accuracy of the measurement. For example, if an experimenter measured all lengths with a "yardstick" which was actually a "meterstick" — a trap into which unwary carpenters sometimes fall when working in a chemistry laboratory — it is evident that all the measurements would be systematically in error.

Indeterminate (random) errors, on the other hand, are those that are present in spite of the experimenter's best efforts. It is difficult, for example, to repeat a procedure in exactly the same way and under the same conditions that it was carried out the first time. In the simple operation of heating a crucible, which is then to be cooled and weighed, for example, the heating time may be inadvertently changed by a few minutes, or the temperature in the flame may be changed because of unrecognized changes in the composition of the gas-air mixture. These differences may be reflected in a slightly changed weight for the crucible. One must not be disturbed, therefore, if the results of two measurements are not exactly the same. On the other hand, the results should not vary so much that the differences cannot be reasonably accounted for on the basis of chance.

Normal Distribution Curve. If a large number of measurements are performed on a quantity having a true value of T, all individual measurements will not be identical to T, but will be scattered about T, owing to random error. If we plot the absolute error of each measurement against the frequency with which a given absolute error and sign occurs, a *normal distribution curve (or error curve)* (Fig. 35-2) results. Two important consequences are apparent from the curve of Fig. 35-2. (1) Both positive and negative errors are equally probable. Thus, the arithmetical mean (or average) of a series of single measurements represents the most reliable value, and the reliability (precision) increases as the number of measurements in the series increases. However, in doing an experiment one must consider the amount of time required to perform a large number of measurements. In the

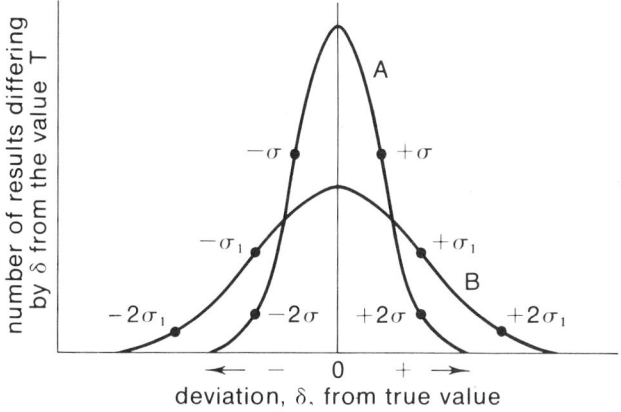

FIGURE   35-2

Normal distribution curves. Curves A and B represent the determination of the same quantity by two methods with inherently different reliability; method A is much more precise than method B.

interest of progress the experimenter must make some rational judgment between the time required to collect a large number of measurements and the reliability. Except in the most precise work, the average of three or four measurements is usually sufficient. (2) Small errors occur frequently, but the occurrence of a large error is relatively improbable.

The breadth or spread of the error curve indicates the precision of the measurements. A precise method gives a curve that is sharply peaked and drops off rapidly for large errors, e.g., curve *A* in Fig. 35-2; for imprecise measurements the curve is broad, e.g., curve *B* in Fig. 35-2. For a normal distribution, 68.3 percent of the results will differ from the arithmetical mean by less than one standard deviation ($\pm 1\sigma$), 95.5 percent by less than two ($\pm 2\sigma$), and 99.7 percent by less than three ($\pm 3\sigma$).

SIGNIFICANT FIGURES

All measuring instruments have limits to their suitability and precision. *Consequently, all measurements and calculations reported in scientific work must reflect the limitations of the instruments used.* This is done by expressing the measurement to the correct number of significant figures.

The meaning of significant figures may be illustrated by considering the following data for the weight of a crucible:

A	B	C
12.8 g.	12.86 g.	12.864 g.

The first result tells the reader that the crucible was weighed with an accuracy of 0.1 g.; the weight is closer to 12.8 g. than to 12.7 or 12.9 g. Hopefully, it lies between 12.75 and 12.85 g., but the reader cannot be sure, because the balance used may not have enabled one to distinguish values in the second decimal place. Similarly, in result B the weight is placed between 12.85 and 12.87 g., and in weight C it is between 12.863 and 12.865 g. Result A is recorded to three significant figures, result B is recorded to four significant figures, and result C is recorded to five significant figures. The third result is, of course, much more precise than either of the others. The weighing in result C is made to within 1 mg., or 0.001 g., in a total weight of approximately 13 g. Thus, the *absolute uncertainty* of the experimental weight is 0.001 g. and the *relative uncertainty* is 0.001/12.864, which may be expressed as 1 part in 12,864 or approximately 1 in 13,000. Weight A has a relative uncertainty of 1 part in only 128 (0.1/12.8), and weight B has a relative uncertainty of 1 part in 1286 or approximately 1 in 1300.

The relative error is often more important than the absolute error, as shown in the following example. The absolute error in result C is 0.001 g., and the value 12.864 g. is very precise (to 1 part in $\sim$ 13,000) for the object weighing about 13 g. However, a weighing made with the same absolute error (0.001 g.) for an object weighing only 0.013 g. would have a relative uncertainty of 1 part in 13.

In reporting the result of a set of measurements we should give the best value for the set, i.e., the arithmetical mean, *and* an indication of the precision observed between the individual values. We have already indicated that the latter may be expressed in terms of the range, average deviation, or standard deviation. One should determine and report all numbers that are known with certainty and, in addition, the first uncertain digit should also be reported.

For example, assume the values reported for an experiment were 21.24, 21.01, 21.22, and 21.15. The mean is 21.155 and the average deviation from the mean is ±0.075. Clearly, the number in the second decimal place is uncertain and we should round the mean value accordingly. Next, we must decide between 21.15 and 21.16, the value 21.155 being spaced equally between them. In a case of a remaining 5, round the number to the nearest even number. Thus, we should report these results as 21.16 ± 0.08 .

The *significant figures* of a number are all digits having values known with certainty plus the first digit having uncertain value. The position of the decimal point is irrelevant. Thus, 0.43214, 4.3214, 432.14, and 43,214 all contain five significant figures.

In some cases, the uncertainty in the final significant figure may be more than one unit. Suppose, for example, that the balance scale is divided by lines into 0.01 g. units, and the needle comes to rest between two lines. The experimenter then visually divides each space into 10 parts, and estimates the position of the needle in terms of these fictitious marks. In Fig. 35-3, for example, the recorded value would be 0.164, because the experimenter would estimate the needle to be four-tenths of the way from 16 to 17. Such an estimate cannot be made with a precision greater than about 0.2 unit, and the result might be recorded as 0.164 ± 0.002 g. to indicate that the uncertainty in the final significant figure was greater than one unit.

Zero as a Significant Figure. The zero occupies an ambiguous position with regard to significant figures. In the numbers 10.7, 17.03, and 17.033 the zero is a significant figure. Zeros enclosed by digits other than zero *are always significant.* Thus, 10.01 contains four significant figures.

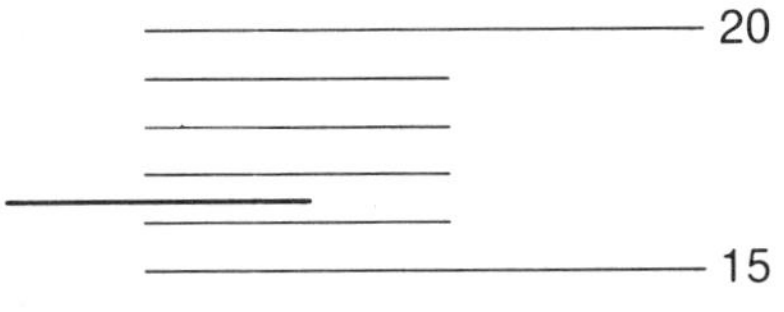

FIGURE 35-3

Estimating the space between graduations.

The zero is not a significant figure in the number .033 since it merely serves to locate the position of the decimal point. A zero is *never* a significant figure when it appears between the decimal and the first number on the right side of the decimal. In the number 0.033 neither zero is a significant figure. It may be helpful to note that when the number is written as a fraction $\left(\dfrac{33}{1000}\right)$, the zero is not a significant figure but simply designates the magnitude.

When the zero comes last in a whole number but before the decimal point, it may or may not be a significant figure. In the number 730, for example, the zero is not a significant figure if the precision of the measurement is ± 10 units, and the intention of the information is that the number lies between 720 and 740. But if 730 represents a number between 729 and 731, the zero is a significant figure. In the first case the number is expressed to only two significant figures, and the zero merely indicates that the decimal point is to be placed one place farther to the right than the last significant figure, thus, 730 and not 73. *Modern scientific notation requires that a number like 730 but containing only two significant figures be written as 7.3 X 10^2.* This removes the ambiguity previously associated with numbers ending in zero.

Propagation of Errors. It is often important to estimate the error of a final result that is calculated from several different measurements, each possessing a degree of uncertainty. It is evident that the precision of the final result can be no greater than that of the least precise measurements from which it was calculated.

Addition and Subtraction. The preceding point may be illustrated by the following determinations of the weight of a salt, based on the gain in weight of a crucible on the balance.

	A	B	C	D
Weight of crucible and salt	17.13 g.	17.033 g.	17.0 g.	17.033 g.
Weight of crucible	12 g.	12.1 g.	12.13 g.	17.020 g.
Weight of salt	5 g.	4.9 g.	4.9 g.	.013 g.

In each of the examples, the number of reliable, i.e., significant, figures that can be carried to the answer is determined by the least precise measured values.

This generalization is important in two ways. In the first place, it shows that it is wasteful of laboratory time to make one measurement with extreme precision when the precision of the result is fixed by a second measurement that is inherently less precise. In the determination of the density of a common liquid, for example, there would be no point in weighing the liquid to the nearest milligram if the volume could be measured only in a cylinder graduated in 2-ml. units (0, 2, 4, 6, 8, 10, etc.). Even if the volume could be measured with the same precision as the weight, there would be little point in doing so unless the temperature measurement could be made with a similar precision. Since the volume of a liquid changes with change in temperature, the experimenter might have a very precise value of the density, but not know to what temperature that value corresponds. Thus, it is often useful to consider what precision is necessary or possible for the several measurements before starting an experiment.

The second way in which the preceding sums are important is in showing that in subtractions the absolute error is often more important than the relative error. For result A, since the crucible was weighed only to the nearest gram, the weight of the salt is also known only to the nearest gram, or the relative error is 1 part in 5, a very imprecise result. This effect is particularly marked when there is a small difference between two large numbers. Thus, the result of the two weighings of D has a precision of only 1 part in 13, although each weighing was reliable to 1 part in ~17,000.

Multiplication and Division. In calculated results involving multiplication or division, the relative error is more important than the absolute error, and the final result will be in error by the amount of the largest relative error. Suppose, for example, you wish to determine the value for the product of pressure and volume for a gas in a case in which the pressure can be measured to the nearest 0.1 cm. of mercury and the volume to the nearest 0.1 ml. Assume the experimental values were 74.5 cm. and 15.5 ml; the numerical product would be 1154.75 ml. × cm. But the volume is measured to 1 part in 155 (0.1 in 15.5) and the pressure to 1 part in 745 (0.1 in 74.5). Since the result is precise only to within the amount of the least precise measurement, it will be reliable to 1 part in 155, or about 1 in the third digit. The value of the pressure-volume product must then be reported with only three significant figures; the numerical answer must be rounded off to 1150 ml. × cm. or, less ambiguously, to 11.5×10^2 ml. × cm.

These examples illustrate two important rules for determining the number of significant figures in the result derived from experimental data:

1. In addition or subtraction, the *absolute* uncertainty of the component is considered. Round off the answer so that it has the same precision as the least precise component.
2. In multiplication or division, the *relative* uncertainty of the component quantities governs the number of significant figures in the product or quotient. The answer should reflect a relative uncertainty that is of the same order as that of the least precise component.

Calculation of Percentage Error. The percentage error is calculated by dividing the difference between the accepted value and the experimenter's measured value by the accepted value and multiplying by 100. For example, an experimenter determines the percentage of oxygen in pure potassium chlorate to be 38.7 percent. The correct value is 39.6 percent. The percentage error is calculated as

$$\frac{39.6 - 38.7}{39.6} \times 100 = 2.3\%$$

In calculations such as these, the difference in the numerator is always considered to be positive. Had the measured value been 40.5 percent the percentage error would have been calculated as

$$\frac{40.5 - 39.6}{39.6} \times 100 = 2.3\%$$

UNITS AND DIMENSIONS

Every measurement involves relating the particular quantity measured to some reference scale. The reference scale is usually composed of arbitrarily chosen standards of measurement called units. A given dimension, such as length, may be expressed in a variety of units, such as angstroms, millimeters, centimeters, meters, kilometers, or light-years. The unit used is usually determined by its convenience for the kind and magnitude of measurement to be made. For example, if a small length, such as the diameter of an atom, is to be measured, it is more convenient to express that value in terms of angstroms than in terms of centimeters or meters. For most objects used in the laboratory the convenient unit of length is the centimeter. For greater distances, kilometers or miles are used and in considering distances in interstellar space, the light-year is the unit usually employed. A light-year is about 10^{26} Å.

Scientific convention requires that units be indicated for every measurement or derived result reported. Students in science should follow this rule not only in the laboratory but also in problem-solving and on examinations.

Carrying Units in Mathematical Operations. When two or more measured quantities are used in mathematical operations, attention must be given to the proper handling of units. The following rules apply to such manipulations:

1. In addition and subtraction, all quantities must be expressed in the same units before the operation is performed. For example, in adding 1.22 cm. and 10.3 mm., both quantities must be expressed in centimeters or in millimeters before the addition is performed. Converting 1.22 cm. to 12.2 mm. and adding this to 10.3 mm. gives 22.5 mm. If both numbers are expressed in centimeters, the answer would be 2.25 cm. which is, of course, the same length as 22.5 mm.
2. In multiplication and division, the units are multiplied or divided in the same way as the numbers are treated. For example, in multiplying 10.2 mm. by 4.00 ml., the product is 40.8 mm. × ml. (expressed as millimeter milliliters); in dividing 14.4 g. by 1.20 ml., the quotient is 12.0 $\frac{g.}{ml.}$ (expressed as grams per milliliter); in multiplying 10.2 cm. by 5.0 cm., the product is 51 cm. × cm. or 51 cm.^{2} (expressed as centimeters squared or square centimeters).

Conversion of Units. Often a quantity expressed in one set of units must be expressed in a different set as in Item 1 preceding, in which 1.22 cm. is changed 12.2 mm. This change is effected by using the conversion factor, which indicates that 1 cm. is equivalent to 10 mm. It is convenient to write such conversion factors as fractions and to use them as such, *retaining the unit labels.* Thus, the conversion of 1.22 cm. to 12.2 mm. can be formally expressed as

$$1.22 \text{ cm.} \times \frac{10 \text{ mm.}}{1 \text{ cm.}} = \frac{1.22 \text{ cm.} \times 10 \text{ mm.}}{1 \text{ cm.}} = \frac{1.22 \times 10 \text{ mm.}}{1} = 12.2 \text{ mm.}$$

Note that the centimeter units are cancelled out, just as if they were numerical factors. For a less trivial case, consider converting a density of 340 mg. per cubic centimeter to the corresponding value in pounds per cubic foot. We need the conversion factors for milligrams to grams, grams to pounds, centimeters to inches, and inches to feet, as shown in the following formal arrangement:

$$\frac{340 \text{ mg.}}{1 \text{ cm.}^3} \times \frac{1 \text{ g.}}{1000 \text{ mg.}} \times \frac{1 \text{ lb.}}{454 \text{ g.}} \times \left(\frac{2.54 \text{ cm.}}{1 \text{ in.}}\right)^3 \times \left(\frac{12 \text{ in.}}{1 \text{ ft.}}\right)^3 =$$

$$\frac{340 \times 1 \times 1 \times (2.54)^3 \times (12)^3 \times \text{mg.} \times \text{g.} \times \text{lb.} \times \text{cm.}^3 \times \text{in.}^3}{1 \times 1000 \times 454 \times 1^3 \times 1^3 \times \text{cm.}^3 \times \text{mg.} \times \text{g.} \times \text{in.}^3 \times \text{ft.}^3} = 21.2 \text{ lb./ft.}^3$$

In writing the conversion factor, the ratio of one unit to another can always be written two (i.e., correct and incorrect) ways. It is very important that fractions representing conversion factors be written in such a way that the unwanted units cancel. The factor for changing grams to milligrams is written $\frac{1 \text{ g.}}{1000 \text{ mg.}}$ so that "mg." in the denominator of the conversion factor can cancel the "mg." that appears in the numerator in the given data. Had the factor been written in the reciprocal manner, i.e., $\frac{1000 \text{ mg.}}{1 \text{ g.}}$, "mg." would not cancel, and "mg.^{2}" would have appeared in the answer. If properly arranged, all unit labels except those desired cancel out in the answer. Thus, one can feel confident that he has worked the problem in a correct manner if he carries through the units and arrives at the answer with the desired units on the numerical part of the answer.

As a further example of the use of conversion factors, consider the following problem: How many liters of oxygen at standard temperature and pressure (STP) can be obtained by heating 61.25 g. $KClO_3$, which decomposes according to the equation

$$2KClO_3 \longrightarrow 2KCl + 3O_2?$$

Here one of the conversion factors comes from the (calculated or known) formula weight of potassium chlorate and another comes from the chemical equation which gives the relation between the number of moles of oxygen formed and the number of moles of potassium chlorate decomposed. The problem is set up with the given information as follows:

$$6.25 \text{ g. } KClO_3 \times \frac{1 \text{ mole } KClO_3}{122.5 \text{ g. } KClO_3} \times \frac{3 \text{ moles } O_2}{2 \text{ moles } KClO_3} \times \frac{22.4 \text{ liters } O_2 \text{ at STP}}{1 \text{ mole } O_2} =$$

$$\frac{61.25 \times 1 \times 3 \times 22.4}{122.5 \times 2 \times 1} \times \frac{\text{g. } KClO_3 \times \text{moles } KClO_3 \times \text{moles } O_2 \times \text{liters } O_2 \text{ at STP}}{\text{g. } KClO_3 \times \text{moles } KClO_3 \times \text{moles } O_2} =$$

$$16.8 \text{ liters } O_2 \text{ at STP}$$

Note again that the conversion factors are arranged so that only the units of the desired answer "liters O_2 at STP" remain uncanceled.

Experiment 35
REPORT SHEET

Name _____   Section No. _____

1. Perform the following operations, observing significant figures.

    a. $(0.071)(16.000) =$

    b. $\dfrac{0.32}{6.023 \times 10^{23}} =$

    c. $\left(\dfrac{3.00}{1.25}\right)(20) =$

    d. $(250)\left(\dfrac{1}{1000}\right)\left(\dfrac{1}{2.50}\right) =$

    e. $\left(\dfrac{2000}{40}\right)\left(\dfrac{4000}{60}\right) =$

    f. $203.1 + 7.21 + 0.3734 =$

    g. $1.0046 - 0.0031 =$

    h. $1.0 - 0.0031 =$

2. Perform the following operations, observing significant figures.

    a. $1.0 \times 10^2 \times 2.0 \times 10^3 \times 3.0 \times 10^{-6} =$

    b. $\dfrac{1.00 \times 10^{-2} \times 3.0 \times 10^{-6}}{4.00 \times 10^8} =$

    c. $\dfrac{1.00 \times 10^{-2} \times 3.00 \times 10^{-6}}{4.0 \times 10^3 \times 6.00 \times 10^{-2}} =$

    d. $\dfrac{7.2 \times 10^{12}}{2.4 \times 10^{-12}} + \dfrac{1.3 \times 10^6}{2.6 \times 10^{-18}} =$

    e. $4.0 \times 10^{-6} + 3.7 \times 10^{-7} - 6.2 \times 10^{-5} =$

    f. $4 \times 10^{-6} \times 9.34 \times 10^4 =$

3. Perform the following operations, carrying units and the proper number of significant figures in the answer.

    a. 3.68 cm. $+ 2.43$ cm. $+ 10.44$ mm. $=$

    b. $\dfrac{1.852 \text{ g.}}{9.26 \text{ ml.}} =$

c.   18.2 cm. × 10.0 cm. =

d. 2.3×10^{-16} cm. + 2.0×10^{-18} m. =

e. $\dfrac{(4/3) \times \pi \times (1.23 \text{ Å})^3}{1 \text{ atom}} \times \dfrac{6.023 \times 10^{23} \text{ atoms}}{1 \text{ mole}} =$

f. 1.8×10^8 coulombs/g. × 1.64×10^{-18} g./sec. =

g. $\dfrac{16.5 \text{ feet}}{8.3 \text{ feet/year}} =$

h.   16.5 amperes × 6.0 volts =

4. Make the following conversions, observing significant figures.

a.   8.63 ergs to joules

b. 4.184×10^7 joules to calories

c.   10.0 a.m.u. to grams

d. 16.7×10^{-22} g. to a.m.u.

e.   28 g. of nitrogen to moles of nitrogen atoms (1 mole of nitrogen atoms is 14.0067 g.)

f.   8.63 ergs to kilocalories

g.   10 a.m.u. to pounds (1 lb. equals 453.59 g.)

h.   10.71 p.s.i. × cubic feet/degrees Rankine × pound mole to torr × cubic centimeter/ degrees Kelvin × gram mole.

i. 8.3143×10^7 ergs × mole^{-1} to calories per mole.

5. From a table of logarithms, determine the logarithms of

a.   6.924   e. 7.86×10^2

b.   710.3   f. 8.692×10^{-3}

c.   0.492   g. 1.63×10^{-11}

d.   0.00821   h. 4.296×10^{22}

6. From a table, determine the numbers y described as indicated

   a. log y = 4.2863   c. log y = – 2.4688

   b. log y = 0.6924   d. log y = – 0.8622

7. Obtain from the instructor the copper to sulfur ratio found (in Exp. 4) by each member of your laboratory section.
   a. Using these values as the experimental data, determine
       (1). The mean of the results
       (2). The deviation, δ, of each
       (3). The average deviation, $\overline{\delta}$
       (4). The standard deviation, σ
   b. Construct a graph similar to Fig. 35-2, assuming the mean value just calculated to be the true value. Does your curve more closely resemble curve A or curve B of Fig. 35-2?
   c. Cite as many sources of error in determining the copper:sulfur ratio as possible, and designate whether they are determinate or indeterminate errors.
   d. Calculate the percentage relative error [Equation (3)] of *your* copper:sulfur ratio by assuming that the true value is the mean value of the class results.

8. a. From your results of Experiment 7 calculate the percentage relative error of the three hydrogen lines. Are all relative percentage errors the same? How do you rationalize the results?
   b. Again using all the values obtained by your laboratory classmates for one of the lines of the hydrogen atomic spectrum, determine:
       (1). The mean value of the results
       (2). The average deviation, $\overline{\delta}$
       (3). The standard deviation, σ
   c. From the results of your calculations, how does the precision of the measurement in Experiment 4 compare with that of Experiment 7?

THE SINGLE-PAN ANALYTICAL BALANCE

Single-pan analytical balances are marketed by a number of firms, but all are quite similar in construction. The weighing mechanism consists of a beam resting near its middle upon a sapphire knife-edge. The end of the beam toward the front of the balance carries a stirrup and hanger device which supports the balance pan, contact again being made on a jewelled knife-edge. Also at the front and rigidly attached to the beam is a rod which carries weights of several denominations adding to a total corresponding to the maximum load of the balance, 99 g. for a nominal maximum of 100 g., for example. The end of the beam toward the back of the balance carries a counter weight equal to the combined mass of weights, rod and balance pan so that when all the weights are on the rod and nothing is on the pan the beam balances horizontally on the central knife-edge (Figure AI-1). The mass of an object placed on the balance pan is determined by counting the weights which must be lifted off the rod to compensate for the object's mass and restore the beam to its horizontal, balanced position. Also attached to the beam is a dash-pot to damp motion of the beam so that it does not rock back and forth indefinitely but comes quickly to rest, and a marker that, through an optical device, registers and translates into milligrams and tenths of milligrams the small differences from absolute horizontal that are caused by fractional differences of these magnitudes between the mass of the object and the mass of the weights removed.

Single-arm substitution balance

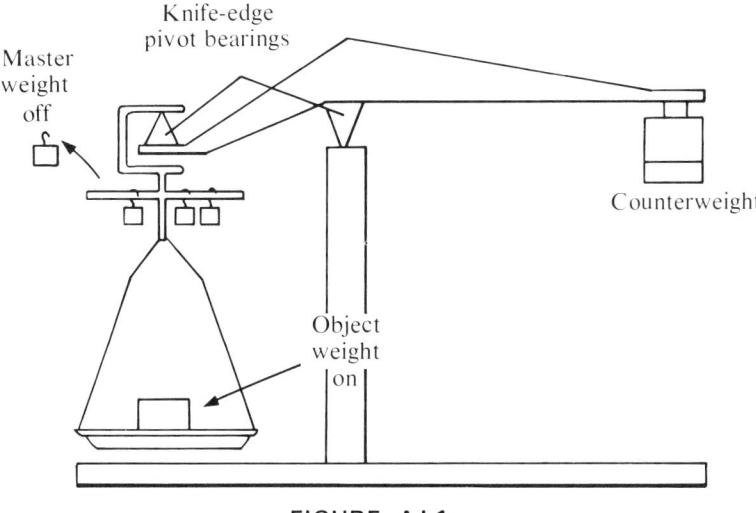

FIGURE AI-1.

Diagram for the beam of a single-pan analytical balance.

For reproducible weighings the beam and knife-edges must always be in the same relation to each other, so each balance carries a leveling mechanism and devices to prevent rubbing of the knife-edges and their opposite bearing plates, with consequent dulling of the knife-edges, when the balance is not in use, and which will drop the bearing plates onto the knife-edges in the same position each time the balance is used. Balances of different manufacture differ in the position on the balance case of the knobs or levers which operate these knife-edge-protecting devices, and the fingers which lift the weights from the rod attached to the beam. They also differ in the optical mechanism and the nature of the read-out which reports the third and fourth decimal place; sometimes this appears as a reference mark which points to a number and sometimes as a number which appears in a gap.

Your instructor will give you directions for operating the particular make of balance in your laboratory, but you should note and look for the following features.

1. **The Level Indicator.** This is usually a small disc on the floor or top of the balance which carries a bubble. When the balance is level, the bubble should be directly in the circle at the top of the disc. It is usually not necessary to adjust the level more than once or twice a day to compensate for temperature changes which change the position of the building or change unequally the length of the legs on the balance table.

2. **The Weight Manipulation Knobs.** These knobs, usually two, control the fingers which lift weights from the beam to balance the mass of the object on the pan. One knob will remove weights in 10-g. increments, the other removes weights in 1-g. increments. Turning the knobs also turns a dial so that the magnitude of the weight removed can be read. It is important that these knobs be turned slowly and gently and only when the beam and balance pan are supported free of the knife-edges. Rapid turning may cause the weights to bounce off their supports, and adding or removing weights when the beam and pan are resting on the knife-edges will dull the knife-edges and make the balance useless.

3. **The Zero Line Adjustment Knob.** This knob changes the position of the optical scale so that it rests on zero when the pan is empty and the beam is at equilibrium. This control also should need only infrequent adjustment, but should be checked before each weighing.

4. **The Optical Scale, Micrometer, or Vernier Adjustment Device.** This control positions the optical scale relative to a reference line or pointer so that a reading to 0.0001 g. can be obtained. On some makes it is not necessary to make this adjustment, the final reading appears automatically when the weights are set in the optical range.

5. **The Pan and Beam Control Knob.** This knob keeps the knife-edges separated from the beam when the balance is not in use and, if turned *gently,* will lower the beam and pan reproducibly when weighing. It should be emphasized that contact with the knife-edges should be made softly and smoothly by gentle turning of this knob. Turning this knob also switches on the light source for the optical scale. There are commonly three positions:

a) *Arrest position:* The beam is completely supported and the knife-edges are not in contact. The balance should be put in the arrest position whenever the operator is placing objects on or removing them from the pan and between weighings.

b) *Partial arrest position:* In this position, the beam is free to move but contact with the fragile knife-edges is restricted. Weights can be removed (with care) when the beam is in the partial arrest position without harm to the balance mechanism. The partial arrest position is used in obtaining the approximate weight needed to balance the mass of the object being weighed.

c) *Fully released position:* This is the position in which the beam is freely swinging and it and the pan are supported only on the knife-edges. The balance is particularly vulnerable in this position and should never be touched except for careful and gentle adjustment of the zero line or optical scale. Final weighings are taken in the fully released position.

d) Figure AI-2 shows a diagram of a representative single-pan balance, and instructions for the use of a typical balance follow.

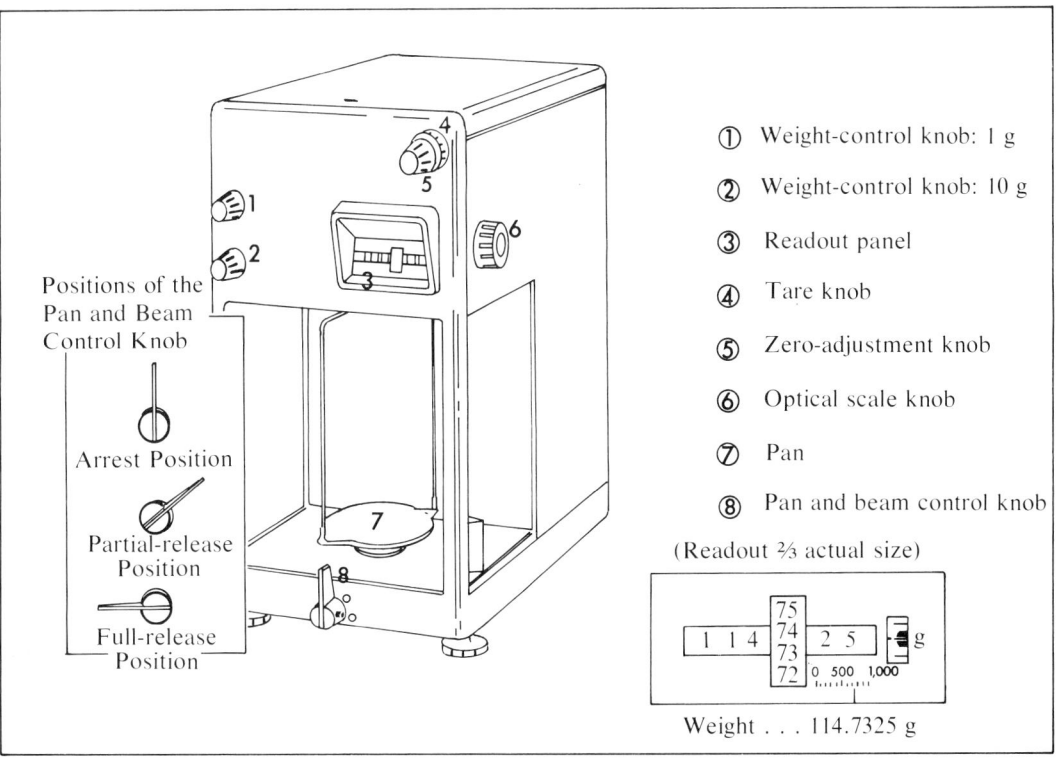

FIGURE AI-2

A typical single-pan balance.

I. Before weighing, be sure that:
 a. The balance is arrested.
 b. There is nothing on the pan, the pan is clean, and the doors are closed.
 c. The weight-control knobs, optical-scale knob, and tare knob are set at zero.

II. Setting the zero point:
 The zero point must be rechecked before each weighing.
 a. Turn the pan and beam control knob to the full-release position.
 b. When the numbered scale stops moving, turn the zero-adjustment knob until the line is perfectly centered at the zero position.
 c. Arrest the balance by turning the pan and beam control.

III. Weighing the sample:
 a. Place sample on pan with forceps. Use a container or weighing paper.
 b. Close the doors.
 c. Turn the pan and beam control to the partial-release position.
 d. Turn the 10-g. weight-control knob until the beam is nearly balanced.
 e. Repeat with the other weight-control knob.
 f. Turn the pan and beam-control knob to the arrest position and after a slight pause, turn it (gently!) to the full-release position.
 g. Turn the optical-scale knob (after the numbered scale stops moving) until a line is perfectly centered in the indicator slit or aligned with the pointer.

IV. Completing the weighing:
 a. Record the result.
 b. Arrest the balance.
 c. Remove sample from pan with forceps.
 d. Return all knobs to zero.
 e. If you spilled anything on the pan or in the balance, clean it up.

DIAGRAMS AND BRIEF INSTRUCTIONS FOR USING SIMPLE SPECTROPHOTOMETERS

BAUSCH & LOMB SPECTRONIC 20
(Bausch & Lomb, Rochester, New York)

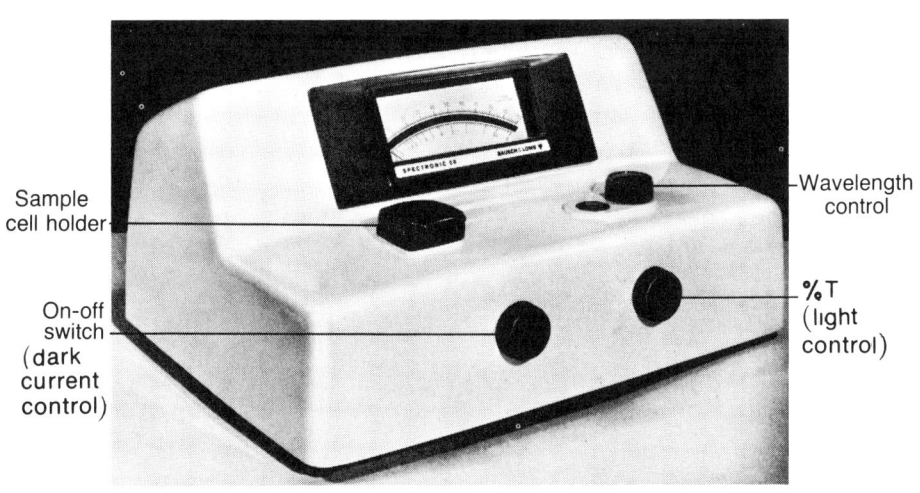

FIGURE AII-1

(Courtesy of Bausch & Lomb.)

Whenever a tube is removed from the sample holder of this spectro-photometer, the shutter automatically falls into the light beam. In this condition, with zero light from the lamp reaching the phototube and with the top to the cell compartment closed, adjust the meter to read zero percent transmittance, using the on-off switch (dark current control). *All settings and readings must be made with the cover of the sample holder closed.* As soon as a cuvette containing solvent is placed in the sample holder, the shutter automatically swings out of the light path; set the meter at 100 percent transmittance by means of the %T control (light control). When the sample solution is substituted for the reference solution in the light beam, the percent transmittance of the sample may be read directly on the meter.

Whenever the wavelength is changed, a cuvette containing solvent must be placed in the sample holder and the %T control readjusted until the meter records 100 on the %T scale. On the other hand, the dark current is adjusted without any light striking the photocell and, therefore, is independent of the wavelength setting.

The optical pathway and the components of the Spectronic 20 are shown in Figure AII-2. Any Spectronic 20 that covers the range of 340 to

Check 0%T, 100%T, and %T of sample in sequence.

Note which knobs control 0%T and 100%T readings.

LIGHT PATH AS VIEWED FROM A POINT DIRECTLY ABOVE THE INSTRUMENT.

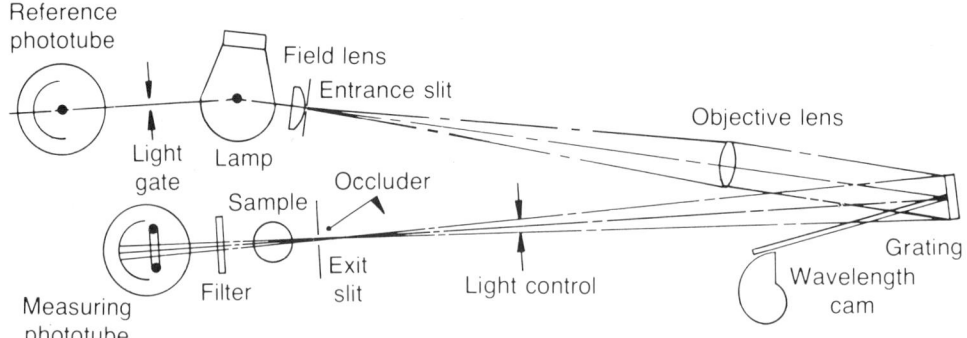

FIGURE AII-2

(Courtesy of Bausch & Lomb.)

625 millimicrons can be extended easily to 700 millimicrons by ordering the Wide Range Accessory. With infrared phototube and filter, the range can be extended to 950 millimicrons.*

The Spectronic 20 has three convenient control knobs, and the readings are made directly from the meter in either absorbance or transmittance.

Operating Instructions

1. Set the wavelength dial to the desired value.
2. Check the zero setting with the dark current control.
3. Set the 100 percent transmittance against the reference with the %T control.
4. Insert the sample.
5. Read the %T or absorbance value.

FIGURE AII-3

(Courtesy of Bausch & Lomb.)

---

*A millimicron is equivalent to a nanometer, which is becoming widely used for measurement of wavelengths.

6 0.3
ABSORBANCE
CONCENTRATION
TRANSMISSION
nm 5 4 7
BAUSCH & LOMB

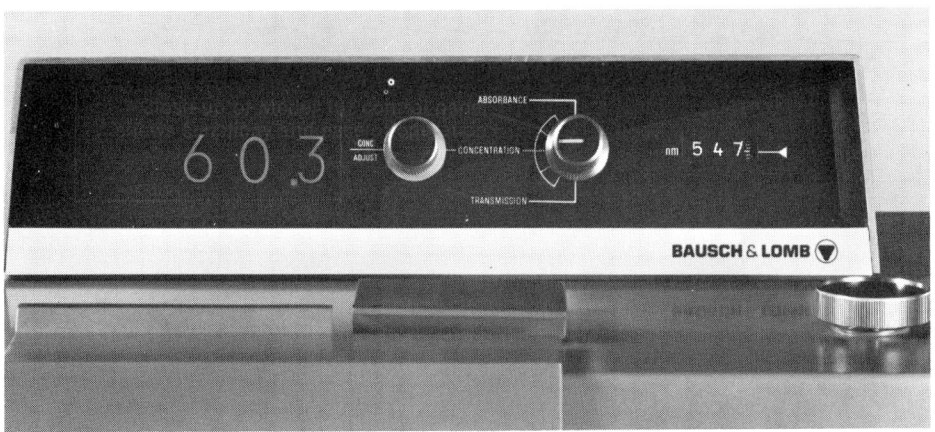

FIGURE AII-4

(Courtesy of Bausch & Lomb.)

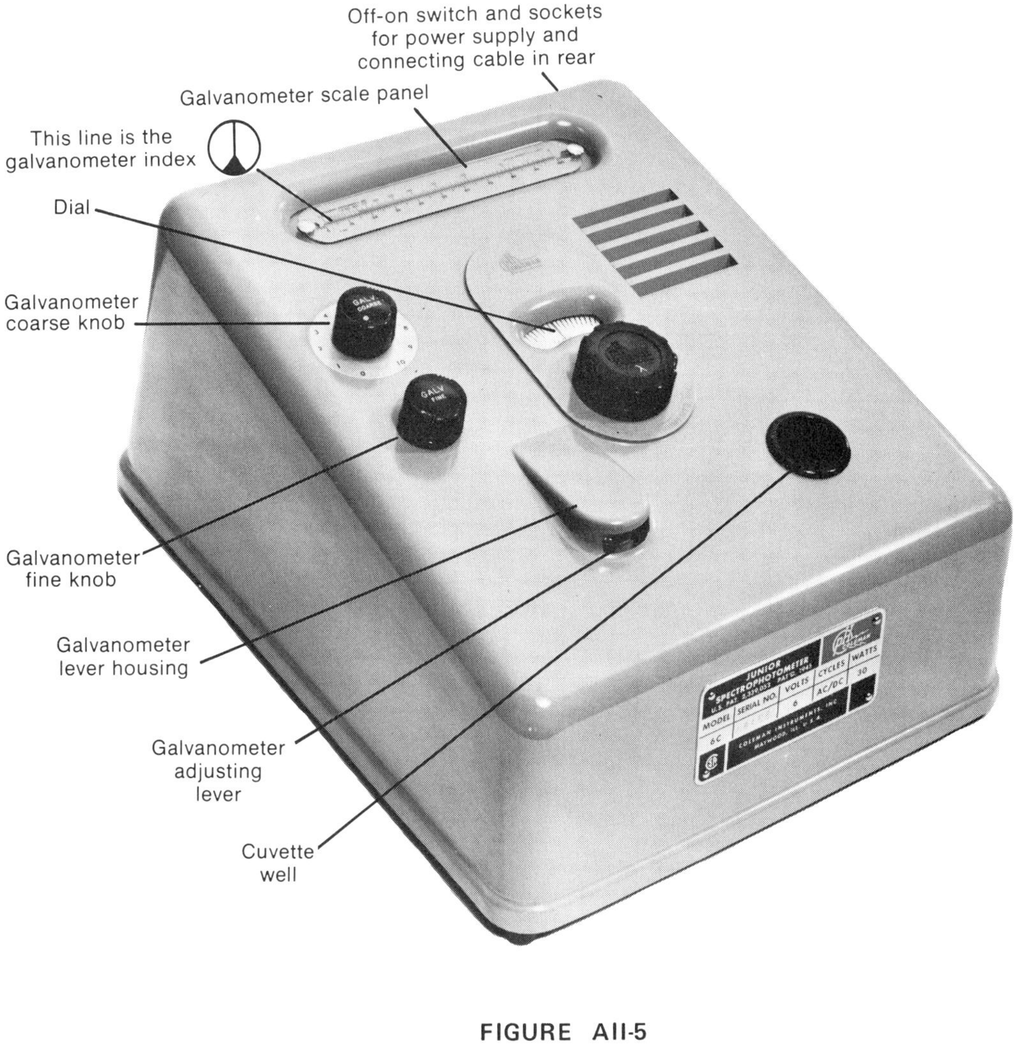

Off-on switch and sockets for power supply and connecting cable in rear
Galvanometer scale panel
This line is the galvanometer index
Dial
Galvanometer coarse knob
Galvanometer fine knob
Galvanometer lever housing
Galvanometer adjusting lever
Cuvette well

FIGURE AII-5

(Courtesy of Coleman Instruments Corporation.)

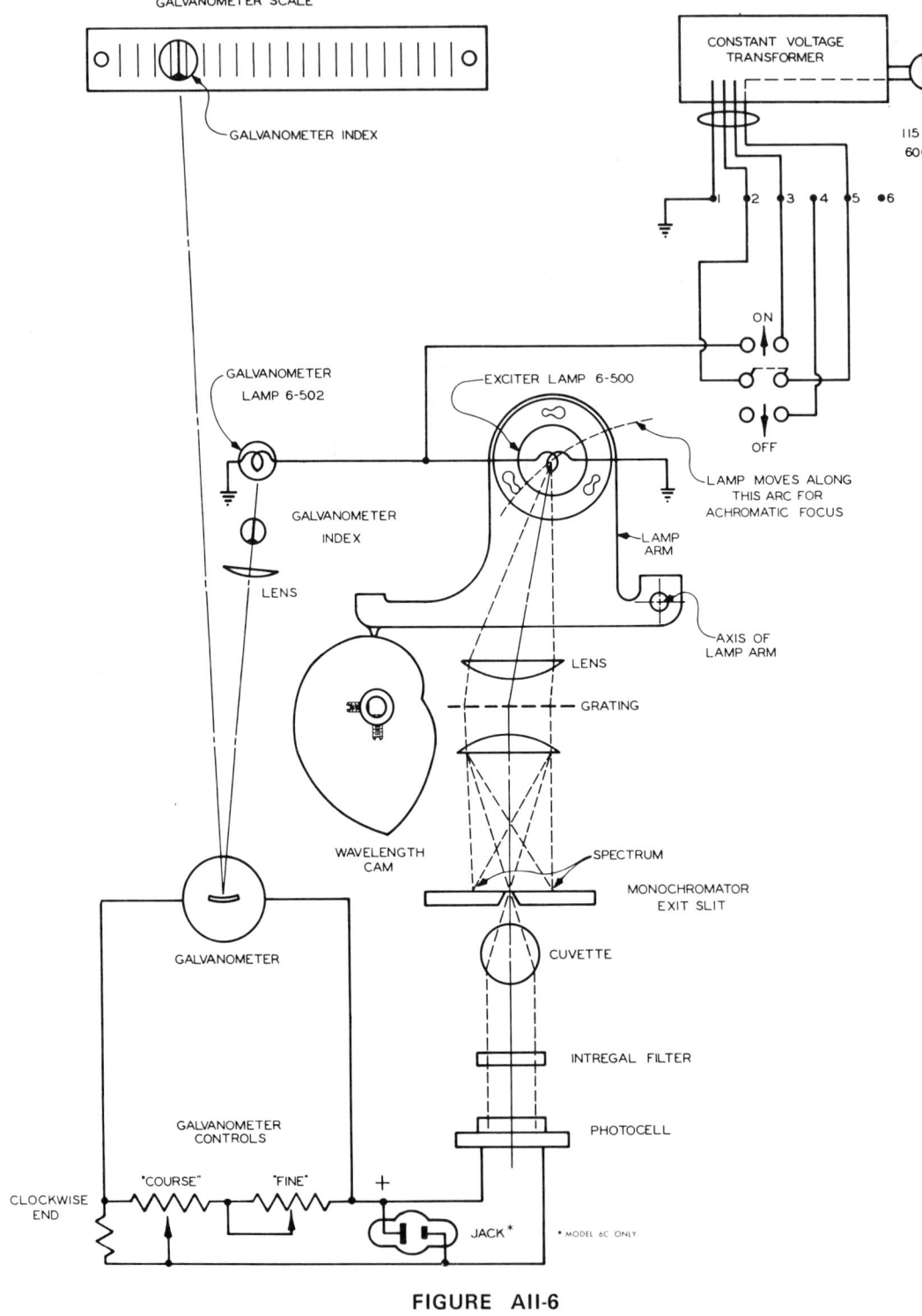

FIGURE AII-6

Schematic diagram of the optical and electrical components of the Coleman Models 6A and 6C Junior Spectrometers. (Courtesy of Coleman Instruments Corporation.)

BAUSCH & LOMB SPECTRONIC 70 AND 100

These newer (and somewhat more expensive) models have automatically interchanged phototubes and cover the range of 325 to 925 millimicrons. The photometric readout on the Spectronic 70 is on an 8-inch, mirror-scale meter, whereas the Spectronic 100 has a digital readout. The operational steps for taking a measurement are similar to those for the Spectronic 20. The design of the two instruments is similar; the Spectronic 70 is shown in Figure AII-3. The digital read-out of the Spectronic 100 is shown in Figure AII-4.

COLEMAN MODELS 6A AND 6C JUNIOR SPECTROPHOTOMETERS
(Coleman Instruments Corporation, Maywood, Illinois)

A photograph of the Coleman Model 6C spectrophotometer is given in Figure AII-5. A schematic diagram of the optical and electrical systems in both models is given in Figure AII-6.

Operating Instructions

1. Mount the selected scale panel in the galvanometer window.
2. Insert into the cuvette well a cuvette adapter of the proper size to accept the type of cuvette specified in the contemplated analytical method. (See detailed instruction booklet for specific dimensions and cuvettes.)
3. Turn on the switch located on the back of the instrument.
4. Verify the galvanometer zero setting and readjust it if necessary.
5. Adjust the λ dial to the wavelength specified.
6. Carefully wipe the lower third of a cuvette containing at least the minimum volume of reference solution and properly position it in the cuvette well.
7. Adjust the galvanometer "Coarse" and "Fine" knobs until the galvanometer index indicates the correct value. Usually this will be 100 percent transmittance on the black transmittance scale of the general purpose scale panel.
8. Remove the cuvette of reference solution and replace it with a similar cuvette containing sample solution, wiped clean and properly positioned as before.
9. Read the position of the galvanometer index on the scale panel.

SCHEMATIC DIAGRAMS OF RELATIVELY INEXPENSIVE INFRARED SPECTROPHOTOMETERS

PERKIN-ELMER INFRACORD MODEL 137
(Perkin-Elmer Corporation, Norwalk, Connecticut)

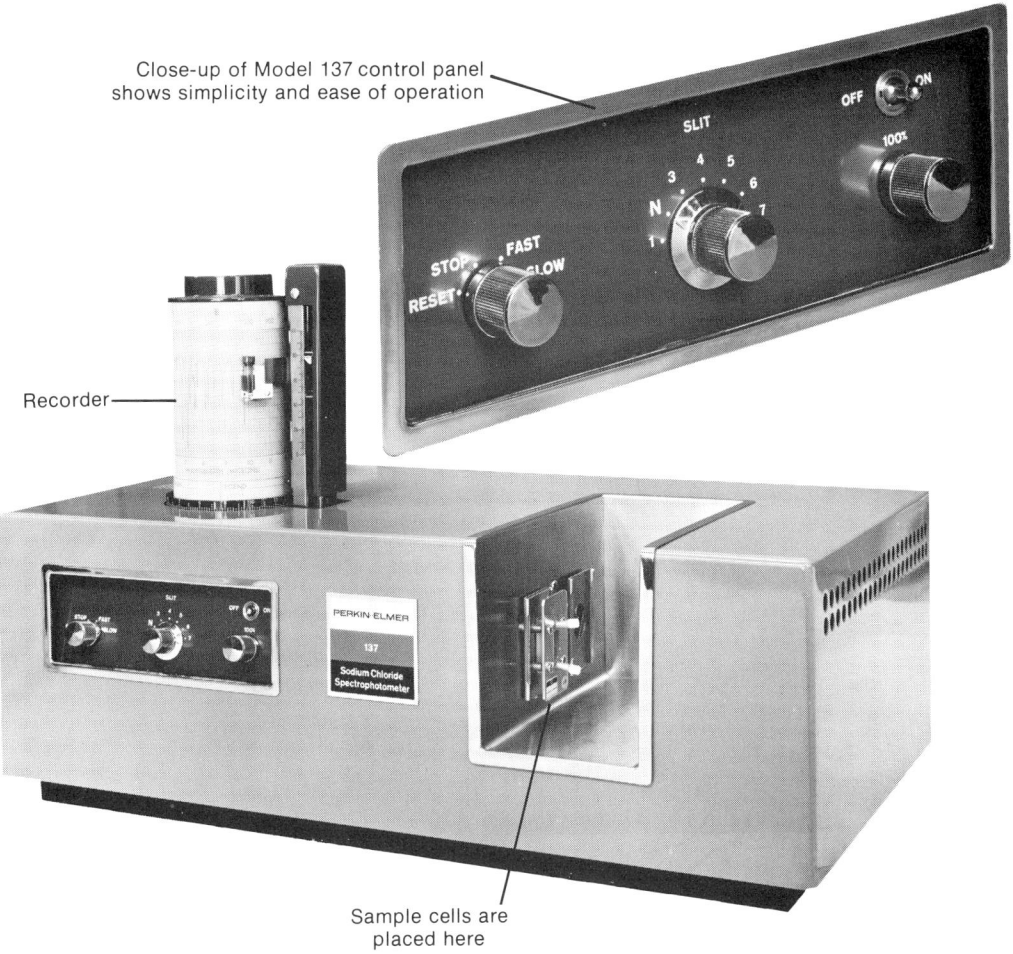

Close-up of Model 137 control panel shows simplicity and ease of operation

Recorder

Sample cells are placed here

FIGURE AIII-1

(Courtesy of Perkin-Elmer Corporation.)

Low cost, automatic recording, double-beam infrared spectrophotometers that have quality performance capabilities and that are relatively easy to operate are now available for instructional uses. Infrared instruments are designed to measure the characteristic vibrations of a sample by passing infrared radiation through it and recording the wavelengths that have been absorbed and the extent of absorption.

To accomplish this function, a typical infrared spectrophotometer has three basic components: a *source* of infrared energy (commonly a globar) to emit radiation over the entire frequency range covered by the spectrophotometer, a *monochromator* (prism or diffraction grating) to separate the energy into its component frequencies, and a *detector* (thermocouple) to measure the intensity of radiation before and after passing through the sample and to actuate the indicating-recording system.

Obtain operation instructions from your instructor before using the infrared spectrophotometer that is available.

In the Infracord 137-B spectrophotometer the radiant energy from the source is split into two beams by a plane mirror and two toroid mirrors. (Fig. AIII-2). Toroid mirror *1* focuses one beam through the sampling area onto the 100 percent adjustment attenuator; toroid mirror *2* focuses the reference beam onto the measuring attenuator. The sample is placed just in front of the 100 percent attenuator. Sample and reference beams, after passing through the sampling area, are recombined by the semicircular, rotating sector mirror. Oriented in relation to the three plane mirrors shown, the sector mirror alternately reflects the reference beam and passes the sample beam through its open half; thus, both beams are the same size and follow identical paths through the remainder of the system.

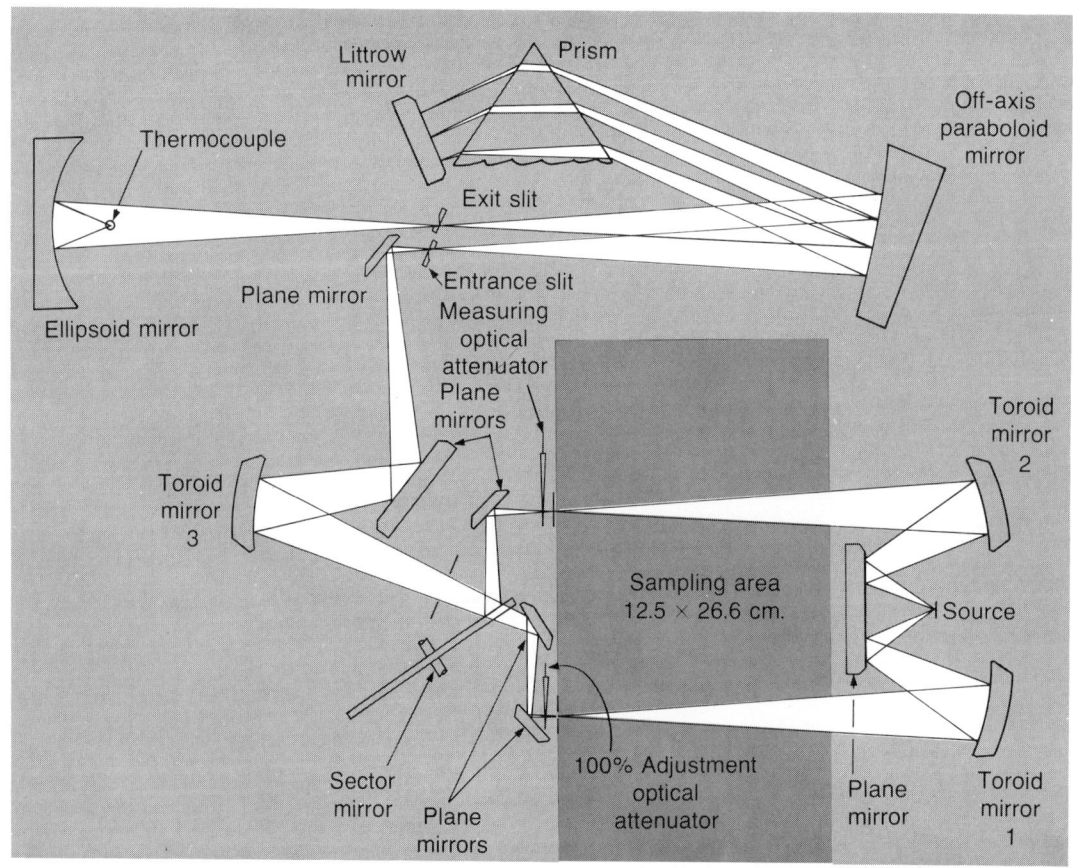

FIGURE AIII-2

Schematic diagram of the optical system of the Infracord spectrophotometer. (Courtesy of Perkin-Elmer Corporation.)

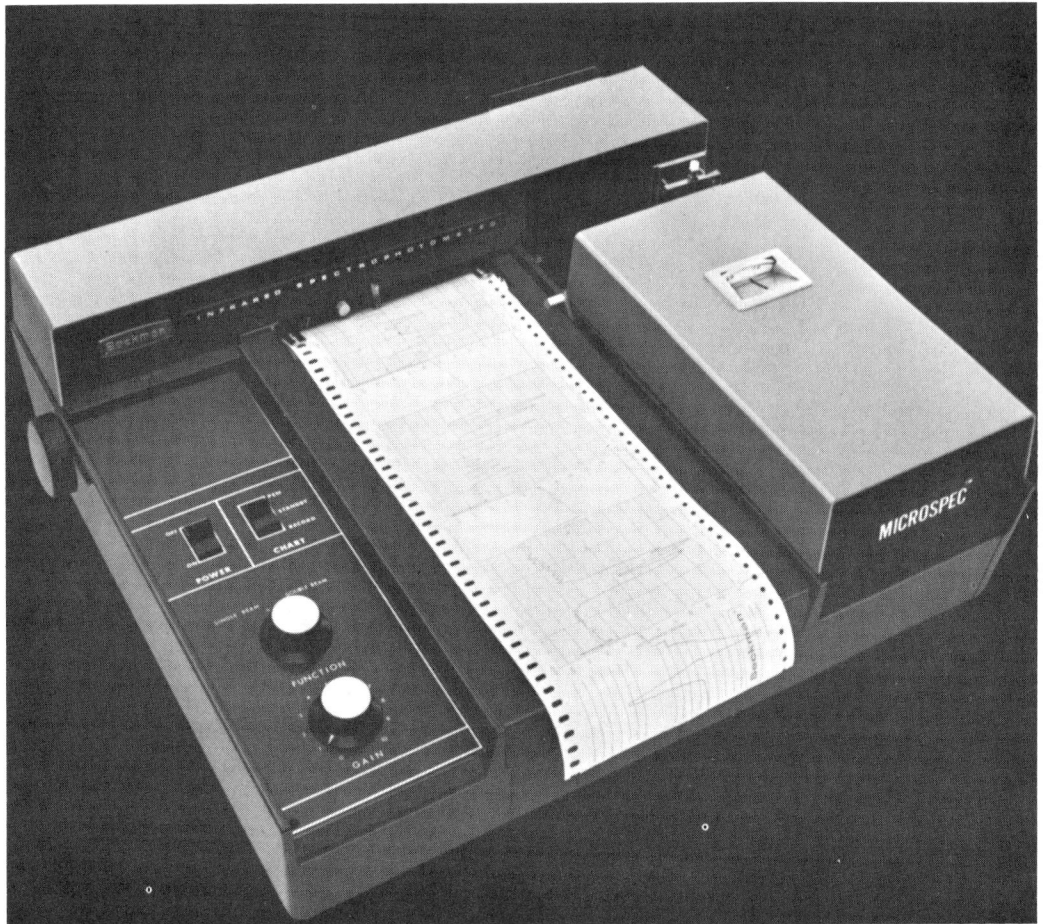

FIGURE   AIII-3

(Courtesy of Beckman Instruments.)

The signal beam, which now consists of alternate pulses of sample and reference radiation, is focused by toroid mirror *3* and reflected by the plane mirrors through the entrance window of the monochromator and onto the entrance slit (at the focal point of the toroid mirror). Leaving the slit, the beam diverges until the off-axis paraboloid reflects it as collimated light (parallel rays) onto the prism. Passing through the prism, the component wavelengths are dispersed and reflected by the Littrow mirror back through the prism where they are further dispersed. The dispersed radiation is then focused on the plane of the exit slit by the paraboloid as a band of individual wavelengths falling across the slit. The particular spectral wavelength that emerges from the slit strikes the ellipsoid mirror, which focuses it on the thermocouple.

If the energy in both sample and reference beams is equal, a direct current voltage is produced by the thermocouple and is not amplified by the alternating current amplifier of the instrument. When the sample absorbs radiation at the characteristic wavelengths, the intensity of the sample beam is reduced. This produces an unequal signal at the detector as the sample and reference beams are alternately compared. This inequality is converted by the thermocouple into an alternating voltage, which is amplified and used to drive a servo motor, which moves the measuring optical attenuator in the right direction to equalize or null the beam intensities. Because the recorder pen is coupled directly to the measuring optical attenuator, its movements provide a precise record of the attenuator position, and therefore of sample transmittance, at each individual wavelength.

The Perkin-Elmer Models 237B and 337 have gratings instead of the prism to disperse infrared energy into its component frequencies. Gratings offer the advantages of more efficient dispersion of energy, better signal-to-noise ratios, better resolution of closely spaced absorptions, and nonsensitivity to moisture in the air. The physical design and operating controls are very similar to those on the Infracord.

BECKMAN IR-33 (Beckman Instruments, Fullerton, California)

The Beckman IR-33 is a low priced infrared spectrophotometer. This instrument features three scanning speeds (2.5, 7.5 and 22 minutes), a continuous chart recorder system and gives a scan of the entire wavelength range of 4000 to 600 cm.$^{-1}$. The IR-33 is shown in Figure AIII-3 and its optical diagram is given in Figure AIII-4.

BECKMAN IR-18A (Beckman Instruments, Fullerton, California)

The Beckman IR-18A features two precision gratings for high resolution throughout the 4000 to 600 cm.$^{-1}$ range. The complete spectrum is recorded on one continuous chart in contrast to the two chart ranges of the Perkin-Elmer instrument. A picture of the IR-18A is given in Figure AIII-5.

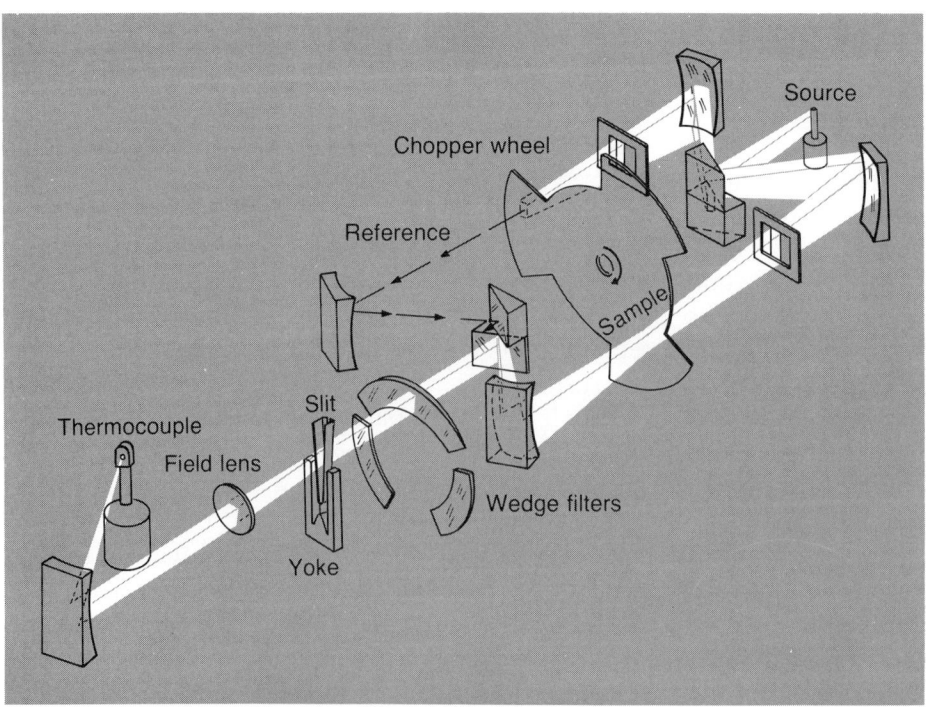

FIGURE AIII-4

(Courtesy of Beckman Instruments.)

FIGURE   AIII-5

(Courtesy of Beckman Instruments.)

CONVERSION FACTORS

THE METRIC SYSTEM

Basic Units		Prefixes Used with Basic Units		
Length:	meter	kilo	=	1000
Weight:	kilogram	deci	=	0.1
Volume:	liter	centi	=	0.01
		milli	=	0.001
		micro	=	0.000001

Length

1 kilometer (km.)	=	1000 m.
1 meter (m.)	=	100 cm. = 1000 mm.
1 centimeter (cm.)	=	10 mm.
1 millimeter (mm.)	=	10,000,000 Å
1 angstrom (Å)	=	1×10^{-8} cm. = 1×10^{-10} m.

Weight

1 kilogram (Kg.)	=	1000 g.
1 gram (g.)	=	1000 mg.
1 milligram (mg.)	=	0.001 g.

Volume

1 liter (l)	=	1000 ml. = 1000.027 cm.^{3}
1 milliliter (ml.)	=	0.001 l
1 ml.	=	1 cm.^{3} = 1 cc. (for all usual purposes)

THE SYSTEM INTERNATIONAL (SI)

SI BASE UNITS

Physical quantity	Name of unit	Symbol
Length	meter	m
Mass	kilogram	kg
Time	second	s
Electric current	ampere	amp
Thermodynamic temperature	kelvin	K
Luminous intensity	candela	cd
Amount of substance	mole	mol

SPECIAL NAMES AND SYMBOLS FOR CERTAIN SI DERIVED UNITS

Physical quantity	Name of SI unit	Symbol for SI unit	Definition of SI unit
force	newton	N	$kg \cdot m \cdot s^{-2}$
pressure	pascal	Pa	$kg \cdot m^{-1} \cdot s^{-2}$ $(=N \cdot m^{-2})$
energy	joule	J	$kg \cdot m^2 \cdot s^{-2}$
power	watt	W	$kg \cdot m^2 \cdot s^{-3}$ $(=J \cdot s^{-1})$
electric charge	coulomb	C	$amp \cdot s$
electric potential difference	volt	V	$kg \cdot m^2 \cdot s^{-3} \cdot amp^{-1} (=J \cdot amp^{-1} \cdot s^{-1})$
electric resistance	ohm	Ω	$kg \cdot m^2 \cdot s^{-3} \cdot amp^{-2}$ $(=V \cdot amp^{-1})$
electric conductance	siemens	S	$kg^{-1} \cdot m^{-2} \cdot s^3 \cdot amp^2$ $(=amp \cdot V^{-1} = \Omega^{-1})$
electric capacitance	farad	F	$amp^2 \cdot s^4 \cdot kg^{-1} \cdot m^{-2}$ $(=amp \cdot s \cdot V^{-1})$

SOME CONVERSION RELATIONSHIPS

Electric Charge

One Coulomb	$= 2.778 \times 10^{-4}$ amp $\cdot$ h
	$= 1.036 \times 10^{-5}$ Faraday
	$= 2.998 \times 10^9$ statcoul

Electric Dipole Moment

One debye (D)	$= 1 \times 10^{-18}$ statcoulomb $\cdot$ cm
	$= 3.336 \times 10^{-20}$ C $\cdot$ Å
	$= 0.21$ electron $\cdot$ Å

Energy and Work   (Mass units are included as energy equivalents.)

One erg	$= 10^{-7}$ J
	$= 2.389 \times 10^{-8}$ cal
	$= 6.242 \times 10^{11}$ eV
	$= 1.113 \times 10^{-24}$ kg
	$= 670.5$ amu
One calorie (cal)	$= 4.1840 \times 10^7$ erg
	$= 4.184$ J
	$= 2.613 \times 10^{19}$ eV
	$= 4.659 \times 10^{-17}$ kg
	$= 2.807 \times 10^{10}$ amu
One electron volt (eV)	$= 1.602 \times 10^{-12}$ erg
	$= 1.602 \times 10^{-19}$ J
	$= 3.827 \times 10^{-20}$ cal
	$= 1.783 \times 10^{-36}$ kg
	$= 1.074 \times 10^{-19}$ amu

SOME CONVERSION RELATIONSHIPS (Continued)

One kilogram (kg)	$= 8.987 \times 10^{23}$ erg
	$= 8.987 \times 10^{-16}$ J
	$= 2.142 \times 10^{16}$ cal
	$= 5.610 \times 10^{35}$ eV
	$= 6.025 \times 10^{26}$ amu
One atomic mass unit (amu)	$= 1.492 \times 10^{-3}$ erg
	$= 1.492 \times 10^{-10}$ J
	$= 3.564 \times 10^{-11}$ cal
	$= 9.31 \times 10^{8}$ eV
	$= 1.660 \times 10^{-27}$ kg

Force

One newton (N)	$= 10^{5}$ dyn
	$= 0.2248$ lb
One pound (lb)	$= 4.448 \times 10^{5}$ dyn
	$= 4.448$ N

Length

One meter (m)	$= 39.37$ in.
	$= 3.281$ ft
	$= 6.214 \times 10^{-4}$ mi
One inch (in.)	$= 2.540$ cm
One Angstrom (Å)	$= 10^{-10}$ m
One micron	$= 10^{-6}$ m
One light-year	$= 9.4600 \times 10^{12}$ km

Mass and Weight

(Mass-weight equivalents are valid for terrestrial use only.)

One gram (g)	$= 6.852 \times 10^{-5}$ slug
	$= 6.024 \times 10^{23}$ amu
	$= 3.27 \times 10^{-2}$ oz
	$= 2.205 \times 10^{-3}$ lb
One atomic mass unit (amu)	$= 1.6602 \times 10^{-24}$ g
One pound (lb)	$= 453.6$ g
One ton	$= 2000$ lb
	$= 907.2$ kg

SOME CONVERSION RELATIONSHIPS (Continued)

Pressure

One atmosphere (atm)	$= 1.013 \times 10^6$ dyn/cm.^{2}
	$= 1.013 \times 10^5$ N·m^{-2}
	$= 76.0$ cm. Hg
	$= 14.70$ lb./in.^{2}
	$= 2116$ lb./ft.^{2}
	$= 760$ torr
One centimeter mercury (cm. Hg)	$= 1$ torr
	$= 1.316 \times 10^{-2}$ atm
	$= 1.333 \times 10^4$ dyn/cm.^{2}
	$= 5.353$ in. H_2O
	$= 0.1934$ lb./in.^{2}
	$= 27.85$ lb./ft.^{2}

GRAPHICAL DISPLAY OF DATA

Data Tables. The data from an experiment often appear as a pair of quantities giving the value of one measurement when the value of a second has been set at a particular magnitude. A second experiment under another condition gives another pair of values, and other experiments give other pairs, so that a table like Table AV-1 can be made. Here an experimenter measured the number of grams of barium nitrate that 100 g. of H_2O will dissolve when the temperature is $0°C$, and again at $10°$, and so forth.

TABLE AV-1 SOLUBILITY OF BARIUM NITRATE IN WATER

Temperature ($°C$)	Solubility (g./100 g. H_2O)
0	5.0
10	7.0
20	9.2
30	11.6
40	14.7
50	17.1
60	20.3

Graphs. Although a table of this sort reports the data, the trend in the solubility as the temperature is changed can be assessed more readily if the data are plotted on a graph. It is evident that the solubility is a *function* of the temperature, and, as in algebra, we plot the *dependent variable* as *ordinate* (along the vertical axis) and the *independent variable* as *abscissa* (along the horizontal axis) on a pair of lines at right angles to each other (Cartesian coordinate axes). Each axis must be assigned a *scale* so that a certain distance along the axis corresponds to a certain value of the variable. For a plot of the data of Table AV-1, for example, we might choose to let $10°C$ be represented as a distance of 20 mm., and 5 g. of $Ba(NO_3)_2/100$ g. of H_2O be represented also by 20 mm., and prepare a piece of graph paper for plotting as in Figure AV-1. The positive directions of the axes are taken from left to right and from bottom to top; this rule holds also for negative numbers so that -15, for example, lies on the horizontal axis to the left of -10, and similarly -4 is below -2 on the vertical axis.

Independent and Dependent Variables. The independent variable in a scientific measurement can usually be identified as the quantity over which the experimenter has control. In the solubility measurement, for example, the experimenter has control of the temperature; he may choose to measure the solubility at 9 or $22°C$ instead of at 10 and $20°C$. Hence, the temperature

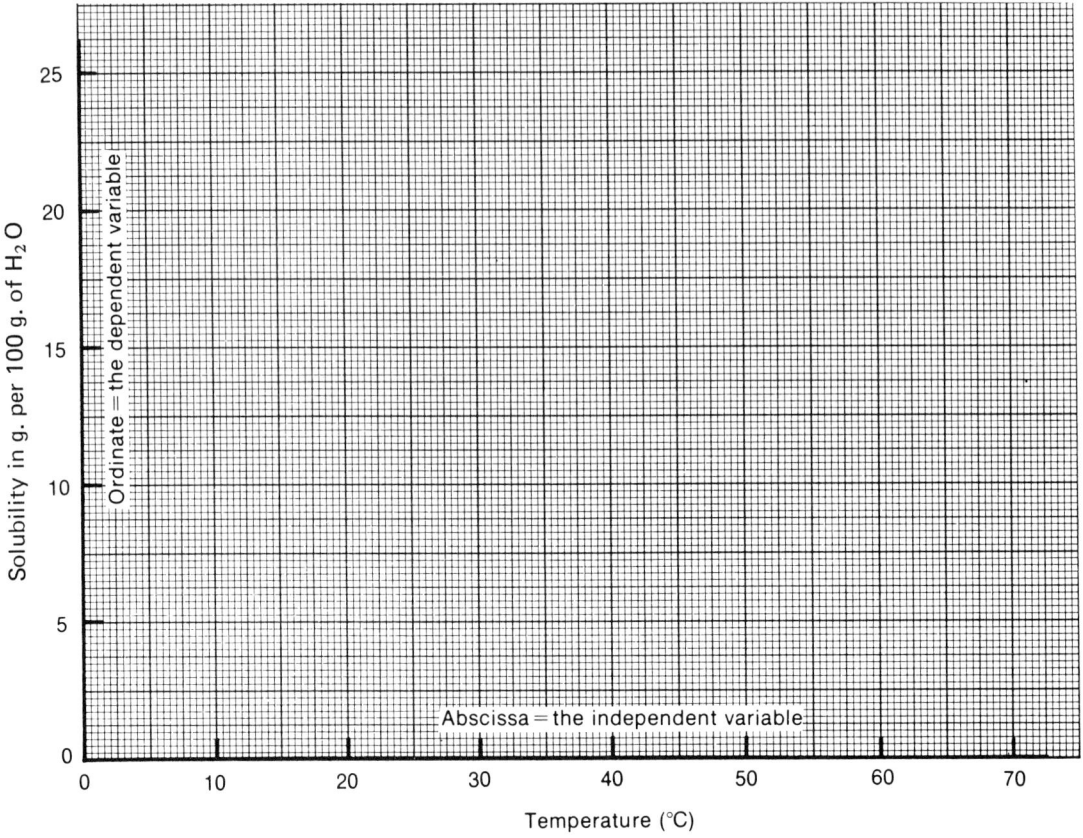

FIGURE AV-1

is the independent variable and is to be plotted horizontally (as abscissa). The solubility at the temperature he has chosen, however, is a fact of nature that he cannot alter; the solubility is thus the dependent variable and is plotted vertically (as ordinate). We speak of plotting the dependent variable *against* the independent variable; therefore, if the directions are to "plot the solubility against the temperature," this means that the temperature is to be the abscissa.

To plot the data of Table AV-1, we plot the paired values (temperature, solubility) as coordinates of experimental points. For the second line of the table, for example, the coordinates are (10, 7.0), and we move horizontally to the scale reading corresponding to 10°C, then vertically to the position corresponding to 7.0 and mark a point. We proceed similarly for other points: over to 20 and up to 9.2, mark a point, and so forth. It is usually convenient to draw a small circle around the point, for better legibility. The plotted data will then look like Figure AV-2.

Note that the points are plotted as exactly as possible, dividing, by eye, the distance between lines on the graph into smaller divisions as necessary. There is no line on the graph paper that corresponds to 9.2 – the nearest line represents 9.25. Hence the distance between 9.00 and 9.25 is considered to be divided into five parts, and the point for 9.2 placed four-fifths of the distance from 9.00 to 9.25. It is not necessary that all graphs start at the origin (0, 0). Had the measurements been made only between 50 and 100°C, for example, it might have been convenient to mark the intersection of the axes as (50, 15) and increase from those values.

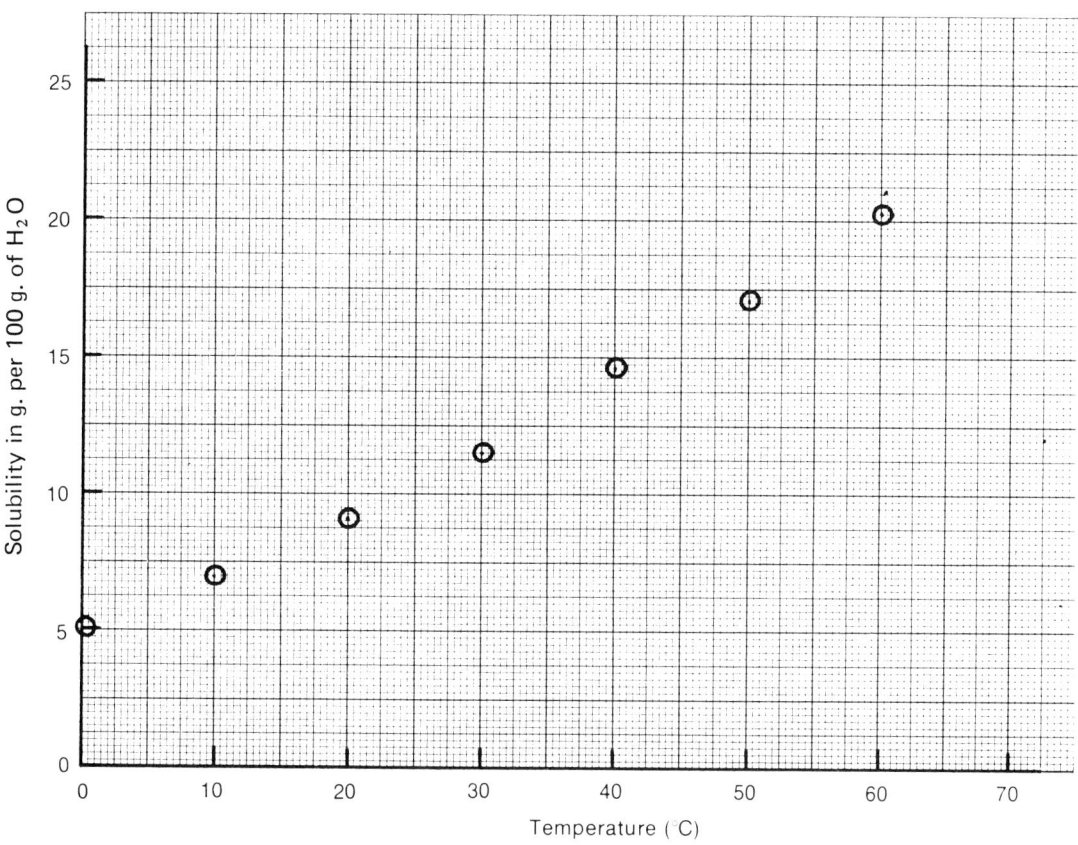

FIGURE   AV-2

Drawing the Curve.   Figure AV-2 properly records the experimental data of Table AV-1. The advantage of a graphical display, however, is that it enables us to predict values of the dependent variable for values of the independent variable at which experimental measurements have not been made. To make this prediction, we rely upon experience, which tells us that things in nature usually change gradually from one value to another. (They do not always change gradually: consider, for example, the quantum jumps in energy that occur in the electronic excitation of an atom.) If the change is gradual, we can connect the experimental points of a graph by a smooth curve representing a continuous function, which will have no abrupt changes of direction. (Note that abrupt changes of direction sometimes do occur, cf. Fig. 13-2). If nature demands a smooth curve, apparent deviations from a smooth, nonbumpy curve must be the result of experimental error and our curve may not actually touch all the experimental points. Thus in completing the graphing of the data in Table AV-1 as in Figure AV-3, we reject exact contact with the points at 40 and $50°C$ in favor of making a smooth curve. Although the curve may be drawn free-hand, it is better to use a ruler (for straight lines) or to draw against the edge of a properly bent flexible spline or a plastic French curve placed at successive positions to fit the points.

All graphs should be titled, the scale numbers marked, and the axes labeled with correct units.

Choice of Scale. In plotting Figures AV-2 and AV-3, the scale was chosen to correspond to the precision of the data. Since the data are

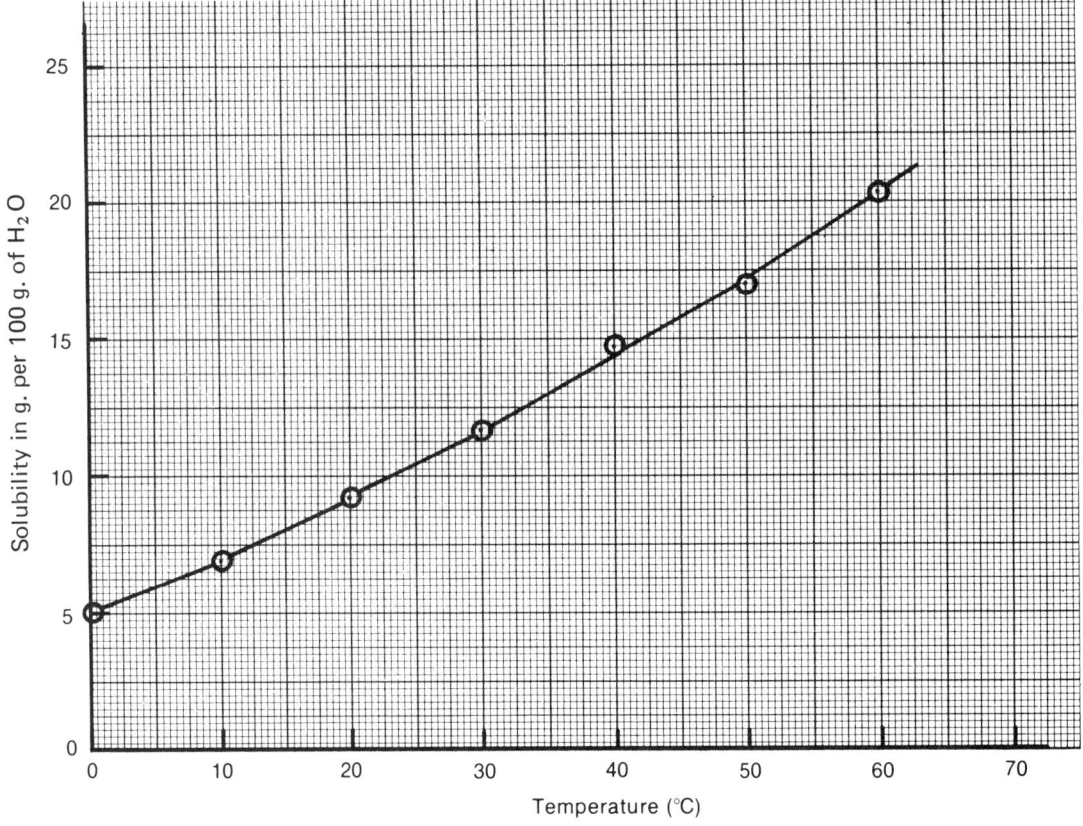

FIGURE AV-3

measured to one decimal place, we want to plot them in such a way that two points differing by a few decimal places will appear at recognizably different places on the graph. It would be possible to plot the data as in Figure AV-4, but this small scale shows only the gross trend of the data and does not do justice to the care with which the measurements were made. The precision of the data for both ordinate and abscissa should be considered in deciding upon a scale.

Comparison with Theory. Graphs of data are often used to compare the data with a prediction from theory. A common procedure is to cast the prediction into a form that can be graphed as a straight line, $y = mx + b$. In this equation y is the dependent variable, x the independent variable, m the

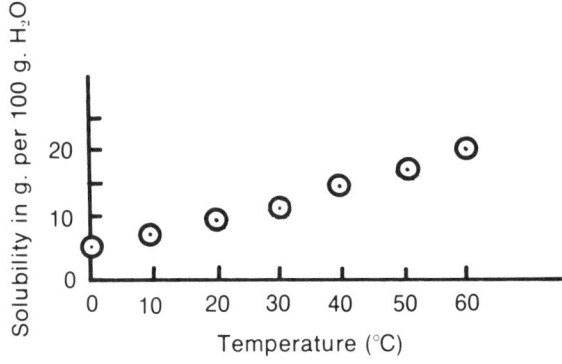

FIGURE AV-4

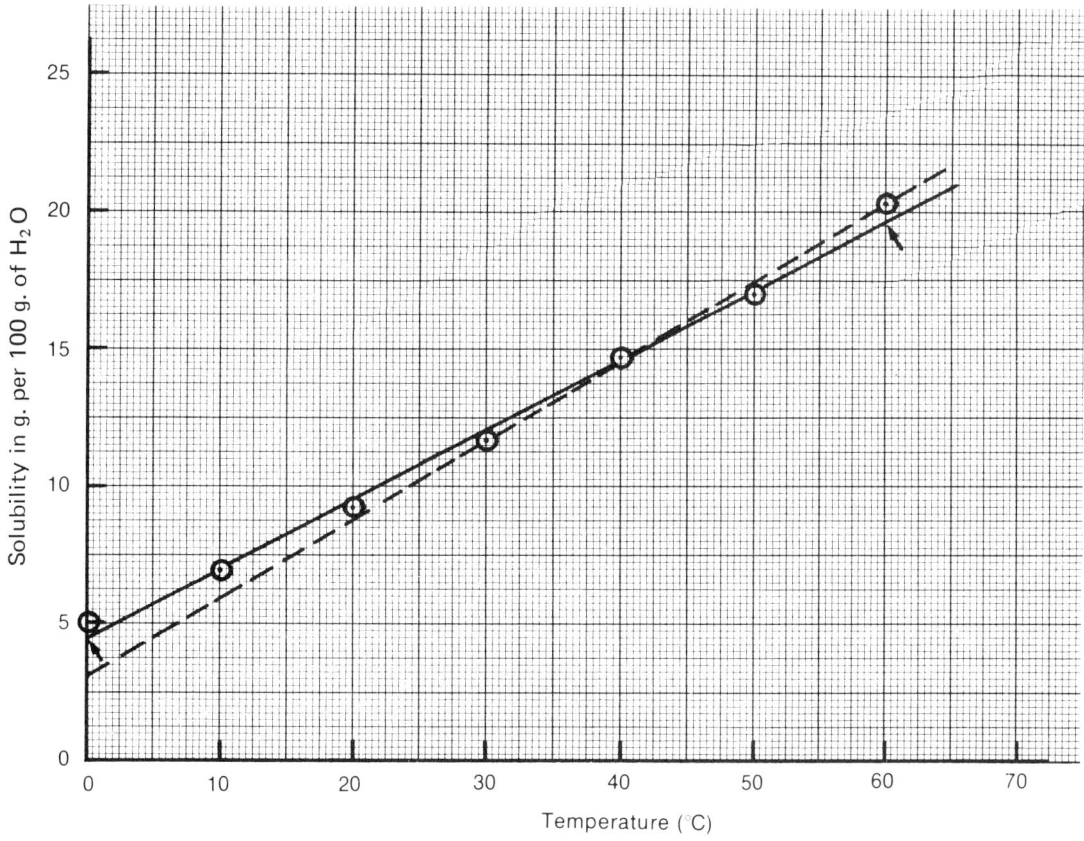

FIGURE   AV-5

slope of the line, and b the *intercept* on the y axis. The intercept is the value of y when x = 0, and the slope of a straight line is obtained from the coordinates of any two points on the line, (x_1, y_1) and (x_2, y_2), by setting up the quotient $\dfrac{y_2 - y_1}{x_2 - x_1}$. The value of this quotient is the slope m. Note that (x_1, y_1) and (x_2, y_2) are points *on the line;* if experimental error causes the measured points to lie off the line, their coordinates should not be used in determining the slope, but new coordinates should be read from the best straight line through the experimental points. The *best* straight line is the one that is drawn as closely as possible to as many of the points as possible. If theory predicted, for example, that the data of Table AV-1 should fit a straight line (the theory does not make such a prediction) this straight line would be drawn as the solid line of Figure AV-5 and not as the dotted line. The slope of the solid line may be obtained from the marked points as

$$\frac{19.75 - 4.35}{60.0 - 0.0} = 0.267 \text{ g. per 100 g. per degree}$$

Derived Functions as Variables.   Often the dependent and independent variables to be plotted are not the quantities that are directly measured, but quantities derived from them by mathematical manipulations. Thus, in Experiment 12, the measured quantities are pressure and temperature. The theoretical equation to be tested, however, is

$$\log p = A - \frac{B}{T}$$

The quantities to be plotted are, therefore, the logarithm of the pressure against the reciprocal of the temperature. The reciprocal of the temperature remains the independent variable and is plotted as abscissa because the temperature is the quantity over which the experimenter has control, while log p is plotted as the ordinate. In plotting logarithmic quantities, it should be remembered that the logarithm of a number less than 1 is negative.

USE OF CELL VOLTAGE TO DETERMINE THE CHOICE OF INDICATOR

The voltage of the cell of Fig. 18-1 is given, using (5) on p. 216, by

$$E = E^{\circ}_{H_2 \rightarrow H^+} + E^{\circ}_{Fe^{3+} \rightarrow Fe^{2+}} - \frac{2.3RT}{F} \log \frac{[Fe^{2+}][H^+_{aq}]}{[Fe^{3+}]} \qquad (16)$$

$$= 0.77 - 0.059 \log \frac{[Fe^{2+}][H^+_{aq}]}{[Fe^{3+}]}$$

Using (7) we have

$$E = E^{\circ}_{H_2 \rightarrow H^+} + E^{\circ}_{Ce^{4+} \rightarrow Ce^{3+}} - \frac{2.3RT}{F} \log \frac{[Ce^{3+}][H^+_{aq}]}{[Ce^{4+}]} \qquad (17)$$

$$= 1.61 - 0.059 \log \frac{[Ce^{3+}][H^+_{aq}]}{[Ce^{4+}]} \qquad (18)$$

As mentioned, it makes no difference whether we use equations (5) and (16) or (7) and (18), since in the reaction solution all four components are at equilibrium, and

$$\frac{[Fe^{3+}][Ce^{3+}]}{[Fe^{2+}][Ce^{4+}]} = K \qquad (19)$$

The value for K in equation (19) is 1.7×10^{14} at 25°C.

If the concentration of hydrogen ion in the left beaker of Figure 18-1 is made 1 molal (an approximation for an activity of unity), $[H^+_{aq}] = 1$ and the equivalent equations (16) and (18) become

$$E = 0.77 - 0.059 \log \frac{[Fe^{2+}]}{[Fe^{3+}]} = 1.61 - 0.059 \log \frac{[Ce^{3+}]}{[Ce^{4+}]} \qquad (20)$$

Note that equations (20) have the same mathematical form as the equation relating the pH of a solution of a weak acid to the relative concentrations of acid and acid anion

$$pH = pK - \log \frac{[HA]}{[A^-]} \qquad (21)$$

as used in Experiment 28. Hence, a graph of E against milliliters of titrant added may be expected to produce a titration curve similar to those shown in Experiment 28.

Near the beginning of the titration, when only a few drops of Ce(IV) solution has been added, $[Fe^{3+}] \ll [Fe^{2+}]$ and E, as calculated from the left-hand form of equation (20), is close to zero. As more Ce(IV) is added, the concentration of Fe^{2+} decreases and the concentration of Fe^{3+} increases, the magnitude of the logarithmic term in equation (8) decreases, and the voltage increases toward 0.77 volt. This value is reached when $[Fe^{2+}] = [Fe^{3+}]$. Continued titration reduces $[Fe^{2+}]$ further while increasing $[Fe^{3+}]$; the ratio $[Fe^{2+}]/[Fe^{3+}]$ becomes less than one; the logarithm becomes negative; and the voltage increases above 0.77 volt.

At the equivalence point, [cf. (10) on p. 218]

$$\frac{[Fe^{2+}]}{[Fe^{3+}]} = \frac{y}{a-y} = \frac{1}{\sqrt{K}}$$

Taking logarithms,

$$\log \frac{[Fe^{2+}]}{[Fe^{3+}]} = - \tfrac{1}{2} \log K = - \tfrac{1}{2} \log (1.7 \times 10^{14}) = -7.12$$

and the voltage at the equivalence point is

$$E_{equiv} = 0.77 - 0.059 \times (-7.12) = 1.19 \text{ volts} \qquad (22)$$

The student may readily check that the same result, 1.19 volts, will be obtained using the ratio $\frac{[Ce^{3+}]}{[Ce^{4+}]}$ in the right-hand form of equations (20). In fact, if the two forms of equations (20) are written for the equivalence point

$$E_{equiv} = 0.77 - 0.059 \times (-\tfrac{1}{2} \log K) \qquad (22)$$

$$E_{equiv} = 1.61 - 0.059 \times (+\tfrac{1}{2} \log K) \qquad (23)$$

added together and divided by two, we find

$$E_{equiv} = \frac{1.61 + 0.77}{2} = 1.19 \text{ volts} \qquad (24)$$

The form of equation (24) written

$$E_{equiv} = \frac{E^{\circ}_{reductant\ couple} + E^{\circ}_{oxidant\ couple}}{2} \qquad (25)$$

is general for an oxidation-reduction titration in which the number of electrons lost by the reductant is the same as the number gained by the oxidant.

Continued addition of Ce(IV) solution displaces the equilibrium in equation (1) farther and farther to the right, decreasing $[Fe^{2+}]$ and increasing $[Fe^{3+}]$, and causing E to increase still more.

The results of the calculation outlined in the last few paragraphs are given in Table AVI-1 and plotted in Figure AVI-1.

TABLE AVI-1. TITRATION OF 50 ml. OF 0.1 M Fe^{2+} WITH 0.1 M Ce(IV)

Ce(IV) Added	Total Volume	E	Ce(IV) Added	Total Volume	E
(ml.)	(ml.)	(volts)	(ml.)	(ml.)	(volts)
10.00	60	0.73	50.00	100	1.19
20.00	70	.76	50.05	100.05	1.43
30.00	80	.78	50.50	100.5	1.49
40.00	90	.81	51.00	101	1.51
45.00	95	.83	55.00	105	1.55
49.00	99	.87	60.00	110	1.57
49.50	99.5	.89	70.00	120	1.59
49.95	99.95	.95	80.00	130	1.60

Note that in Figure AVI-1, a large change in voltage occurs with 2 drops of titrant at the equivalence point. This large change makes the oxidation-reduction titration feasible, if we can find a device that will enable us to recognize the point in the titration at which this large change occurs. One such device would be the voltmeter or potentiometer of Figure 18-1, so that we could follow directly the changes recorded in Table AVI-1. This procedure is the basis for *potentiometric titrations*. An indicator that, when

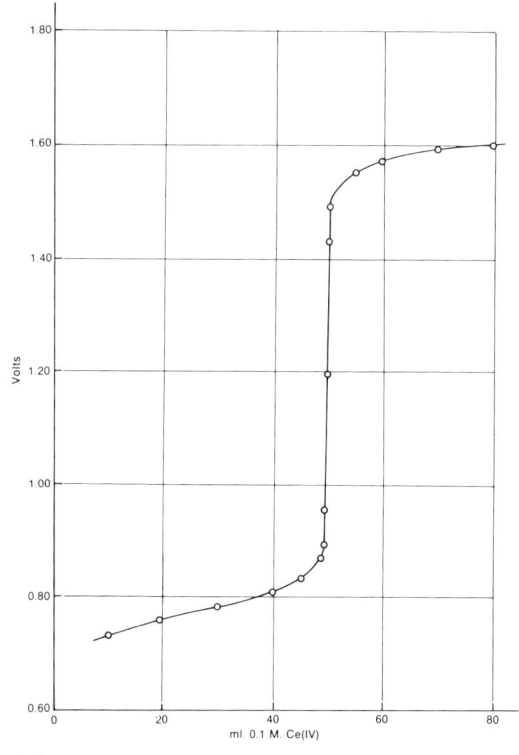

FIGURE AVI-1

Voltage as a function of oxidizing agent added in the titration of 50 ml. 0.1 M Fe(II) with 0.1 M Ce(IV).

present in negligible amount, produces a marked color change as it is oxidized by the titrant solution at the equivalence point, will also be suitable.

Formal Potentials. Before examining indicators, however, we must amend the previous discussion, which is correct for *ideal* systems, in which activities are equal to concentrations and all ions are considered to be simple and uncomplexed, to make it apply to the *real* systems with which we work in the laboratory. We know that activities are not equal to concentrations, and there is evidence to suggest that Ce(III) and Ce(IV) ions, in particular, react with anions, such as nitrate, perchlorate, or sulfate ions, to form complex ions of formulas $(Ce(NO_3)_6^{2-}$, $Ce(ClO_4)_6^{2-}$, $Ce(SO_4)_3^{2-}$, etc. The concentrations of free Ce^{4+} and Ce^{3+} in the solutions and the voltage of cells of Figure 18-1 depend upon these factors also. This is true also of Fe(II) and Fe(III). Analytical chemists have found it convenient to make allowance for this behavior by using what they have called *formal potentials,* which give the voltages, corresponding to the $E°$ values in equations (4) and (5) (p. 216) not for standard states of unit activity, but for 1 M concentrations of the reagents taking part in the cell reaction and specified concentrations for all other ions present in the solution. Thus, whereas the standard oxidation potential for the reaction

$$Ce^{3+} (a=1) \longrightarrow Ce^{4+}(a=1) + e^-$$

is −1.61 volts, the formal potential of a solution 1 M in Ce(III), 1 M in Ce(IV), and 1 M in H_2SO_4 is −1.44 volts, similarly, although the standard oxidation potential for

$$Fe^{2+} (a=1) \longrightarrow Fe^{3+} (a=1) + e^-$$

is −0.77 volt, the value for 1 M Fe(III) and 1 M Fe(II) in 1 M H_2SO_4 is −0.68 volt. The values for the formal potentials, $E_f^\circ = -1.44$ volt and $E_f^\circ = -0.68$ volt, are often more closely related to the actual laboratory conditions, and their use gives better approximations to laboratory results than does the use of standard electrode potentials. Thus, the equivalence point in the Ce(IV) − Fe(II) titration in 1 M H_2SO_4 would be closer to (cf. equation (22))

$$E_{equiv} = 0.68 - 0.059 \times (-7.12) = 1.10 \text{ volts}$$

than to 1.19 volts as calculated from the standard electrode potentials. In 1 M HCl it would come at 0.97 volt. In choosing an indicator that changes color at the equivalence point, therefore, it is advisable to consider the exact composition of the solution at the equivalence point, or more usually, to add materials to the solution to bring it to a previously chosen condition at the equivalence point.

Choice of Indicator. If an indicator is added to the reaction solution in Figure 18-1, the color of the solution depends upon the relative concentrations of Ind and Ind$^+$ as determined by the position of equilibrium in the reaction

$$Ind + Ce^{4+} \rightleftharpoons Ind^+ + Ce^{3+}$$

The voltage of the cell also depends upon the relative concentrations by

$$E = E^{\circ}_{H_2 \to H^+} + E^{\circ}_{Ind^+ \to Ind} - 0.059 \log \frac{[Ind]}{[Ind^+]} \qquad (26)$$

The color change of $\frac{[Ind]}{[Ind^+]}$ in the range from 10 to 0.1 occurs in the voltage range of 0.059 volt on each side of the E° value, for a total range of 0.12 volt. For a useful indicator, the midpoint of the range should come at the voltage corresponding to the equivalence point. For the titration of Fe(II) with Ce(IV), the indicator ferroin, which changes from red in the reduced form to faint blue in the oxidized form at 1.11 volts in 1 M H_2SO_4, is suitable.

QUESTIONS–PROBLEMS

1. Using equation (19) in Appendix VI and the value of K defined by that equation, show that equations (16) and (18) give the same value of E.

2. Remembering that $\Delta G^{\circ} = -n\mathscr{F}E^{\circ}$ and that $\Delta G^{\circ} = -2.303RT \log K$, calculate the equilibrium constant for the reaction

$$Sn^{2+} + 2Ce^{4+} \longrightarrow Sn^{4+} + 2Ce^{3+}$$

at room temperature. E° values are:

$$E^{\circ}_{Sn(IV) \to Sn(II)} = 0.15 \text{ volt}$$

$$E^{\circ}_{Ce(IV) \to Ce(III)} = 1.61 \text{ volts}$$

Be sure to evaluate correctly the number of electrons transferred.

3. Consider the titration of tin(II) ion with cerium(IV) ion. (a) Write equations corresponding to equations (5) and (7) in Experiment 18 and equations (16) and (18) in Appendix VI for this situation and (b) calculate the voltage at the equivalence point for the ideal situation for a cell measured against the standard hydrogen electrode, using the data of Question 2. Be sure to evaluate correctly the number of electrons transferred in the cell reactions written. (c) Select an indicator for this titration from among the following:

Indicator	Oxidized Form	Reduced Form	Transition Voltage
Methylene blue	colorless	blue	0.53
Diphenylamine	colorless	violet	0.76
Diphenylamine sulfonic acid	colorless	purple	0.85
Ferroin	red	light blue	1.11

Explain your selection.

4. Using the information that the E° value for the transition of diphenylamine sulfonic acid is 0.85 volt, calculate the voltage range over which this indicator might be useful.

5. Show that, in general, for reactions of the type

$$bMe^{n+} + aOx^{m+} \longrightarrow bMe^{(n+a)+} + aRe^{(m-b)+}$$

the ideal voltage at the equivalence point for a cell using the standard hydrogen electrode, as in Figure 18-1, is given by

$$E_{equiv} = \frac{aE^\circ_{Me^{(n+a)+} \to Me^{n+}} + bE^\circ_{Ox^{m+} \to Re^{(m-b)+}}}{a + b}$$

6. (a) Write the equations corresponding to equations (5) and (7) in Experiment 18 and equations (16) and (18) in this Appendix for the dichromate oxidation of iron(II) as given by equation (15). Note that for the dichromate oxidation, the voltage depends upon the hydrogen ion concentration in the titrating solution.

(b) The equilibrium constant for the dichromate oxidation written as equation (15) is 9×10^{56}. The values of E° are

$$E^\circ_{Fe^{3+} \to Fe^{2+}} = 0.77 \text{ volt}$$

$$E^\circ_{Cr_2O_7^{2-} \to Cr^{3+}} = 1.33 \text{ volts}$$

Calculate the voltage at the equivalence point for a solution 1 M in hydrogen ion and 0.01 M in Cr(III) ion at the end-point.

(c) Would you suggest ferroin or diphenylamine sulfonic acid (Question 2) as the more satisfactory indicator for this titration, given the preceding data? Explain your choice.

(d) The formal potentials in 1 M HCl for this system have the values:

$$E^\circ_{fFe(III) \to Fe(II)} = 0.70 \text{ volt}$$

$$E^\circ_{fCr(VI) \to Cr(III)} = 1.00 \text{ volt}$$

Would this information affect your choice of indicator? How and why?

7. Calculate the voltages corresponding to those of Table AVI-1 for the range 40 to 60 ml. for the titration of 50 ml. of a 0.1 M solution of a reducing agent Me^{2+} for which the equilibrium constant for the reaction

$$Me^{2+} + Ce^{4+} \rightleftharpoons Me^{3+} + Ce^{3+}$$

is 1.08×10^2 using 0.1 M Ce^{4+} as titrant. In this case $E^\circ_{Me^{3+} \to Me^{2+}} = 1.49$ volts. Plot these data and those from Table AVI-1 on the same graph, and draw conclusions that are warranted by the comparison of the two curves.

TABLES OF DATA

TABLE VII-A IONIZATION CONSTANTS OF WEAK ELECTROLYTES

Acids	Ionization Equation		K_{ion} at 25°C
Acetic	CH_3CO_2H	$\rightleftharpoons H^+ + CH_3CO_2^-$	1.85×10^{-5}
Aqueous carbon	$CO_2 + H_2O$	$\rightleftharpoons H^+ + HCO_3^-$	4.2×10^{-7}
dioxide	HCO_3^-	$\rightleftharpoons H^+ + CO_3^{2-}$	4.8×10^{-11}
Hydrogen cyanide	HCN	$\rightleftharpoons H^+ + CN^-$	4.0×10^{-10}
Hydrogen sulfate	HSO_4^-	$\rightleftharpoons H^+ + SO_4^{2-}$	1.3×10^{-2}
Hydrogen sulfide	H_2S	$\rightleftharpoons H^+ + HS^-$	1.1×10^{-7}
	HS^-	$\rightleftharpoons H^+ + S^{2-}$	1×10^{-14}
Hypochlorous	$HClO$	$\rightleftharpoons H^+ + ClO^-$	3.2×10^{-8}
Nitrous	HNO_2	$\rightleftharpoons H^+ + NO_2^-$	4.5×10^{-4}
Phosphoric	H_3PO_4	$\rightleftharpoons H^+ + H_2PO_4^-$	7.5×10^{-3}
	$H_2PO_4^-$	$\rightleftharpoons H^+ + HPO_4^{2-}$	6.2×10^{-8}
	HPO_4^{2-}	$\rightleftharpoons H^+ + PO_4^{3-}$	1.0×10^{-12}
Phosphorous	H_3PO_3	$\rightleftharpoons H^+ + H_2PO_3^-$	1.6×10^{-2}
	$H_2PO_3^-$	$\rightleftharpoons H^+ + HPO_3^{2-}$	7×10^{-7}
Sulfurous	H_2SO_3	$\rightleftharpoons H^+ + HSO_3^-$	1.3×10^{-2}
	HSO_3^-	$\rightleftharpoons H^+ + SO_3^{2-}$	5.6×10^{-8}
Bases			
Aqueous ammonia	$NH_3 + H_2O$	$\rightleftharpoons NH_4^+ + OH^-$	1.8×10^{-5}

TABLE VII-B INSTABILITY CONSTANTS OF COMPLEX IONS

Ion	K_{inst}
$Ag(NH_3)_2^+$	6.8×10^{-8}
$Cd(NH_3)_4^{2+}$	1×10^{-7}
$Co(NH_3)_6^{2+}$	1.3×10^{-5}
$Co(NH_3)_6^{3+}$	2.2×10^{-34}
$Cu(NH_3)_4^{2+}$	2.6×10^{-13}
$Zn(NH_3)_4^{2+}$	2.6×10^{-10}
$HgBr_4^{2-}$	2.3×10^{-22}
$HgCl_4^{2-}$	8.0×10^{-16}
$Ag(CN)_2^-$	1.7×10^{-19}
$Au(CN)_2^-$	5×10^{-39}
$Cd(CN)_4^{2-}$	1.6×10^{-19}
$Cu(CN)_3^{2-}$	1×10^{-35}
$Fe(CN)_6^{4-}$	1×10^{-35}
$Fe(CN)_6^{3-}$	1×10^{-42}
$Zn(CN)_4^{2-}$	1×10^{-18}
AlF_6^{3-}	1×10^{-21}
FeF_5^{2-}	4×10^{-16}
HgI_4^{2-}	5×10^{-31}
$FeNCS^{2+}$	8×10^{-3}
$Hg(NCS)_4^{2-}$	1×10^{-21}
$Ag(S_2O_3)_2^{3-}$	6×10^{-14}

TABLE VII-C SOLUBILITY PRODUCT CONSTANTS AT 25°C

Substance		K_{sp}
Aluminum hydroxide	$Al(OH)_3$	3×10^{-33}
Antimony(III) sulfide	Sb_2S_3	2.9×10^{-59}
Barium carbonate	$BaCO_3$	5.0×10^{-9}
Barium chromate	$BaCrO_4$	1.8×10^{-10}
Barium fluoride	BaF_2	1.0×10^{-5}
Barium oxalate	BaC_2O_4	1.7×10^{-7}
Barium sulfate	$BaSO_4$	1.1×10^{-10}
Bismuth sulfide	Bi_2S_3	6.8×10^{-97}
Cadmium carbonate	$CdCO_3$	2.2×10^{-13}
Cadmium hydroxide	$Cd(OH)_2$	2.2×10^{-14}
Cadmium sulfide	CdS	7.8×10^{-27}
Calcium carbonate	$CaCO_3$	7.5×10^{-9}
Calcium chromate	$CaCrO_4$	7.1×10^{-4}
Calcium fluoride	CaF_2	9.6×10^{-11}
Calcium oxalate	CaC_2O_4	2.0×10^{-9}
Calcium phosphate	$Ca_3(PO_4)_2$	1×10^{-25}
Calcium sulfate	$CaSO_4$	2.4×10^{-5}
Cobalt(II) hydroxide	$Co(OH)_2$	2.5×10^{-16}
Cobalt(II) sulfide	CoS	5.9×10^{-21}
Cobalt(III) hydroxide	$Co(OH)_3$	3.0×10^{-43}
Chromium(III) hydroxide	$Cr(OH)_3$	7×10^{-31}
Copper(I) chloride	$CuCl$	4×10^{-7}
Copper(I) iodide	CuI	1×10^{-12}
Copper(I) sulfide	Cu_2S	1.6×10^{-48}
Copper(II) iodate	$Cu(IO_3)_2$	1.4×10^{-7}
Copper(II) sulfide	CuS	8.7×10^{-36}
Iron(II) hydroxide	$Fe(OH)_2$	2.0×10^{-15}
Iron(II) sulfide	FeS	4.9×10^{-18}
Iron (III) hydroxide	$Fe(OH)_3$	6.0×10^{-38}
Lead(II) bromide	$PbBr_2$	5.0×10^{-6}
Lead(II) carbonate	$PbCO_3$	1.2×10^{-13}
Lead(II) chloride	$PbCl_2$	1.6×10^{-5}
Lead(II) chromate	$PbCrO_4$	2×10^{-15}
Lead(II) fluoride	PbF_2	3.0×10^{-8}

All solubility products for sulfides are from Waggoner, *J. Chem. Educ.,* 35, 339 (1958). Averaged values from the literature are used for other compounds.

TABLE VII-C SOLUBILITY PRODUCT CONSTANTS AT $25°C$ (Continued)

Substance		K_{sp}
Lead(II) hydroxide	$Pb(OH)_2$	9×10^{-15}
Lead(II) iodate	$Pb(IO_3)_2$	1.9×10^{-13}
Lead(II) iodide	PbI_2	9.6×10^{-9}
Lead(II) phosphate	$Pb_3(PO_4)_2$	3×10^{-44}
Lead(II) sulfate	$PbSO_4$	1.4×10^{-8}
Lead(II) sulfide	PbS	8.4×10^{-28}
Lithium phosphate	Li_3PO_4	3.5×10^{-13}
Magnesium ammonium phosphate	$MgNH_4PO_4$	2.5×10^{-13}
Magnesium carbonate	$MgCO_3$	4×10^{-5}
Magnesium fluoride	MgF_2	8×10^{-8}
Magnesium hydroxide	$Mg(OH)_2$	1.2×10^{-11}
Magnesium oxalate	MgC_2O_4	8.6×10^{-5}
Manganese(II) carbonate	$MnCO_3$	8.8×10^{-11}
Manganese(II) hydroxide	$Mn(OH)_2$	6×10^{-14}
Manganese(II) sulfide	MnS	5.1×10^{-15}
Mercury(I) bromide	Hg_2Br_2	1.0×10^{-22}
Mercury(I) chloride	Hg_2Cl_2	1.3×10^{-18}
Mercury(I) iodide	Hg_2I_2	3.3×10^{-25}
Mercury(I) sulfide	Hg_2S	5.8×10^{-44}
Mercury(II) sulfide	HgS	8.6×10^{-53}
Nickel carbonate	$NiCO_3$	1.4×10^{-7}
Nickel hydroxide	$Ni(OH)_2$	2×10^{-16}
Nickel sulfide	NiS	1.8×10^{-21}
Silver acetate	AgO_2CCH_3	2.5×10^{-3}
Silver arsenate	Ag_3AsO_4	1×10^{-22}
Silver bromide	$AgBr$	4.8×10^{-13}
Silver carbonate	Ag_2CO_3	8×10^{-12}
Silver chloride	$AgCl$	1.2×10^{-10}
Silver chromate	Ag_2CrO_4	2.2×10^{-12}
Silver cyanide	$AgCN$	1.5×10^{-14}
Silver iodide	AgI	1.4×10^{-16}
Silver nitrite	$AgNO_2$	2.5×10^{-4}
Silver phosphate	Ag_3PO_4	1.0×10^{-18}
Silver sulfide	Ag_2S	6.8×10^{-50}
Silver thiocyanate	$AgSCN$	1×10^{-12}
Strontium carbonate	$SrCO_3$	1.0×10^{-9}
Strontium chromate	$SrCrO_4$	3.6×10^{-5}
Strontium fluoride	SrF_2	7.9×10^{-10}
Strontium oxalate	SrC_2O_4	5.6×10^{-8}
Strontium sulfate	$SrSO_4$	3.0×10^{-7}
Tin(II) hydroxide	$Sn(OH)_2$	5×10^{-26}
Tin(II) sulfide	SnS	1.2×10^{-25}
Zinc carbonate	$ZnCO_3$	1×10^{-10}
Zinc hydroxide	$Zn(OH)_2$	4×10^{-17}
Zinc sulfide	ZnS	1.1×10^{-21}

All solubility products for sulfides are from Waggoner, *J. Chem. Educ.*, 35, 339 (1958). Averaged values from the literature are used for other compounds.

TABLE VII-D STANDARD OXIDATION POTENTIALS, ACID SOLUTIONS, 25°C
(AFTER LATIMER)

Potential, $\mathcal{E}^\circ$ volts	Half-Reaction		
	Reducing Agents	Oxidizing Agents	
3.045	Li	$\rightleftharpoons Li^+$	$+ e^-$
2.925	K	$\rightleftharpoons K^+$	$+ e^-$
2.90	Ba	$\rightleftharpoons Ba^{2+}$	$+ 2e^-$
2.87	Ca	$\rightleftharpoons Ca^{2+}$	$+ 2e^-$
2.714	Na	$\rightleftharpoons Na^+$	$+ e^-$
2.37	Mg	$\rightleftharpoons Mg^{2+}$	$+ 2e^-$
1.66	Al	$\rightleftharpoons Al^{3+}$	$+ 3e^-$
1.18	Mn	$\rightleftharpoons Mn^{2+}$	$+ 2e^-$
0.763	Zn	$\rightleftharpoons Zn^{2+}$	$+ 2e^-$
0.74	Cr	$\rightleftharpoons Cr^{3+}$	$+ 3e^-$
0.440	Fe	$\rightleftharpoons Fe^{2+}$	$+ 2e^-$
0.41	Cr^{2+}	$\rightleftharpoons Cr^{3+}$	$+ e^-$
0.403	Cd	$\rightleftharpoons Cd^{2+}$	$+ 2e^-$
0.356	$Pb \quad + SO_4^{2-}$	$\rightleftharpoons PbSO_4$	$+ 2e^-$
0.250	Ni	$\rightleftharpoons Ni^{2+}$	$+ 2e^-$
0.136	Sn	$\rightleftharpoons Sn^{2+}$	$+ 2e^-$
0.126	Pb	$\rightleftharpoons Pb^{2+}$	$+ 2e^-$
0.000*	H_2*	$\rightleftharpoons 2H^+$	$+ 2e^-$
−0.141	H_2S	$\rightleftharpoons S$	$+ 2H^+ + 2e^-$
−0.15	Sn^{2+}	$\rightleftharpoons Sn^{4+}$	$+ 2e^-$
−0.153	Cu^+	$\rightleftharpoons Cu^{2+}$	$+ e^-$
−0.337	Cu	$\rightleftharpoons Cu^{2+}$	$+ 2e^-$
−0.45	$S \quad + 3H_2O$	$\rightleftharpoons H_2SO_3$	$+ 4H^+ + 4e^-$
−0.521	Cu	$\rightleftharpoons Cu^+$	$+ e^-$
−0.535	$2I^-$	$\rightleftharpoons I_2$	$+ 2e^-$
−0.682	H_2O_2	$\rightleftharpoons O_2$	$+ 2H^+ + 2e^-$
−0.771	Fe^{2+}	$\rightleftharpoons Fe^{3+}$	$+ e^-$
−0.789	$2Hg$	$\rightleftharpoons Hg_2^{2+}$	$+ 2e^-$
−0.799	Ag	$\rightleftharpoons Ag^+$	$+ e^-$
−0.854	Hg	$\rightleftharpoons Hg^{2+}$	$+ 2e^-$
−0.920	Hg_2^{2+}	$\rightleftharpoons 2Hg^{2+}$	$+ 2e^-$
−0.94	$HNO_2 + H_2O$	$\rightleftharpoons NO_3^-$	$+ 3H^+ + 2e^-$
−1.065	$2Br^-$	$\rightleftharpoons Br_2(l)$	$+ 2e^-$
−1.229	$2H_2O$	$\rightleftharpoons O_2$	$+ 4H^+ + 4e^-$
−1.33	$2Cr^{3+} + 7H_2O$	$\rightleftharpoons Cr_2O_7^{2-}$	$+ 14H^+ + 6e^-$
−1.360	$2Cl^-$	$\rightleftharpoons Cl_2$	$+ 2e^-$
−1.42	Au	$\rightleftharpoons Au^{3+}$	$+ 3e^-$
−1.45	$Cl^- \quad + 3H_2O$	$\rightleftharpoons ClO_3^-$	$+ 6H^+ + 6e^-$
−1.456	$Pb^{2+} + 2H_2O$	$\rightleftharpoons PbO_2$	$+ 4H^+ + 2e^-$
−1.51	$Mn^{2+} + 4H_2O$	$\rightleftharpoons MnO_4^-$	$+ 8H^+ + 5e^-$
−1.52	$Br^- \quad + 3H_2O$	$\rightleftharpoons BrO_3^-$	$+ 6H^+ + 6e^-$
−1.685	$PbSO_4 + 2H_2O$	$\rightleftharpoons PbO_2$	$+ SO_4^{2-} + 4H^+ + 2e^-$
−1.695	$MnO_2 + 2H_2O$	$\rightleftharpoons MnO_4^-$	$+ 4H^+ + 3e^-$
−1.77	$2H_2O$	$\rightleftharpoons H_2O_2$	$+ 2H^+ + 2e^-$
−2.01	$2SO_4^{2-}$	$\rightleftharpoons S_2O_8^{2-}$	$+ 2e^-$
−2.07	$O_2 \quad + H_2O$	$\rightleftharpoons O_3$	$+ 2H^+ + 2e^-$
−2.87	$2F^-$	$\rightleftharpoons F_2$	$+ 2e^-$

*Taken as standard.

INFRARED SPECTRA OF ALCOHOLS, ALDEHYDES, KETONES AND CARBOXYLIC ACIDS

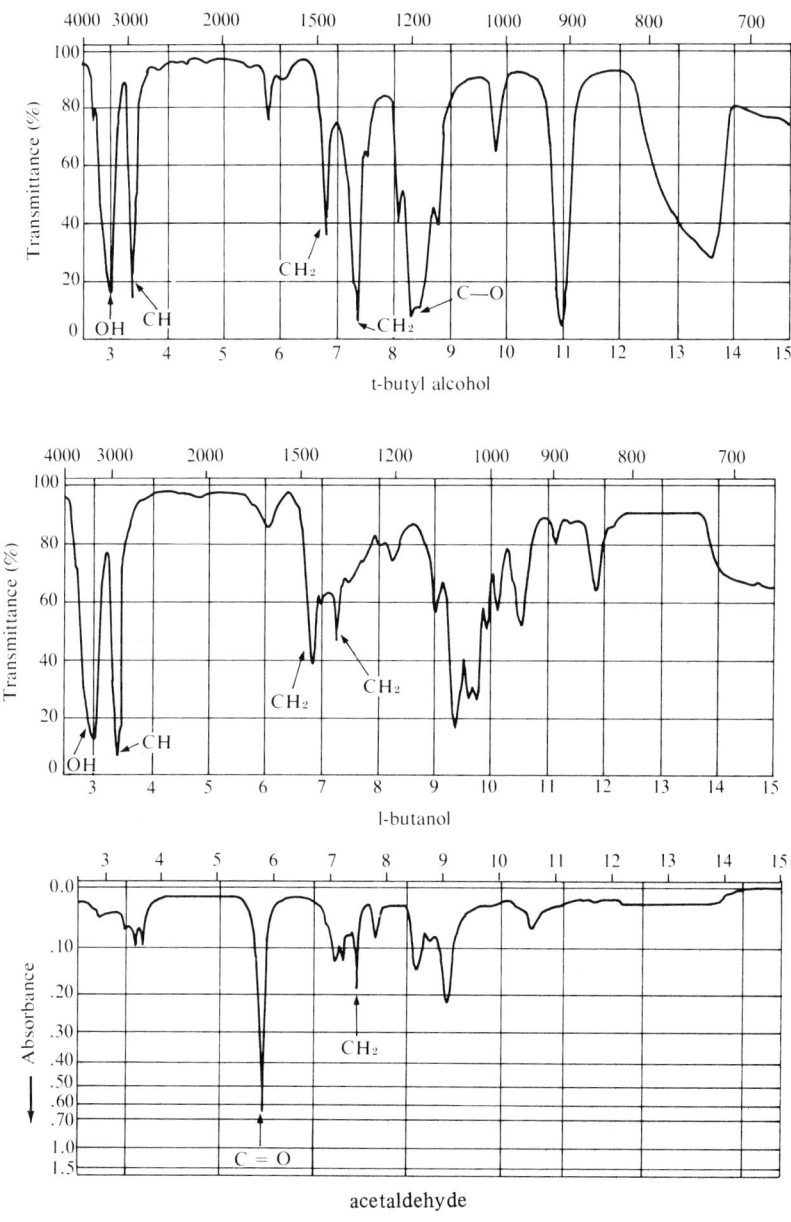

t-butyl alcohol

l-butanol

acetaldehyde

(Adapted from Sadtler Research Laboratory Spectra.)

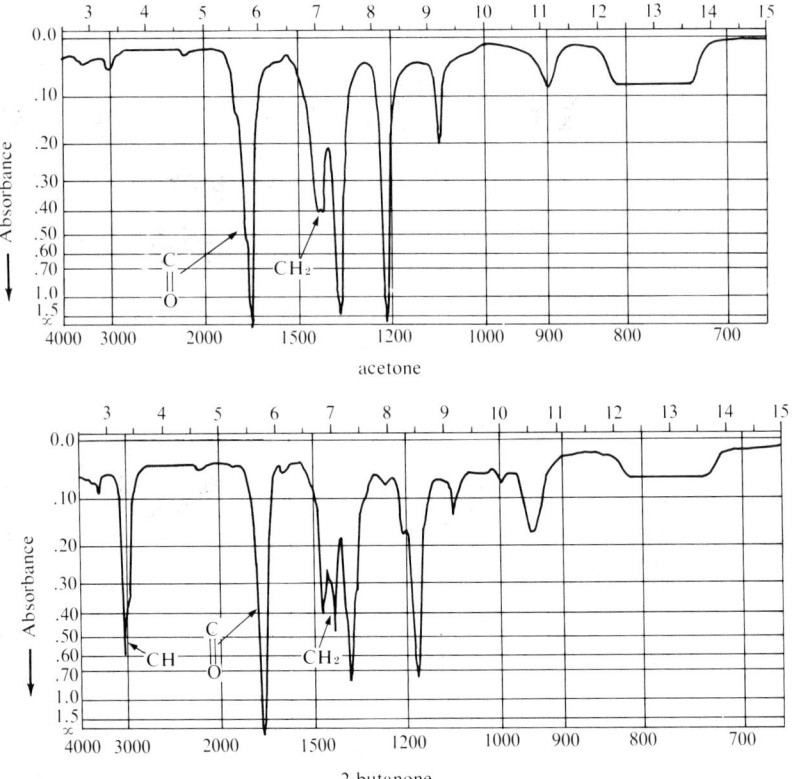

acetone

2-butanone

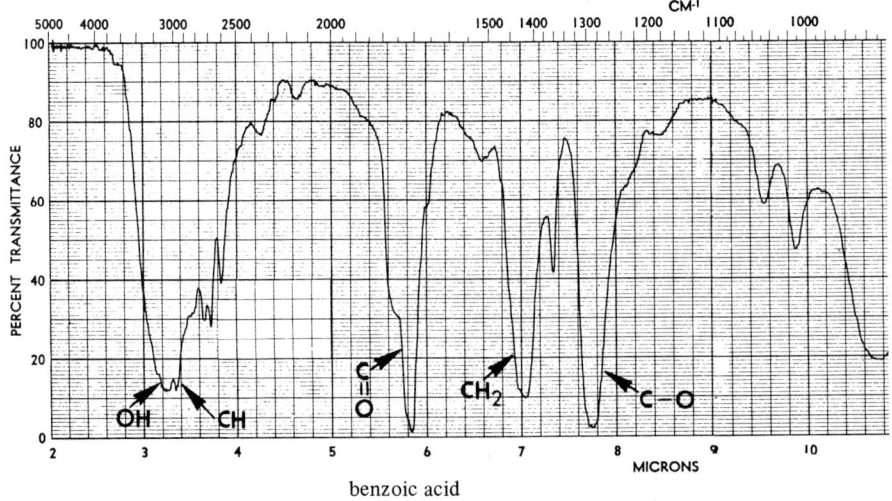

benzoic acid

TABLES OF BOILING OR MELTING POINTS OF ALCOHOLS, ALDEHYDES, KETONES, CARBOXYLIC ACIDS AND DERIVATIVES

ACIDS-LIQUID

Compound	B.p.	M.p. of Amide
Acetic	118	82
Propionic	141	81
Isobutyric	155	128
n-Butyric	163	115
Crotonic (cis)	169	101
Ethylmethylacetic	176	112
Isovaleric	176	135
n-Valeric	186	106
Dichloroacetic	194	98 subl.
Diethylacetic	195	112
Isocaproic	199	120
n-Caproic	205	100
Ethoxyacetic	206	81

ACIDS-SOLID

Compounds	M.p.	M.p. of Amide
Oleic	16	76
Methacrylic	16	106
dl-Lactic	18	74
Caproic	31.5	108
Levulinic	33	107
Lauric	43	99
β-Phenylpropionic (Hydrocinnamic)	48	105
Myristic	54	103
Chloroacetic	61	121
Stearic	70	109
Phenylacetic	76.5 subl.	156
o-Toluic	104	143
Azelaic	106.5	175
m-Toluic	112	97
p-Isopropylbenzoic	117	133
Benzoic	122	130
o-Benzoylbenzoic	128	165

ACIDS-SOLID (Continued)

Compounds	M.p.	M.p. of Amide
2, 5-Dimethylbenzoic	132	186
Sebacic	133	210
Acetylsalicylic	135	138
Diphenylacetic	145	168
p-Hydroxyphenylacetic	148	175
Adipic	152	220
2, 5-Dichlorobenzoic	153	155
Salicylic	158	139
p-Toluic	179	158
β-Naphthoic	185	195
Succinic	188	242
Phthalic	206	149

ALCOHOLS-LIQUID

Compounds	B.p.	M.p. of 3,5-Dinitrobenzoate
Methyl	65	108
Ethyl	78	93
Isopropyl	82	123
n-Propyl	97	74
Isobutyl	108	87
3-Pentanol	116	101
n-Butyl	117	64
2,3-Dimethyl-2-butanol	118	111
3,3-Dimethyl-2-butanol	120	107
2-Methyl-2-pentanol	121	72
3-Methyl-3-pentanol	123	96.5
2-Methylbutanol	129	70
4-Methyl-2-pentanol	132	65
Isoamyl (3-Methylbutanol)	132	61
2-Ethoxyethanol	135	75
3-Hexanol	136	77
n-Amyl (1-Pentanol)	138	46
Cyclopentanol	141	115
2,3-Dimethylbutanol	145	51.5
2-Methylpentanol	148	50.5
2-Ethylbutanol	148	51.5

ALCOHOLS-SOLID

Compound	M.p.	M.p. of 3,5-Dinitrobenzoate
α-Methylbenzyl (α-Phenylethyl)	20	95
2-Methylcyclohexanol (trans or α)	21	115
Dodecyl (Lauryl)	24	60
Cyclohexanol	25	113
tert-Butyl	25	142
Cinnamyl	33	121
α-Terpineol	35	79
L-Menthol	44	153

ALCOHOLS-SOLID (Continued)

Compound	M.p.	M.p. of 3,5-Dinitrobenzoate
Cetyl (Hexadecanol)	49	66
Octadecanol (Stearyl)	60	66
p-Methylbenzyl (p-Tolylcarbinol)	60	117
Benzhydrol (Diphenylcarbinol)	68	141
Ergosterol (anh.)	165	202

ALDEHYDES-LIQUID

Compounds	B.p.	M.p. of 2,4-Dinitrophenylhydrazone
Acetaldehyde (Ethanal)	20	168
Propionaldehyde (Propanal)	48	148
Isovaleraldehyde	64	187
n-Butyraldehyde (Butanal)	75	123
Pivaldehyde (Trimethylacetaldehyde)	75	209
Isovaleraldehyde	92.5	123
2-Methyl-1-butanal (α-Methyl-n-butyraldehyde)	92	120
Chloral	98	131
n-Pentanal (Valeraldehyde)	103	98
n-Hexanal (Caproaldehyde)	131	104
n-Heptanal	153	108

ALDEHYDES-SOLID

Compounds	M.p.	M.p. of 2,4-Dinitrophenylhydrazone
Palmitaldehyde	34	108
Phenylacetaldehyde	34	121
Piperonal (Heliotropin)	37	266
o-Methoxybenzaldehyde	38	253
Stearaldehyde	38	101
Lauraldehyde	44	106
p-Chlorobenzaldehyde	48	254
2-Napthaldehyde	60	270
Vanillin	81	271
p-Nitrobenzaldehyde	106	320
p-Hydroxybenzaldehyde	116	280

KETONES-LIQUID

Compound	B.p.	M.p. of 2,4-Dinitrophenylhydrazone
Acetone	56	126
2-Butanone (Ethyl methyl)	80	116
2-Methyl-3-butanone	94	120
3-Pentanone (Diethyl)	102	156
2-Pentanone (Methyl propyl)	102	143
4-Methyl-2-pentanone	117	95
2,4-Dimethyl-3-pentanone	124	88

KETONES-LIQUID (Continued)

Compound	B.p.	M.p. of 2,4-Dinitrophenylhydrazone
3-Hexanone	125	130
2-Hexanone	128	106
Cyclopentanone	131	146
4-Heptanone (Dipropyl)	144	75
Cyclohexanone	156	160

KETONES-SOLID

Compound	M.p.	M.p. of 2,4-Dinitrophenylhydrazone
Acetophenone	20	238
Benzyl methyl ketone	27	156
p-Methylacetophenone	28	260
p-Methoxyacetophenone	38	220
Benzalacetone	41	227
Benzophenone	48	238
Phenyl p-tolyl ketone	55	199
β-Naphthyl methyl ketone	56	262
Benzalacetophenone	58	244
p-Methoxybenzophenone	63	180
p-Toluquinone	69	269
p-Chlorobenzophenone	78	185
m-Nitroacetophenone	81	228
Fluorenone	83	283
Benzil	95	189
p-Hydroxyacetophenone	96	261

CALIBRATION OF THE
THERMOMETER

Thermometers, like other measuring devices, must be calibrated if their temperature measurements are to be considered reliable. Common chemical thermometers are often in error by a degree or more. This error may be caused by inaccuracy in graduation, irregularity in the bore of the tubing, or shrinkage of the glass after heating to a high temperature.

Procedures for calibrating thermometers involve comparing the readings on the thermometer with those on a thermometer calibrated at the National Bureau of Standards, or comparing the thermometer readings at the melting or boiling temperatures of a number of pure substances with the known values. Since nearly all pure substances melt at one characteristic temperature, it is possible to use this temperature to check the accuracy of the thermometer at that temperature. The latter procedure for calibration is most often used in teaching laboratories.

Calibration of the complete range of the thermometer can be made by checking the accuracy at four or five points. Some convenient standards are:
1. The transition temperature of sodium sulfate decahydrate, 32.4°C.*
2. The melting point of p-dichlorobenzene, 53.2°C.
3. The melting point of naphthalene, 80.2°C.
4. The melting point of oxalic acid hydrate, 101°C.

The results of this calibration should be recorded and the proper correction applied whenever precise temperature measurements are made.

Most thermometers are manufactured for total immersion use; i.e., the whole thermometer body, including the mercury thread in the capillary, is supposed to be at the temperature to be measured. When a considerable portion of the mercury column protrudes from the heated zone, a correction in the thermometer reading must be made because of the contraction of the mercury column in the cooler region. Since most work in the general chemistry laboratory is carried out with only the lower portion in the hot zone and the remainder at approximately room temperature, the thermometer should be calibrated for partial immersion. Your thermometer may have an *immersion line* marked on it 6 or 8 cm. above the bulb. If it does not, choose some fixed point about the same distance up, e.g., – 10 or – 20°, to represent the immersion line. Always immerse the thermometer to this immersion line during calibration and in later use.

---

*Sodium sulfate decahydrate, $Na_2SO_4 \cdot 10H_2O$, decomposes sharply at 32.38°C to anhydrous sodium sulfate, Na_2SO_4, and water; these form a saturated solution of sodium sulfate. This process is reversible and the temperature remains constant as long as both solids are present.

LOGARITHMS OF NUMBERS

TABLE I.

$$y = \log_{10} x$$

x	0	1	2	3	4	5	6	7	8	9
10	0000	0043	0086	0128	0170	0212	0253	0294	0334	0374
11	0414	0453	0492	0531	0569	0607	0645	0682	0719	0755
12	0792	0828	0864	0899	0934	0969	1004	1038	1072	1106
13	1139	1173	1206	1239	1271	1303	1335	1367	1399	1430
14	1461	1492	1523	1553	1584	1614	1644	1673	1703	1732
15	1761	1790	1818	1847	1875	1903	1931	1959	1987	2014
16	2041	2068	2095	2122	2148	2175	2201	2227	2253	2279
17	2304	2330	2355	2380	2405	2430	2455	2480	2504	2529
18	2553	2577	2601	2625	2648	2672	2695	2718	2742	2765
19	2788	2810	2833	2856	2878	2900	2923	2945	2967	2989
20	3010	3032	3054	3075	3096	3118	3139	3160	3181	3201
21	3222	3243	3263	3284	3304	3324	3345	3365	3385	3404
22	3424	3444	3464	3483	3502	3522	3541	3560	3579	3598
23	3617	3636	3655	3674	3692	3711	3729	3747	3766	3784
24	3802	3820	3838	3856	3874	3892	3909	3927	3945	3962
25	3979	3997	4014	4031	4048	4065	4082	4099	4116	4133
26	4150	4166	4183	4200	4216	4232	4249	4265	4281	4298
27	4314	4330	4346	4362	4378	4393.	4409	4425	4440	4456
28	4472	4487	4502	4518	4533	4548	4564	4579	4594	4609
29	4624	4639	4654	4669	4683	4698	4713	4728	4742	4757
30	4771	4786	4800	4814	4829	4843	4857	4871	4886	4900
31	4914	4928	4942	4955	4969	4983	4997	5011	5024	5038
32	5051	5065	5079	5092	5105	5119	5132	5145	5159	5172
33	5185	5198	5211	5224	5237	5250	5263	5276	5289	5302
34	5315	5328	5340	5353	5366	5378	5391	5403	5416	5428
35	5441	5453	5465	5478	5490	5502	5514	5527	5539	5551
36	5563	5575	5587	5599	5611	5623	5635	5647	5658	5670
37	5682	5694	5705	5717	5729	5740	5752	5763	5775	5786
38	5798	5809	5821	5832	5843	5855	5866	5877	5888	5899
39	5911	5922	5933	5944	5955	5966	5977	5988	5999	6010
40	6021	6031	6042	6053	6064	6075	6085	6096	6107	6117
41	6128	6138	6149	6160	6170	6180	6191	6201	6212	6222
42	6232	6243	6253	6263	6274	6284	6294	6304	6314	6325
43	6335	6345	6355	6365	6375	6385	6395	6405	6415	6425
44	6435	6444	6454	6464	6474	6484	6493	6503	6513	6522
45	6532	6542	6551	6561	6571	6580	6590	6599	6609	6618
46	6628	6637	6646	6656	6665	6675	6684	6693	6702	6712
47	6721	6730	6739	6749	6758	6767	6776	6785	6794	6803
48	6812	6821	6830	6839	6848	6857	6866	6875	6884	6893
49	6902	6911	6920	6928	6937	6946	6955	6964	6972	6981
50	6990	6998	7007	7016	7024	7033	7042	7050	7059	7067
51	7076	7084	7093	7101	7110	7118	7126	7135	7143	7152
52	7160	7168	7177	7185	7193	7202	7210	7218	7226	7235
53	7243	7251	7259	7267	7275	7284	7292	7300	7308	7316
54	7324	7332	7340	7348	7356	7364	7372	7380	7388	7396
x	0	1	2	3	4	5	6	7	8	9

From *Rinehart Mathematical Tables, Formulas, and Curves,* Enl. Ed. Copyright, 1948, 1953, by Harold D. Larsen.

TABLE I.

$$y = \log_{10} x$$

x	0	1	2	3	4	5	6	7	8	9
55	7404	7412	7419	7427	7435	7443	7451	7459	7466	7474
56	7482	7490	7497	7505	7513	7520	7528	7536	7543	7551
57	7559	7566	7574	7582	7589	7597	7604	7612	7619	7627
58	7634	7642	7649	7657	7664	7672	7679	7686	7694	7701
59	7709	7716	7723	7731	7738	7745	7752	7760	7767	7774
60	7782	7789	7796	7803	7810	7818	7825	7832	7839	7846
61	7853	7860	7868	7875	7882	7889	7896	7903	7910	7917
62	7924	7931	7938	7945	7952	7959	7966	7973	7980	7987
63	7993	8000	8007	8014	8021	8028	8035	8041	8048	8055
64	8062	8069	8075	8082	8089	8096	8102	8109	8116	8122
65	8129	8136	8142	8149	8156	8162	8169	8176	8182	8189
66	8195	8202	8209	8215	8222	8228	8235	8241	8248	8254
67	8261	8267	8274	8280	8287	8293	8299	8306	8312	8319
68	8325	8331	8338	8344	8351	8357	8363	8370	8376	8382
69	8388	8395	8401	8407	8414	8420	8426	8432	8439	8445
70	8451	8457	8463	8470	8476	8482	8488	8494	8500	8506
71	8513	8519	8525	8531	8537	8543	8549	8555	8561	8567
72	8573	8579	8585	8591	8597	8603	8609	8615	8621	8627
73	8633	8639	8645	8651	8657	8663	8669	8675	8681	8686
74	8692	8698	8704	8710	8716	8722	8727	8733	8739	8745
75	8751	8756	8762	8768	8774	8779	8785	8791	8797	8802
76	8808	8814	8820	8825	8831	8837	8842	8848	8854	8859
77	8865	8871	8876	8882	8887	8893	8899	8904	8910	8915
78	8921	8927	8932	8938	8943	8949	8954	8960	8965	8971
79	8976	8982	8987	8993	8998	9004	9009	9015	9020	9025
80	9031	9036	9042	9047	9053	9058	9063	9069	9074	9079
81	9085	9090	9096	9101	9106	9112	9117	9122	9128	9133
82	9138	9143	9149	9154	9159	9165	9170	9175	9180	9186
83	9191	9196	9201	9206	9212	9217	9222	9227	9232	9238
84	9243	9248	9253	9258	9263	9269	9274	9279	9284	9289
85	9294	9299	9304	9309	9315	9320	9325	9330	9335	9340
86	9345	9350	9355	9360	9365	9370	9375	9380	9385	9390
87	9395	9400	9405	9410	9415	9420	9425	9430	9435	9440
88	9445	9450	9455	9460	9465	9469	9474	9479	9484	9489
89	9494	9499	9504	9509	9513	9518	9523	9528	9533	9538
90	9542	9547	9552	9557	9562	9566	9571	9576	9581	9586
91	9590	9595	9600	9605	9609	9614	9619	9624	9628	9633
92	9638	9643	9647	9652	9657	9661	9666	9671	9675	9680
93	9685	9689	9694	9699	9703	9708	9713	9717	9722	9727
94	9731	9736	9741	9745	9750	9754	9759	9763	9768	9773
95	9777	9782	9786	9791	9795	9800	9805	9809	9814	9818
96	9823	9827	9832	9836	9841	9845	9850	9854	9859	9863
97	9868	9872	9877	9881	9886	9890	9894	9899	9903	9908
98	9912	9917	9921	9926	9930	9934	9939	9943	9948	9952
99	9956	9961	9965	9969	9974	9978	9983	9987	9991	9996
x	0	1	2	3	4	5	6	7	8	9